AF388966

New Frontiers in Parallel Robots

New Frontiers in Parallel Robots

Editors

Zhufeng Shao
Dan Zhang
Stéphane Caro

MDPI • Basel • Beijing • Wuhan • Barcelona • Belgrade • Manchester • Tokyo • Cluj • Tianjin

Editors

Zhufeng Shao
Tsinghua University
China

Dan Zhang
York University
Canada

Stéphane Caro
National Centre for Scientific
Research (CNRS) and the
Laboratory of Digital Sciences
of Nantes (LS2N)
France

Editorial Office
MDPI
St. Alban-Anlage 66
4052 Basel, Switzerland

This is a reprint of articles from the Special Issue published online in the open access journal *Machines* (ISSN 2075-1702) (available at: https://www.mdpi.com/journal/machines/special_issues/Parallel_Robots).

For citation purposes, cite each article independently as indicated on the article page online and as indicated below:

LastName, A.A.; LastName, B.B.; LastName, C.C. Article Title. *Journal Name* **Year**, *Volume Number*, Page Range.

ISBN 978-3-0365-7252-9 (Hbk)
ISBN 978-3-0365-7253-6 (PDF)

Cover image courtesy of Dan Zhang and Stéphane Caro

Contents

About the Editors

Zhufeng Shao

Zhufeng Shao, Ph.D., is an associate professor in the Department of Mechanical Engineering at Tsinghua University, a member of the National Industrial Foundation Expert Committee, a special expert of the Chinese Academy of Engineering for foundation strategy research, a senior member of the Chinese Mechanical Engineering Society, an Editorial Board member of the journal *Defence Technology* and a member of the National Professional Standards Technical Committee. He has won the Silver Award of China Patent, the First Prize of Technical Invention of the Science and Technology Award of the China Instrument Society, the Special Prize and Second Prize of Scientific and Technological Progress of the China Machinery Industry Science and Technology Award and the Second Prize of Natural Science of the Ministry of Education, as well as others. Dr. Shao received his Ph.D. in Mechanical Engineering from Tsinghua University, China, in June 2011. He is mainly engaged in the research of high-performance robotic equipment and intelligent manufacturing. Dr. Shao's research interests include parallel mechanisms, cable-driven parallel robots, soft robots, the innovative design of robots, macro and micro robot systems, humanoid and rehabilitation robots, autonomous systems and intelligent manufacturing. He has published 38 SCI-indexed papers, 32 EI-indexed papers and two books, and has been granted 45 Chinese invention patents and 12 software copyrights. He has developed an intelligent painting system for large civil aircraft and an automatic painting system for the non-structural surface of large ship segments, and has invented a new type of cable-driven high-speed parallel robot. Dr. Shao led the construction of a remote operation and maintenance system for CNC metal-cutting machine tools and has released six national standards related to intelligent manufacturing.

Dan Zhang

Dan Zhang, Ph.D., is a Kaneff Professor and Tier 1 York Research Chair in Advanced Robotics and Mechatronics in the Department of Mechanical Engineering at York University. Dr. Zhang was a Canada Research Chair in Advanced Robotics and Automation and was a founding Chair of the Department of Automotive, Mechanical and Manufacturing Engineering with the Faculty of Engineering and Applied Science at Ontario Tech University. He received his Ph.D. in Mechanical Engineering from Laval University, Canada, in June 2000. Dr. Zhang's research interests include the synthesis and optimization of parallel and hybrid mechanisms, generalized parallel mechanisms research, reconfigurable robots, the innovative design of parallel robots, the parallelization of serial robots, micro/nano manipulation and mems devices (e.g., sensors), rescue robots, smart biomedical instruments (e.g., exoskeleton robots and rehabilitation robotics), AI/robotics/autonomous systems, aerial and underwater robotics, artificial intelligence for robotics and intelligent reconfigurable adaptive landing gear and manipulators (manipulanders). Dr. Zhang has published 241 journal papers and 187 conference papers, 12 books, nine book chapters and numerous other technical publications. Dr. Zhang has served as a General Chair for 67 International Conferences and delivered 117 keynote speeches. Dr. Zhang was listed in the World's Top Two Percent Researchers by Stanford's Standardized Citation Indicators in 2020 and 2021. Dr. Zhang is a Fellow of the Canadian Academy of Engineering (CAE), a Fellow of the Engineering Institute of Canada (EIC), a Fellow of the American Society of Mechanical Engineers (ASME), a Fellow of the Canadian Society for Mechanical Engineering (CSME), a Senior Member of the Institute of Electrical and Electronics Engineers (IEEEs) and a Senior Member of SME.

Stéphane Caro

Stéphane Caro, Ph.D., is the Director of Research at the National Centre for Scientific Research (CNRS) and at the Laboratory of Digital Sciences of Nantes (LS2N), UMR CNRS 6004, France. Dr. Caro received his Engineering and M.Sc. degrees in mechanical engineering from Ecole Centrale Nantes, Nantes, France, in 2001, and his Doctorate degree in mechanical engineering from the University of Nantes in 2004. He was a Post-Doctoral Fellow in the Centre for Intelligent Machines, McGill University, Montreal, QC, Canada, from 2005 to 2006. He was awarded the accreditation to supervise research (HDR) in 2014. Dr. Caro's research focuses on the design, modeling and control of cable-driven parallel robots and reconfigurable parallel robots. Dr. Caro is the head of the "Robots and Machines for Manufacturing Society and Services" (RoMaS) Team at LS2N. He is also a part-time researcher at IRT Jules Verne, a mutualized industrial research institute, and a lecturer at Ecole Centrale de Nantes. Dr. Caro is a member of ASME and IEEE and was one of the two recipients of the 2018 Reviewers of the Year for the ASME *Journal of Mechanisms and Robotics*. He was also one of two recipients of the 2019 Crossley Award for Mechanism and Machine Theory. Dr. Caro is Associate Editor for the ASME *Journal of Mechanisms and Robotics* and the *IEEE Robotics and Automation Letters*. He is a member of the IFToMM Computational Kinematics and ASME Mechanisms and Robotics Technical Committees.

machines

MDPI

Editorial

New Frontiers in Parallel Robots

Zhufeng Shao [1,*], **Dan Zhang** [2] **and Stéphane Caro** [3]

1   Department of Mechanical Engineering, Tsinghua University, Beijing 100084, China
2   Department of Mechanical Engineering, Lassonde School of Engineering, York University,
    Toronto, ON M3J 1P3, Canada
3   Nantes Université, École Centrale Nantes, CNRS, LS2N, UMR 6004, F-44000 Nantes, France
*   Correspondence: shaozf@mail.tsinghua.edu.cn

Citation: Shao, Z.; Zhang, D.; Caro, S. New Frontiers in Parallel Robots. *Machines* **2023**, *11*, 386. https://doi.org/10.3390/machines11030386

Received: 13 March 2023
Accepted: 13 March 2023
Published: 15 March 2023

In the field of parallel robots, marked by the birth and application of the Gough–Stewart parallel mechanism [1] in the 1960s, great progress has been made in the past 60 years. The most notable feature of a parallel robot is that there are multiple closed-loop branch chains jointly connecting and driving the moving platform [2], which gives great flexibility in its configurations, creating a new way to change performance through robot configuration. Parallel robots usually have outstanding advantages of high stiffness, high precision, and high speed [3], which make up for the performance shortcomings of serial robots. The abundant configurations and complex mechanisms of parallel robots also present challenges in terms of configuration synthesis, performance evaluation, modeling, calibration, control, etc. Opportunities coexist with challenges, and parallel robots attract attention from both academia and industry. Today, parallel robots are constantly enriched, and new types of parallel robots, such as cable-driven parallel robots (CDPRs) [4], soft parallel robots [5], and hybrid robots [6], are constantly emerging. In particular, while inheriting the abovementioned advantages, a CDPR has the advantages of low cost, high energy efficiency, easy reconfiguration, and light weight, showing great application potential in scenarios such as large working spaces, heavy loads, high speeds, and bionics [7,8]. Parallel robotics research and applications show continued vitality and are expected to transform the industry in the future.

This Special Issue provides an international forum for professionals, academics, and researchers to present the latest developments from theoretical studies and applications of parallel robots. It includes 10 selected papers, covering important aspects of parallel robots such as modeling and control, error analysis and calibration, singularity analysis, and trajectory planning. The contents of these studies are briefly described here.

In [9], an evaluation model is established to analyze the influence of geometric errors on limbs' comprehensive deformations for an over-constrained parallel manipulator. The evaluation model is established based on kinematics and verified through simulations. Two global sensitivity indices are proposed, and a sensitivity analysis is conducted using the Monte Carlo method throughout the reachable workspace. The geometric errors that have greater effects on the average angular comprehensive deformation are identified.

In [10], a consistent solution strategy for static equilibrium workspaces of different types of under-constrained cab-driven robots is presented. The dynamic models and parameters that are applied to make the system stable for point-to-point movements are introduced. The constraints of the dynamics model are incorporated into the trajectory planning process to achieve point-to-point trajectory planning for the under-constrained cable-driven robots.

In [11], the authors present the singularity analysis and the geometric optimization of a 6-DOF (Degrees of Freedom) parallel robot for SILS (Single-Incision Laparoscopic Surgery). Based on a defined set of input/output constraint equations, the singularities of the parallel robotic system are determined and geometrically interpreted. Then, the geometric parameters for the 6-DOF parallel robot are optimized to make the operational workspace singularity-free.

Paper [12] focuses on pick-and-place trajectory planning and tracking control of a cable-based gangue-sorting robot in the operation space. A four-phase pick-and-place trajectory planning scheme based on an S-shaped acceleration/deceleration algorithm and the quantic polynomial trajectory planning method is proposed. A robust adaptive fuzzy tracking control strategy is presented against inevitable uncertainties and unknown external disturbances. The proposed method guarantees a stable and accurate pick-and-place trajectory tracking process.

Paper [13] proposes a fractional-order impedance control scheme, named KDHD, in which additional damping is added, proportional to the half-order derivatives of the end-effector position errors according to the half-derivative damping matrix, HD. The proposed impedance controller represents an extension to multi-input multi-output robotic systems of the $PDD^{1/2}$ controller for single-input single-output systems, which over performs the PD scheme in the transient behavior.

Paper [14] focuses on the dynamic modeling, workspace analysis, and multi-objective structural optimization of a large-span, high-speed, cable-driven parallel camera robot. The curved cable, due to the self-weight, is modeled as a catenary, and the dynamic model is derived by decomposing the motion of the cable into an in-plane motion and an out-plane motion. An optimization model is presented to simultaneously improve the workspace volume, anti-wind disturbance ability, and impulse of tensions on the camera and pan-tilt device system (CPTDS).

In [15], the authors present kinematic and dynamic modeling and workspace analysis for a novel suspended CDPR which generates Schönflies motions. The kinematics of the CDPR are solved through a geometrical approach. The dynamic feasible workspace of the robot is determined. Experiments are performed on a prototype of the robot to demonstrate the correctness of the derived models and workspace.

Paper [16] proposes a new method for the kinematic calibration of parallel robots to strict pose error bounds. The new method includes a new pose error model with 60 error parameters and a different kinematic parameter error identification algorithm based on L-infinity parameter estimation. Parameter errors are identified by using linear programming. The feasibility and validity of the proposed kinematic calibration are verified through both simulations and experiments.

Paper [17] studies the 3-DOF cutting stability and surface quality optimization of a parallel kinematic manipulator (PKM). A prediction model for the 3-DOF stability of helical milling based on the PKM is established through a semi-discrete method based on the natural frequency analysis of the PKM and a cutting force model of titanium alloy helical milling. A step-cutter is used to improve the machining process by enhancing the stability domain. The proposed method can provide a reference for further optimization of the prediction and optimization of the milling of difficult-to-process materials based on a PKM in the future.

In [18], the authors develop a simple model to evaluate the first natural frequencies of over-constrained PKMs. The PKM legs are modeled by beams, and constraint equations between the parameters are determined according to screw theory. The focus of this paper is to determine the global mass and stiffness matrices of the PKM in stationary configurations without the use of Jacobian matrices. The proposed method can be easily used at the conceptual design stage of PKMs.

The Guest Editors thank all of our colleagues who have taken interest in this Special Issue, especially the authors of the papers published in this Special Issue. All the of papers underwent a rigorous review process to ensure the high quality of the publications. We are grateful to the reviewers who evaluated these papers and provided valuable comments based on their professional perspectives. We also would like to thank the editors from MDPI for their support and effort in the organization and publication of this Special Issue.

It is hoped that the papers published in this Special Issue can be used as vehicles to promote knowledge sharing in the field of parallel robots. More importantly, we hope more people will be informed about and understand parallel robots and their latest technologies and actively participate in the innovation, research, development, and application promotion of parallel robots.

Author Contributions: Conceptualization, Z.S., D.Z. and S.C.; writing—original draft preparation, Z.S.; writing—review and editing, D.Z. and S.C. All authors have read and agreed to the published version of the manuscript.

Funding: This work was supported by the National Natural Science Foundation of China (No. U19A20101) and the National High-tech Ship Research Project of China (No. MC-202003-Z01).

Acknowledgments: We express great thanks to the editors of MDPI for their excellent support for this Special Issue, and it would have been impossible without their persistence and help. Many thanks to all our colleagues who are interested in these research topics and submitted their research for our Special Issue. A special thanks to the reviewers for their efforts and time spent in order to maintain the high quality of all contributions.

Conflicts of Interest: The authors declare no conflict of interest.

References

1. Gough, V.; Whitehall, S. Universal Tire Test Machine. In Proceedings of the 9th International Congress FISITA, London, UK, 30 April–5 May 1962; pp. 117–137.
2. Zhang, Z.; Wang, L.; Shao, Z. Improving the kinematic performance of a planar 3-RRR parallel manipulator through actuation mode conversion. *Mech. Mach. Theory* **2018**, *130*, 86–108. [CrossRef]
3. Pierrot, F.; Reynaud, C.; Fournier, A. DELTA: A simple and efficient parallel robot. *Robotica* **1990**, *8*, 105–109. [CrossRef]
4. Zhang, Z.; Shao, Z.; You, Z.; Tang, X.; Zi, B.; Yang, G.; Gosselin, C.; Caro, S. State-of-the-art on theories and applications of cable-driven parallel robots. *Front. Mech. Eng.* **2022**, *17*, 37. [CrossRef]
5. Huang, X.; Zhu, X.; Gu, G. Kinematic modeling and characterization of soft parallel robots. *IEEE Trans. Robot.* **2022**, *38*, 3792–3806. [CrossRef]
6. Wu, J.; Yu, G.; Gao, Y.; Wang, L. Mechatronics modeling and vibration analysis of a 2-DOF parallel manipulator in a 5-DOF hybrid machine tool. *Mech. Mach. Theory* **2018**, *121*, 430–445. [CrossRef]
7. Zhang, Z.; Shao, Z.; Wang, L. Optimization and implementation of a high-speed 3-DOFs translational cable-driven parallel robot. *Mech. Mach. Theory* **2018**, *145*, 103693. [CrossRef]
8. Shao, Z.; Li, T.; Tang, X.; Tang, L.; Deng, H. Research on the dynamic trajectory of spatial cable-suspended parallel manipulators with actuation redundancy. *Mechatronics* **2018**, *49*, 26–35. [CrossRef]
9. Du, X.; Wang, B.; Zheng, J. Geometric Error Analysis of a 2UPR-RPU Over-Constrained Parallel Manipulator. *Machines* **2022**, *10*, 990. [CrossRef]
10. Duan, Q.; Zhao, Q.; Wang, T. Consistent Solution Strategy for Static Equilibrium Workspace and Trajectory Planning of Under-Constrained Cable-Driven Parallel and Planar Hybrid Robots. *Machines* **2022**, *10*, 920. [CrossRef]
11. Pisla, D.; Birlescu, I.; Crisan, N.; Pusca, A.; Andras, I.; Tucan, P.; Radu, C.; Gherman, B.; Vaida, C. Singularity Analysis and Geometric Optimization of a 6-DOF Parallel Robot for SILS. *Machines* **2022**, *10*, 764. [CrossRef]
12. Liu, P.; Tian, H.; Cao, X.; Qiao, X.; Gong, L.; Duan, X.; Qiu, Y.; Su, Y. Pick–and–Place Trajectory Planning and Robust Adaptive Fuzzy Tracking Control for Cable–Based Gangue–Sorting Robots with Model Uncertainties and External Disturbances. *Machines* **2022**, *10*, 714. [CrossRef]
13. Bruzzone, L.; Polloni, A. Fractional Order KDHD impedance control of the Stewart Platform. *Machines* **2022**, *10*, 604. [CrossRef]
14. Su, Y.; Qiu, Y.; Liu, P.; Tian, J.; Wang, Q.; Wang, X. Dynamic modeling, workspace analysis and multi-objective structural optimization of the large-span high-speed cable-driven parallel camera robot. *Machines* **2022**, *10*, 565. [CrossRef]
15. Wang, R.; Xie, Y.; Chen, X.; Li, Y. Kinematic and dynamic modeling and workspace analysis of a suspended cable-driven parallel robot for Schönflies motions. *Machines* **2022**, *10*, 451. [CrossRef]
16. Yu, D. Kinematic Calibration of Parallel Robots Based on L-Infinity Parameter Estimation. *Machines* **2022**, *10*, 436. [CrossRef]
17. Qin, X.; Shi, M.; Hou, Z.; Li, S.; Li, H.; Liu, H. Analysis of 3-DOF Cutting Stability of Titanium Alloy Helical Milling Based on PKM and Machining Quality Optimization. *Machines* **2022**, *10*, 404. [CrossRef]
18. Chanal, H.; Guichard, A.; Blaysat, B.; Caro, S. Elasto-Dynamic Modeling of an Over-Constrained Parallel Kinematic Machine Using a Beam Model. *Machines* **2022**, *10*, 200. [CrossRef]

Article

Geometric Error Analysis of a 2UPR-RPU Over-Constrained Parallel Manipulator

Xu Du [1], Bin Wang [1] and Junqiang Zheng [2,*]

[1] School of Mechanical Engineering, Zhejiang Sci.-Tech. University, Hangzhou 310018, China
[2] School of Mechanical Engineering, Hangzhou Dianzi University, Hangzhou 310018, China
* Correspondence: zhengjunqiang@hdu.edu.cn

Abstract: For a 2UPR-RPU over-constrained parallel manipulator, some geometric errors result in internal forces and deformations, which limit the improvement of the pose accuracy of the moving platform and shorten the service life of the manipulator. Analysis of these geometric errors is important for restricting them. In this study, an evaluation model is established to analyse the influence of geometric errors on the limbs' comprehensive deformations for this manipulator. Firstly, the nominal inverse and actual forward kinematics are analysed according to the vector theory and the local product of the exponential formula. Secondly, the evaluation model of the limbs' comprehensive deformations is established based on kinematics. Thirdly, 41 geometric errors causing internal forces and deformations are identified and the results are verified through simulations based on the evaluation model. Next, two global sensitivity indices are proposed and a sensitivity analysis is conducted using the Monte Carlo method throughout the reachable workspace of the manipulator. The results of the sensitivity analysis indicate that 10 geometric errors have no effects on the average angular comprehensive deformation and that the identified geometric errors have greater effects on the average linear comprehensive deformation. Therefore, the distribution of the global sensitivity index of the average linear comprehensive deformation is more meaningful for accuracy synthesis. Finally, simulations are performed to verify the results of sensitivity analysis.

Keywords: 2UPR-RPU parallel manipulator; over-constrained parallel manipulator; geometric error; deformation; sensitivity analysis

Citation: Du, X.; Wang, B.; Zheng, J. Geometric Error Analysis of a 2UPR-RPU Over-Constrained Parallel Manipulator. *Machines* **2022**, *10*, 990. https://doi.org/10.3390/machines10110990

Academic Editors: Zhufeng Shao, Dan Zhang and Stéphane Caro

Received: 18 September 2022
Accepted: 27 October 2022
Published: 29 October 2022

1. Introduction

Parallel mechanisms with three DOFs have been successfully applied to hybrid serial–parallel machine tools, such as the well-known Eco-speed series, Tricept, and Exechon [1–6], owing to their high stiffness, large payload, and good dynamics. To achieve a simpler structure, Li et al. [3] designed a 2R1T (R denotes a rotational DOF, and T denotes a translational DOF) parallel mechanism named 2UPR-RPU. This mechanism is not only easier to control but also suitable for many operations along the surfaces. However, it is an over-constrained parallel mechanism with common constraints and over-constraints [7,8]. Some geometric errors in a manipulator based on this mechanism break the common constraints and over-constraints, resulting in internal forces and deformations. The internal forces and deformations not only limit the further improvement of the pose accuracy of the moving platform but also shorten the service life of the manipulator [9,10]. Therefore, it is necessary to restrict the internal-force-and-deformation-related geometric errors in the 2UPR-RPU parallel manipulator.

The accuracy design [11–13] can be applied to restrict geometric errors by determining the tolerances of the fabrication and assembly of machines. It consists of three components: error modelling [14–16], sensitivity analysis [17–19], and accuracy synthesis [20–22], where error modelling is the basis of sensitivity analysis and accuracy synthesis. Zhang et al. [13] applied the closed-loop vector and first-order perturbation methods to establish a geometric error model for a 2UPR-RPS over-constrained manipulator, and they identified the geometric errors that affected the pose errors of the moving platform. Zhang et al. [15] utilised the screw theory to establish a geometric error model for a 4RSR-SS over-constrained parallel tracking machine. With the use of the geometric error model, 53 geometric errors that had a significant influence on the pose errors of the moving platform were identified after sensitivity analysis. However, neither of the above two methods considers the deformations caused by internal forces in over-constrained parallel manipulators. Taking parameter uncertainties into account, Tang et al. [23] built a general interval kinetostatic model for a 2UPR-SPR over-constrained parallel machine to perform sensitivity analysis and tolerance allocation. To predict the pose errors of an over-constrained extendible support structure, Yu et al. [24] proposed a comprehensive model that simultaneously considered geometric errors, joint gaps, and link flexibility. In spite of good accuracy, these two models are complicated for the stiffness matrix needs to be derived and the stiffness coefficients of parts need to be obtained via finite element software.

Affected by geometric errors, the end poses of different limbs of a parallel manipulator should be theoretically inconsistent. However, they can be consistent in non-overconstrained parallel manipulators due to the existence of the moving platform and the motion deviations of passive joints. On this basis, a numerical iterative algorithm [25,26] was proposed to analyse the kinematics of non-overconstrained parallel manipulators with kinematic errors. Inspired by this algorithm, this study aims to establish an evaluation model based on kinematics to analyse the influence of geometric errors on the limbs' comprehensive deformations for the 2UPR-RPU over-constrained parallel manipulator.

Based on the established evaluation model, sensitivity analysis can help reveal the influence of different internal-force-and-deformation-related geometric errors on the limbs' comprehensive deformations. The interval analysis method and probabilistic method have been commonly used for sensitivity analysis of the moving platform's pose error in literature. The interval analysis method treats geometric errors as interval variables and can get a balance between calculation speed and accuracy [11,18]. Treating geometric errors as random variables with a normal distribution, the probabilistic method can be divided into the Monte Carlo method and the probability modelling method. The Monte Carlo method calculates the moving platform's pose errors according to the geometric error model and lots of random values of a geometric error [22,27]. It has good accuracy and low computational efficiency. The probability modelling method establishes an analytical model between the standard deviation of each geometric error and that of the moving platform's pose error based on the geometric error model [28]. In spite of high computational efficiency, this method needs prior knowledge about probability distributions. Considering that the interval analysis method and probability modelling method are not suitable when the geometric error model is iterative, the Monte Carlo method is utilised to analyse the influence of geometric errors on the limbs' comprehensive deformations in this paper.

The remainder of this paper is organised as follows. In Section 2, the 2UPR-RPU parallel mechanism is briefly introduced. Section 3 presents an analysis of the nominal inverse kinematics and actual forward kinematics. Section 4 establishes an evaluation model of the limbs' comprehensive deformations caused by geometric errors. Based on the evaluation model, the internal-force-and-deformation-related geometric errors are identified and the results are verified through simulations in Section 5. In Section 6, two global sensitivity indices are proposed and sensitivity analysis is conducted. Simulations are also performed to verify the results of sensitivity analysis. Finally, the conclusions are drawn in Section 7.

2. 2UPR-RPU Parallel Mechanism

As shown in Figure 1, the 2UPR-RPU parallel mechanism mainly consists of a moving platform, two UPR limbs, one RPU limb, and one fixed base, where the moving platform and fixed base are represented by the isosceles right triangles $\Delta A_1 A_2 A_3$ and $\Delta B_1 B_2 B_3$. U, P, and R denote universal, prismatic, and revolute joints, respectively. B_1, B_2 and A_3 are the centres of U, and A_1, A_2 and B_3 are the centres of R. Because each universal joint is equivalent to two mutually perpendicular revolute joints, the UPR limb is equivalent to the RRPR limb, and the RPU limb is equivalent to the RPRR limb. The axis of the *j*th joint of the *i*th limb is denoted by $\mathbf{s}_{i,j}$. A fixed coordinate system $\{\mathbf{o}_B; \mathbf{x}, \mathbf{y}, \mathbf{z}\}$ is established at the midpoint between $\mathbf{B}_1$ and $\mathbf{B}_2$, where **x** points from $\mathbf{B}_2$ to $\mathbf{B}_1$ and **y** points from $\mathbf{o}_B$ to $\mathbf{B}_3$. Similarly, a moving coordinate system $\{\mathbf{o}_A; \mathbf{u}, \mathbf{v}, \mathbf{w}\}$ is also established, where **u** points from $\mathbf{A}_2$ to $\mathbf{A}_1$ and **v** points from $\mathbf{o}_A$ to $\mathbf{A}_3$. The coordinate axes **z** and **w** are determined using the right-hand rule. For the 2UPR-RPU parallel mechanism, each limb exerts a force and a couple on the moving platform [8], where the two forces from the UPR limbs are parallel to **v**, and the three couples from the UPR and RPU limbs rotate around **w**. It is worth mentioning that the two forces parallel to **v** will lead to over-constraint, and the three couples rotating around **w** will lead to common constraints. Thus, the 2UPR-RPU parallel mechanism is an over-constrained parallel mechanism.

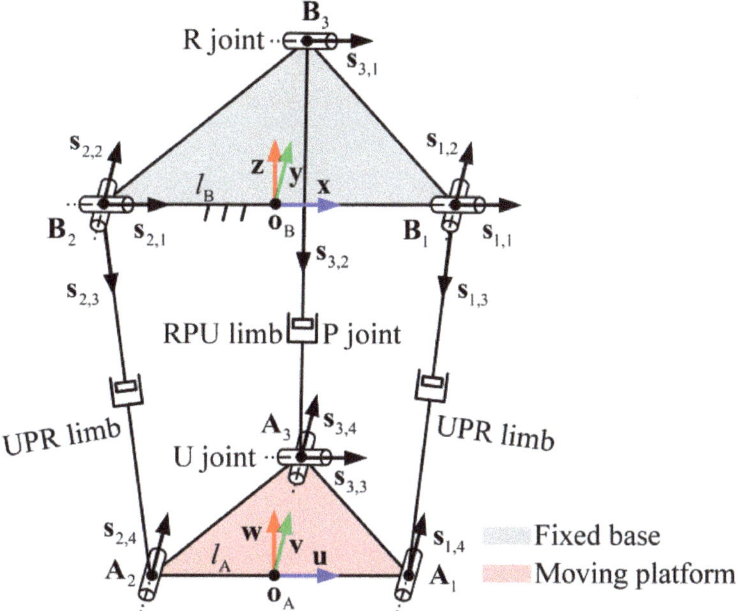

Figure 1. Schematic diagram of the 2UPR-RPU parallel mechanism.

3. Kinematics

Inverse kinematics aims to calculate the displacements of all joints relative to their initial positions or angles according to a given target pose of the moving platform. Forward kinematics is the reverse operation of inverse kinematics. Inverse kinematics without considering geometric errors is called nominal inverse kinematics. In this section, the nominal inverse kinematics of actuated joints and passive joints is first introduced. Then, the actual forward kinematics of the limbs is derived.

3.1. Nominal Inverse Kinematics

The position and orientation of the moving platform shown in Figure 1 can be described by $\begin{bmatrix} x & y & z \end{bmatrix}^T$ and $\begin{bmatrix} \alpha & \beta & \gamma \end{bmatrix}^T$, respectively, where $\begin{bmatrix} x & y & z \end{bmatrix}^T$ denotes the position coordinates of $\mathbf{o}_A$ with respect to $\{\mathbf{o}_B; \mathbf{x}, \mathbf{y}, \mathbf{z}\}$ and $\begin{bmatrix} \alpha & \beta & \gamma \end{bmatrix}^T$ denotes the Euler angle with respect to **z-x-v**. Because only the translation motion along $\mathbf{o}_B\mathbf{o}_A$ and the rotations around **x** and **v** can be achieved by the moving platform [8], $\begin{bmatrix} z & \beta & \gamma \end{bmatrix}^T$ is sufficient to represent the poses. For a given target pose of the moving platform, the nominal displacements of actuated P-joints can be derived using the closed-loop vector method [10] as follows:

$$\begin{cases} q_{1,3} = \|\mathbf{B}_1\mathbf{A}_1\| - \|\mathbf{B}_1\tilde{\mathbf{A}}_1\| \\ q_{2,3} = \|\mathbf{B}_2\mathbf{A}_2\| - \|\mathbf{B}_2\tilde{\mathbf{A}}_2\| \\ q_{3,2} = \|\mathbf{B}_3\mathbf{A}_3\| - \|\mathbf{B}_3\tilde{\mathbf{A}}_3\| \end{cases} \tag{1}$$

where $\|\cdot\|$ represents the Euclidean norm. $\tilde{\mathbf{A}}_i$ denotes the initial position of $\mathbf{A}_i$, which is determined by

$$\begin{cases} \mathbf{B}_1\mathbf{A}_1 = \begin{bmatrix} l_A\cos\gamma - l_B & l_A\sin\beta\sin\gamma - z\tan\beta & -l_A\cos\beta\sin\gamma + z \end{bmatrix}^{\mathrm{T}} \\ \mathbf{B}_2\mathbf{A}_2 = \begin{bmatrix} -l_A\cos\gamma + l_B & -l_A\sin\beta\sin\gamma - z\tan\beta & l_A\cos\beta\sin\gamma + z \end{bmatrix}^{\mathrm{T}} \\ \mathbf{B}_3\mathbf{A}_3 = \begin{bmatrix} 0 & l_A\cos\beta - z\tan\beta - l_B & l_A\sin\beta + z \end{bmatrix}^{\mathrm{T}} \end{cases} \tag{2}$$

where $l_A = \|\mathbf{A}_1\mathbf{A}_2\|/2$ and $l_B = \|\mathbf{B}_1\mathbf{B}_2\|/2$.

For the first UPR limb, the first, second, and fourth joints are passive. The nominal displacement of the first joint can be expressed as

$$q_{1,1} = \beta \tag{3}$$

The nominal displacement of the second joint can be expressed as

$$q_{1,2} = \arccos\left(\frac{\mathbf{e}_1^{\mathrm{T}}\mathbf{B}_1\mathbf{A}_1}{\|\mathbf{B}_1\mathbf{A}_1\|}\right) - \arccos\left(\frac{\mathbf{e}_1^{\mathrm{T}}\mathbf{B}_1\tilde{\mathbf{A}}_1}{\|\mathbf{B}_1\tilde{\mathbf{A}}_1\|}\right) \tag{4}$$

where $\mathbf{e}_1$ is the unit vector along $\mathbf{x}$.

The nominal displacement of the fourth joint can be expressed as

$$q_{1,4} = \arccos\left(\frac{-\left(\mathbf{A}_1\tilde{\mathbf{A}}_2\right)^{\mathrm{T}}\left(\mathbf{B}_1\tilde{\mathbf{A}}_1\right)}{\|\mathbf{A}_1\tilde{\mathbf{A}}_2\|\|\mathbf{B}_1\tilde{\mathbf{A}}_1\|}\right) - \arccos\left(\frac{-(\mathbf{A}_1\mathbf{A}_2)^{\mathrm{T}}(\mathbf{B}_1\mathbf{A}_1)}{\|\mathbf{A}_1\mathbf{A}_2\|\|\mathbf{B}_1\mathbf{A}_1\|}\right) \tag{5}$$

Because the two UPR limbs are symmetrically distributed with respect to $\mathbf{o}_A\mathbf{o}_B$, we have

$$q_{2,1} = \beta \tag{6}$$

$$q_{2,2} = \arccos\left(\frac{\mathbf{e}_1^{\mathrm{T}}\mathbf{B}_2\mathbf{A}_2}{\|\mathbf{B}_2\mathbf{A}_2\|}\right) - \arccos\left(\frac{\mathbf{e}_1^{\mathrm{T}}\mathbf{B}_2\tilde{\mathbf{A}}_2}{\|\mathbf{B}_2\tilde{\mathbf{A}}_2\|}\right) \tag{7}$$

$$q_{2,4} = \arccos\left(\frac{(\mathbf{A}_1\mathbf{A}_2)^{\mathrm{T}}(\mathbf{B}_2\mathbf{A}_2)}{\|\mathbf{A}_1\mathbf{A}_2\|\|\mathbf{B}_2\mathbf{A}_2\|}\right) - \arccos\left(\frac{\left(\mathbf{A}_1\tilde{\mathbf{A}}_2\right)^{\mathrm{T}}\left(\mathbf{B}_2\tilde{\mathbf{A}}_2\right)}{\|\mathbf{A}_1\tilde{\mathbf{A}}_2\|\|\mathbf{B}_2\tilde{\mathbf{A}}_2\|}\right) \tag{8}$$

Similarly, the nominal displacements of the first, third, and fourth joints of the RPU limb can be expressed as

$$q_{3,1} = \arccos\left(\frac{-\mathbf{e}_2^{\mathrm{T}}\mathbf{B}_3\mathbf{A}_3}{\|\mathbf{B}_3\mathbf{A}_3\|}\right) - \arccos\left(\frac{-\mathbf{e}_2^{\mathrm{T}}\mathbf{B}_3\tilde{\mathbf{A}}_3}{\|\mathbf{B}_3\tilde{\mathbf{A}}_3\|}\right) \tag{9}$$

$$q_{3,3} = \arccos\left(\frac{(\mathbf{Re_2})^{\mathrm{T}}(\mathbf{B_3 A_3})}{\|\mathbf{B_3 A_3}\|}\right) - \arccos\left(\frac{(\tilde{\mathbf{R}}\mathbf{e_2})^{\mathrm{T}}(\mathbf{B_3 \tilde{A}_3})}{\|\mathbf{B_3 \tilde{A}_3}\|}\right) \tag{10}$$

$$q_{3,4} = \gamma \tag{11}$$

Here, $\mathbf{e_2}$ is the unit vector along $\mathbf{y}$, and $\tilde{\mathbf{R}}$ denotes the initial state of $\mathbf{R}$, which is given as follows:

$$\mathbf{R} = \begin{bmatrix} \cos\gamma & 0 & \sin\gamma \\ \sin\beta\sin\gamma & \cos\beta & -\sin\beta\cos\gamma \\ -\cos\beta\sin\gamma & \sin\beta & \cos\beta\cos\gamma \end{bmatrix} \tag{12}$$

3.2. Actual Forward Kinematics

The nominal inverse kinematics described above does not consider geometric errors. However, geometric errors exist in the 2UPR-RPU parallel manipulator. In this section, the actual forward kinematics of the limbs in the manipulator is derived in detail.

As shown in Figure 2, four local coordinate systems $\{\mathbf{F}_{i,j}; \mathbf{x}_{i,j}, \mathbf{y}_{i,j}, \mathbf{z}_{i,j}\}$ are assigned to each limb to describe the geometric errors of the 2UPR-RPU parallel manipulator, where the initial pose of the moving platform is $\begin{bmatrix} z_0 & \beta_0 & \gamma_0 \end{bmatrix}^{\mathrm{T}} = \begin{bmatrix} -0.2\text{m} & 0 & 0 \end{bmatrix}^{\mathrm{T}}$ under the home configuration. The coordinate systems $\{\mathbf{o}_{B}; \mathbf{x}, \mathbf{y}, \mathbf{z}\}$ and $\{\mathbf{o}_{A}; \mathbf{u}, \mathbf{v}, \mathbf{w}\}$ are identical to those in Figure 1. For brevity, we use $\{\mathbf{F}_{i,j}\}$ instead of $\{\mathbf{F}_{i,j}; \mathbf{x}_{i,j}, \mathbf{y}_{i,j}, \mathbf{z}_{i,j}\}$. It is worth mentioning that this figure only shows $\mathbf{x}_{i,j}$ and $\mathbf{z}_{i,j}$ of the local coordinate systems, and $\mathbf{y}_{i,j}$ can be determined according to the right-hand rule, which will not be illustrated in detail here. The definitions of the local coordinate systems for the two UPR limbs and the RPU limb are listed in Tables 1–3.

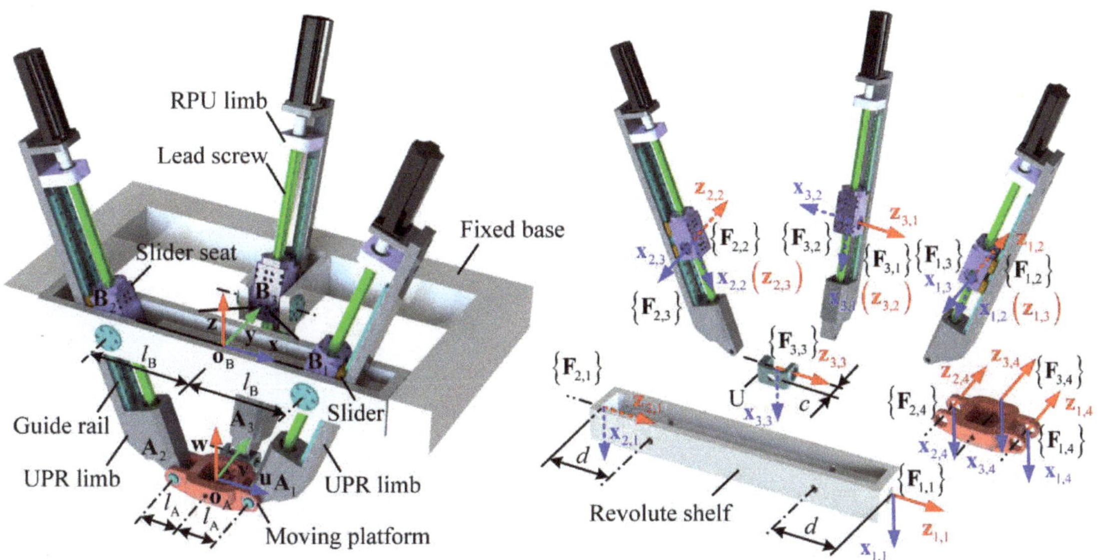

Figure 2. 2UPR-RPU parallel manipulator and its local coordinate systems.

Table 1. Definitions of local coordinate systems for the first UPR limb.

$\{F_{i,j}\}$	The Location	$F_{i,j}$	$x_{i,j}$	$z_{i,j}$
$\{F_{1,1}\}$	On the revolute shelf	The intersection of the right hole axis of the revolute shelf and the right end face of the revolute shelf	Parallel to the intersection of the front and rear symmetry plane of the right hole of the revolute shelf and the vertical plane of the right hole axis	Coincide with the right hole axis of the revolute shelf
			Point down	Point outwards
$\{F_{1,2}\}$	On the slider seat	The midpoint of the hole axis of the slider seat	Parallel to the intersection of the slider mounting plane and the vertical plane of the hole axis of the slider seat	Coincide with the hole axis of the slider seat
			Point to the moving platform	Point to the RPU limb
$\{F_{1,3}\}$	On the lead screw	The intersection of the lead screw axis and the plane passing through $z_{1,2}$ and perpendicular to the slider mounting plane	Parallel to the intersection of the guide rail plane and the vertical plane of the lead screw axis	Coincide with the lead screw axis
			Point in the direction opposite to the RPU limb	Point to the moving platform
$\{F_{1,4}\}$	On the moving platform	The midpoint of the right hole axis of the moving platform	Parallel to the intersection of the vertical plane of the right hole axis of the moving platform and the plane constructed with **v** and **w**	Coincide with the right hole axis of the moving platform
			Point down	Point to the RPU limb

Table 2. Definitions of local coordinate systems for the second UPR limb.

$\{F_{i,j}\}$	The Location	$F_{i,j}$	$x_{i,j}$	$z_{i,j}$
$\{F_{2,1}\}$	On the revolute shelf	The intersection of the left hole axis of the revolute shelf and the left end face of the revolute shelf	Parallel to the intersection of the front and rear symmetry plane of the left hole of the revolute shelf and the vertical plane of the left hole axis	Coincide with the left hole axis of the revolute shelf
			Point down	Point inwards
$\{F_{2,2}\}$	On the slider seat	The midpoint of the hole axis of the slider seat	Parallel to the intersection of the slider mounting plane and the vertical plane of the hole axis of the slider seat	Coincide with the hole axis of the slider seat
			Point to the moving platform	Point to the RPU limb
$\{F_{2,3}\}$	On the lead screw	The intersection of the lead screw axis and the plane passing through $z_{2,2}$ and perpendicular to the slider mounting plane	Parallel to the intersection of the guide rail plane and the vertical plane of the lead screw axis	Coincide with the lead screw axis
			Point in the direction opposite to the RPU limb	Point to the moving platform
$\{F_{2,4}\}$	On the moving platform	The midpoint of the left hole axis of the moving platform	Parallel to the intersection of the vertical plane of the left hole axis of the moving platform and the plane constructed with **v** and **w**	Coincide with the left hole axis of the moving platform
			Point down	Point to the RPU limb

Table 3. Definitions of local coordinate systems for the RPU limb.

$\{F_{i,j}\}$	The Location	$F_{i,j}$	$x_{i,j}$	$z_{i,j}$
$\{F_{3,1}\}$	On the slider seat	The midpoint of the hole axis of the slider seat	Parallel to the intersection of the slider mounting plane and the vertical plane of the hole axis of the slider seat	Coincide with the hole axis of the slider seat
			Point to the moving platform	Point to the first UPR limb
$\{F_{3,2}\}$	On the lead screw	The intersection of the lead screw axis and the plane passing through $z_{3,1}$ and perpendicular to the slider mounting plane	Parallel to the intersection of the guide rail plane and the vertical plane of the lead screw axis	Coincide with the lead screw axis
			Point to the second UPR limb	Point to the moving platform
$\{F_{3,3}\}$	On the U joint	The midpoint of the hole axis of the U joint	Parallel to the intersection of the vertical planes of the two hole axes of the U joint	Coincide with the hole axis of the U joint
			Point down	Point to the first UPR limb
$\{F_{3,4}\}$	On the moving platform	The intersection of the rear hole axis of the moving platform and the rear end face of the moving platform	Parallel to the intersection of the vertical plane of the rear hole axis of the moving platform and the plane constructed with $\mathbf{v}$ and $\mathbf{w}$	Coincide with the rear hole axis of the moving platform
			Point down	Point to the RPU limb

The end poses of the *i*th limb can be obtained from the local product of the exponential formula [25] as

$$\mathbf{g}_i(\mathbf{q}_i) = \mathbf{g}_{i,0} e^{\hat{\zeta}_{i,1} q_{i,1}} \mathbf{g}_{i,1} e^{\hat{\zeta}_{i,2} q_{i,2}} \mathbf{g}_{i,2} e^{\hat{\zeta}_{i,3} q_{i,3}} \mathbf{g}_{i,3} e^{\hat{\zeta}_{i,4} q_{i,4}} \mathbf{g}_{i,4}, i = 1, 2, 3 \tag{13}$$

where $\mathbf{g}_i$ denotes the homogeneous transformation matrix (HTM) of $\{\mathbf{o}_A; \mathbf{u}, \mathbf{v}, \mathbf{w}\}$ with respect to $\{\mathbf{o}_B; \mathbf{x}, \mathbf{y}, \mathbf{z}\}$ calculated using the *i*th limb. $\zeta_{i,j}$ denotes the screw coordinates of $\mathbf{s}_{i,j}$ with respect to $\{F_{i,j}\}$, which can be written as [25,26]

$$\begin{cases} \zeta_{i,j} = \begin{bmatrix} 0 & 0 & 1 & 0 & 0 & 0 \end{bmatrix}^{\mathrm{T}} \text{ for R joint} \\ \zeta_{i,j} = \begin{bmatrix} 0 & 0 & 0 & 0 & 0 & 1 \end{bmatrix}^{\mathrm{T}} \text{ for P joint} \end{cases} \tag{14}$$

Here, $e^{\hat{\zeta}_{i,j} q_{i,j}}$ denotes the exponential map from the Lie algebra $se(3)$ to the special Euclidean group $SE(3)$, which can be obtained using (A1)–(A4) in Appendix A. $\mathbf{g}_{i,j}$ is the HTM between adjacent coordinate systems when the parallel manipulator is under the home configuration. To be more specific, $\mathbf{g}_{i,0}$ denotes the HTM of $\{F_{i,1}\}$ with respect to $\{\mathbf{o}_B; \mathbf{x}, \mathbf{y}, \mathbf{z}\}$; $\mathbf{g}_{i,4}$ denotes the HTM of $\{\mathbf{o}_A; \mathbf{u}, \mathbf{v}, \mathbf{w}\}$ with respect to $\{F_{i,4}\}$; when $j \neq 0$ and $j \neq 4$, $\mathbf{g}_{i,j}$ is the HTM of $\{F_{i,j+1}\}$ with respect to $\{F_{i,j}\}$. $\mathbf{g}_{i,j}$ can be written as

$$\begin{cases} \mathbf{g}_{1,0} = \mathbf{Trans}(x, l_B + d)\mathbf{Rot}(y, \pi/2) \\ \mathbf{g}_{1,1} = \mathbf{Trans}(z, -d)\mathbf{Rot}(x, -\pi/2)\mathbf{Rot}(z, \widetilde{q}_{1,2} - \pi/2) \\ \mathbf{g}_{1,2} = \mathbf{Rot}(y, \pi/2) \\ \mathbf{g}_{1,3} = \mathbf{Trans}(z, \widetilde{q}_{1,3})\mathbf{Rot}(y, -\pi/2)\mathbf{Rot}(z, -\widetilde{q}_{1,2} + \pi/2) \\ \mathbf{g}_{1,4} = \mathbf{Trans}(y, l_A)\mathbf{Rot}(y, -\pi/2)\mathbf{Rot}(z, -\pi/2) \end{cases} \tag{15}$$

$$\begin{cases} \mathbf{g}_{2,0} = \mathbf{Trans}(x, -l_B - d)\mathbf{Rot}(y, \pi/2) \\ \mathbf{g}_{2,1} = \mathbf{Trans}(z, d)\mathbf{Rot}(x, -\pi/2)\mathbf{Rot}(z, \widetilde{q}_{2,2} - \pi/2) \\ \mathbf{g}_{2,2} = \mathbf{Rot}(y, \pi/2) \\ \mathbf{g}_{2,3} = \mathbf{Trans}(z, \widetilde{q}_{2,3})\mathbf{Rot}(y, -\pi/2)\mathbf{Rot}(z, -\widetilde{q}_{2,2} + \pi/2) \\ \mathbf{g}_{2,4} = \mathbf{Trans}(y, -l_A)\mathbf{Rot}(y, -\pi/2)\mathbf{Rot}(z, -\pi/2) \end{cases} \tag{16}$$

$$\begin{cases} \mathbf{g}_{3,0} = \mathbf{Trans}(y, l_B)\mathbf{Rot}(y, \pi/2)\mathbf{Rot}(z, \widetilde{q}_{3,1} - \pi/2) \\ \mathbf{g}_{3,1} = \mathbf{Rot}(y, \pi/2) \\ \mathbf{g}_{3,2} = \mathbf{Trans}(z, \widetilde{q}_{3,2})\mathbf{Rot}(y, -\pi/2)\mathbf{Rot}(z, -\widetilde{q}_{3,1} + \pi/2) \\ \mathbf{g}_{3,3} = \mathbf{Trans}(y, -c)\mathbf{Rot}(x, -\pi/2) \\ \mathbf{g}_{3,4} = \mathbf{Trans}(z, c - l_A)\mathbf{Rot}(y, -\pi/2)\mathbf{Rot}(z, -\pi/2) \end{cases} \tag{17}$$

where $\mathbf{Trans}(x, l_B)$ denotes the HTM that translates by l_B along x, and $\mathbf{Rot}(y, \pi/2)$ denotes the HTM that rotates by $\pi/2$ around y. $\widetilde{q}_{1,2}$ is the initial angle between $\mathbf{x}$ and $\mathbf{B}_1\mathbf{A}_1$, $\widetilde{q}_{2,2}$ is the initial angle between $\mathbf{B}_2\mathbf{B}_1$ and $\mathbf{B}_2\mathbf{A}_2$, and $\widetilde{q}_{3,1}$ is the initial angle between $\mathbf{B}_3\mathbf{o}_B$ and $\mathbf{B}_3\mathbf{A}_3$, which can be expressed as

$$\widetilde{q}_{1,2} = \arccos\left(\frac{\mathbf{e}_1^T \mathbf{B}_1 \widetilde{\mathbf{A}}_1}{\|\mathbf{B}_1 \widetilde{\mathbf{A}}_1\|}\right) \tag{18}$$

$$\widetilde{q}_{2,2} = \arccos\left(\frac{\mathbf{e}_1^T \mathbf{B}_2 \widetilde{\mathbf{A}}_2}{\|\mathbf{B}_2 \widetilde{\mathbf{A}}_2\|}\right) \tag{19}$$

$$\widetilde{q}_{3,1} = \arccos\left(\frac{-\mathbf{e}_2^T \mathbf{B}_3 \widetilde{\mathbf{A}}_3}{\|\mathbf{B}_3 \widetilde{\mathbf{A}}_3\|}\right) \tag{20}$$

In contrast to $\widetilde{q}_{1,2}$, $\widetilde{q}_{2,2}$, and $\widetilde{q}_{3,1}$, $\widetilde{q}_{1,3}$, $\widetilde{q}_{2,3}$, and $\widetilde{q}_{3,2}$ are the initial positions of the actuated P-joints, and we have

$$\widetilde{q}_{1,3} = \|\mathbf{B}_1 \widetilde{\mathbf{A}}_1\| \tag{21}$$

$$\widetilde{q}_{2,3} = \|\mathbf{B}_2 \widetilde{\mathbf{A}}_2\| \tag{22}$$

$$\widetilde{q}_{3,2} = \|\mathbf{B}_3 \widetilde{\mathbf{A}}_3\| \tag{23}$$

The linear errors $\boldsymbol{\delta}_{i,j}$ of $\{\mathbf{F}_{i,j+1}\}$ along $\mathbf{x}_{i,j}$, $\mathbf{y}_{i,j}$, and $\mathbf{z}_{i,j}$ can be expressed as follows:

$$\boldsymbol{\delta}_{i,j} = \begin{bmatrix} \delta_{i,j}^x & \delta_{i,j}^y & \delta_{i,j}^z \end{bmatrix}^T, i = 1, 2, 3 \text{ and } j = 0, \cdots, 4 \tag{24}$$

In addition to linear errors, angular errors also exist. The angular errors $\boldsymbol{\varepsilon}_{i,j}$ of $\{\mathbf{F}_{i,j+1}\}$ around $\mathbf{x}_{i,j}$, $\mathbf{y}_{i,j}$, and $\mathbf{z}_{i,j}$ can be expressed as follows:

$$\boldsymbol{\varepsilon}_{i,j} = \begin{bmatrix} \varepsilon_{i,j}^x & \varepsilon_{i,j}^y & \varepsilon_{i,j}^z \end{bmatrix}^T, i = 1, 2, 3 \text{ and } j = 0, \cdots, 4 \tag{25}$$

where $\boldsymbol{\delta}_{i,0}$ and $\boldsymbol{\varepsilon}_{i,0}$ denote the linear and angular errors of $\{\mathbf{F}_{i,1}\}$ with respect to $\{\mathbf{o}_B; \mathbf{x}, \mathbf{y}, \mathbf{z}\}$, respectively. $\boldsymbol{\delta}_{i,4}$ and $\boldsymbol{\varepsilon}_{i,4}$ denote the linear and angular errors of $\{\mathbf{F}_{i,4}\}$ with respect to $\{\mathbf{o}_A; \mathbf{u}, \mathbf{v}, \mathbf{w}\}$, respectively. Among the 90 error parameters, $\varepsilon_{1,0}^x$, $\varepsilon_{1,1}^y$, $\delta_{1,3}^z$, $\varepsilon_{1,3}^x$, $\varepsilon_{2,0}^x$, $\varepsilon_{2,1}^y$, $\delta_{2,3}^z$, $\varepsilon_{2,3}^x$, $\varepsilon_{3,0}^x$, $\delta_{3,2}^z$, $\varepsilon_{3,2}^x$, and $\varepsilon_{3,3}^y$ represent the initial displacement errors of the 12 joints. In addition, the values of $\delta_{1,2}^x$, $\delta_{2,2}^x$, $\varepsilon_{1,4}^y$, $\varepsilon_{2,4}^y$, $\delta_{3,1}^x$, $\varepsilon_{3,3}^z$, and $\varepsilon_{3,4}^y$ are zeros since the definitions of local coordinate systems. Therefore, the rest 71 error parameters represent the linear and angular geometric errors.

Setting the values of error parameters other than geometric errors to zeros, the HTM of the geometric errors between adjacent coordinate systems can be written as

$$\Delta \mathbf{g}_{i,j} = \begin{bmatrix} e^{\hat{\boldsymbol{\varepsilon}}_{i,j}} & \boldsymbol{\delta}_{i,j} \\ \mathbf{0}_{1\times 3} & 1 \end{bmatrix}, i = 1,\, 2,\, 3 \text{ and } j = 0,\, \cdots,\, 4 \tag{26}$$

where $e^{\hat{\boldsymbol{\varepsilon}}_{i,j}}$ denotes the exponential map from the Lie algebra $so(3)$ to the special orthogonal group $SO(3)$, which can be determined using (A3) and (A5) in Appendix A.

The end poses of the ith limb that include sthe linear and angular geometric errors can then be obtained as follows:

$$\mathbf{g}_i^{ge}(\mathbf{q}_i) = \Delta\mathbf{g}_{i,0}\mathbf{g}_{i,0}e^{\hat{\zeta}_{i,1}q_{i,1}}\Delta\mathbf{g}_{i,1}\mathbf{g}_{i,1}e^{\hat{\zeta}_{i,2}q_{i,2}}\Delta\mathbf{g}_{i,2}\mathbf{g}_{i,2}e^{\hat{\zeta}_{i,3}q_{i,3}}\Delta\mathbf{g}_{i,3}\mathbf{g}_{i,3}e^{\hat{\zeta}_{i,4}q_{i,4}}\mathbf{g}_{i,4}\Delta\mathbf{g}_{i,4}^{-1}, i = 1,2,3 \tag{27}$$

which can be rewritten as

$$\mathbf{g}_i^{ge}(\mathbf{q}_i) = \mathbf{g}_{i,0}^{ge}e^{\hat{\zeta}_{i,1}q_{i,1}}\mathbf{g}_{i,1}^{ge}e^{\hat{\zeta}_{i,2}q_{i,2}}\mathbf{g}_{i,2}^{ge}e^{\hat{\zeta}_{i,3}q_{i,3}}\mathbf{g}_{i,3}^{ge}e^{\hat{\zeta}_{i,4}q_{i,4}}\mathbf{g}_{i,4}^{ge}, i = 1,2,3 \tag{28}$$

4. Evaluation Model of Deformations

As mentioned previously, the 2UPR-RPU parallel manipulator is over-constrained. Theoretically, the end poses of any two limbs can also be consistent with each other through the motion deviations of passive joints when the internal-force-and-deformation-related geometric errors are zero, which can be expressed as

$$\mathbf{g}_i^{ge} + \Delta\mathbf{g}_i^{ge} = \mathbf{g}_k^{ge} + \Delta\mathbf{g}_k^{ge} \tag{29}$$

where $\Delta\mathbf{g}_i^{ge}$ and $\Delta\mathbf{g}_k^{ge}$ denote the end-pose deviations of the ith and kth limbs caused by the motion deviations of passive joints, respectively. The end-pose deviation between the ith and kth limbs can be written as [25,26]

$$\Delta\boldsymbol{\mu}_{k,i} = \left\{ \log\left[\mathbf{g}_k^{ge}\left(\mathbf{g}_i^{ge}\right)^{-1} \right] \right\}^{\vee} \tag{30}$$

where $\log[\cdot]$ stands for the logarithmic operation from $SE(3)$ to $se(3)$, and it can be obtained using (A6) and (A7) in Appendix A. $\vee$ represents the reverse operation of (A1). The end-pose deviation can be rewritten in screw form as follows:

$$\Delta\boldsymbol{\mu}_{k,i} = \Delta\boldsymbol{\mu}_i - \Delta\boldsymbol{\mu}_k \tag{31}$$

where the screws $\Delta\boldsymbol{\mu}_i$ and $\Delta\boldsymbol{\mu}_k$ denote the end-pose deviations of the ith and kth limbs originating from the motion deviations of passive joints, respectively. Take $\Delta\boldsymbol{\mu}_i$ as an example. Taking the partial differential of (28) with respect to the displacements of the passive joints, $\Delta\boldsymbol{\mu}_i$ can be expressed as follows:

$$\Delta\boldsymbol{\mu}_i = \boldsymbol{\Psi}_i\boldsymbol{\Phi}_i\Delta\mathbf{q}_i, i = 1,2,3 \tag{32}$$

where $\Delta\mathbf{q}_i$ denotes the motion deviation of the passive joints of the ith limb.

When $i = 1$ and $i = 2$, we have

$$\Delta\mathbf{q}_i = \begin{bmatrix} \Delta q_{i,1} & \Delta q_{i,2} & \Delta q_{i,4} \end{bmatrix}^{\mathrm{T}} \tag{33}$$

and when $i = 3$, we have

$$\Delta\mathbf{q}_i = \begin{bmatrix} \Delta q_{i,1} & \Delta q_{i,3} & \Delta q_{i,4} \end{bmatrix}^{\mathrm{T}} \tag{34}$$

For the coefficient matrices $\mathbf{\Psi}_i$ and $\mathbf{\Phi}_i$, when $i = 1$ and $i = 2$, we obtain

$$\mathbf{\Psi}_i = \begin{bmatrix} \mathbf{I}_6 & \mathrm{Ad}\left(e^{\hat{\xi}_{i,1}q_{i,1}} \right) & \mathrm{Ad}\left(e^{\hat{\xi}_{i,1}q_{i,1}} e^{\hat{\xi}_{i,2}q_{i,2}} e^{\hat{\xi}_{i,3}q_{i,3}} \right) \end{bmatrix} \in \mathbb{R}^{6 \times 18} \tag{35}$$

$$\mathbf{\Phi}_i = \mathrm{Blkdiag}(\xi_{i,1}, \xi_{i,2}, \xi_{i,4}) \in \mathbb{R}^{18 \times 3} \tag{36}$$

When $i = 3$, we obtain

$$\mathbf{\Psi}_i = \begin{bmatrix} \mathbf{I}_6 & \mathrm{Ad}\left(e^{\hat{\xi}_{i,1}q_{i,1}} e^{\hat{\xi}_{i,2}q_{i,2}} \right) & \mathrm{Ad}\left(e^{\hat{\xi}_{i,1}q_{i,1}} e^{\hat{\xi}_{i,2}q_{i,2}} e^{\hat{\xi}_{i,3}q_{i,3}} \right) \end{bmatrix} \in \mathbb{R}^{6 \times 18} \tag{37}$$

$$\mathbf{\Phi}_i = \mathrm{Blkdiag}(\xi_{i,1}, \xi_{i,3}, \xi_{i,4}) \in \mathbb{R}^{18 \times 3} \tag{38}$$

where $\mathbf{I}_6$ is an identity matrix of order six. $\mathrm{Ad}(\cdot)$ is an adjoint representation of $SE(3)$ and is given in (A8) in Appendix A. $\mathrm{Blkdiag}(\cdot)$ denotes a block-diagonal matrix. $\xi_{i,j}$ denotes the screw coordinates of $\mathbf{s}_{i,j}$ with respect to $\{\mathbf{o}_B; \mathbf{x}, \mathbf{y}, \mathbf{z}\}$, which can be written as follows [25]:

$$\xi_{i,j} = \mathrm{Ad}\left(\mathbf{g}_{i,0}^{ge}, \mathbf{g}_{i,1}^{ge}, \cdots, \mathbf{g}_{i,j-1}^{ge} \right) \zeta_{i,j} \tag{39}$$

Combining (31) with (32) yields

$$\begin{bmatrix} \Delta\mu_{2,1} \\ \Delta\mu_{3,2} \\ \Delta\mu_{1,3} \end{bmatrix} = \begin{bmatrix} \mathbf{\Psi}_1\mathbf{\Phi}_1 & -\mathbf{\Psi}_2\mathbf{\Phi}_2 & 0 \\ 0 & \mathbf{\Psi}_2\mathbf{\Phi}_2 & -\mathbf{\Psi}_3\mathbf{\Phi}_3 \\ -\mathbf{\Psi}_1\mathbf{\Phi}_1 & 0 & \mathbf{\Psi}_3\mathbf{\Phi}_3 \end{bmatrix} \begin{bmatrix} \Delta\mathbf{q}_1 \\ \Delta\mathbf{q}_2 \\ \Delta\mathbf{q}_3 \end{bmatrix} \tag{40}$$

Let

$$\Delta\mu = \begin{bmatrix} \Delta\mu_{2,1}^{\mathrm{T}} & \Delta\mu_{3,2}^{\mathrm{T}} & \Delta\mu_{1,3}^{\mathrm{T}} \end{bmatrix}^{\mathrm{T}} \in \mathbb{R}^{18 \times 1} \tag{41}$$

$$\Delta\mathbf{q} = \begin{bmatrix} \Delta\mathbf{q}_1^{\mathrm{T}} & \Delta\mathbf{q}_2^{\mathrm{T}} & \Delta\mathbf{q}_3^{\mathrm{T}} \end{bmatrix}^{\mathrm{T}} \in \mathbb{R}^{9 \times 1} \tag{42}$$

$$\mathbf{J} = \begin{bmatrix} \mathbf{\Psi}_1\mathbf{\Phi}_1 & -\mathbf{\Psi}_2\mathbf{\Phi}_2 & 0 \\ 0 & \mathbf{\Psi}_2\mathbf{\Phi}_2 & -\mathbf{\Psi}_3\mathbf{\Phi}_3 \\ -\mathbf{\Psi}_1\mathbf{\Phi}_1 & 0 & \mathbf{\Psi}_3\mathbf{\Phi}_3 \end{bmatrix} \in \mathbb{R}^{18 \times 9} \tag{43}$$

Note that when $\Delta\mu$ is obtained using (30) and (41), the motion deviations of passive joints can be calculated as

$$\Delta\mathbf{q} = \left(\mathbf{J}^{\mathrm{T}}\mathbf{J} \right)^{-1} \mathbf{J}^{\mathrm{T}} \Delta\mu = \mathbf{J}_c \Delta\mu \tag{44}$$

Based on the above work, an iterative model can be proposed to evaluate the deformations caused by geometric errors of the 2UPR-RPU over-constrained manipulator. The detailed processes are described below.

As shown in Figure 3, the proposed evaluation model mainly includes the following steps. Firstly, a target pose of the moving platform is input, and specified values are assigned to some of the 71 geometric errors; secondly, the nominal displacements of all joints are calculated based on the inverse kinematics; thirdly, the displacements of the passive joints are iteratively updated starting with the nominal values and the end condition is given as the maximum number of iterations or the target value of the infinity norm $\|\Delta\mu^j - \Delta\mu^{j-1}\|_\infty$; finally, the latest end-pose deviation $\Delta\mu^j$ for the target pose is output. When the internal-force-and-deformation-related geometric errors are not all zeros, the end poses of the limbs cannot be consistent without deformations. Therefore, the latest $\Delta\mu^j$ and indices based on it can be used to indirectly evaluate the limbs' comprehensive deformations caused by geometric errors of the 2UPR-RPU over-constrained manipulator.

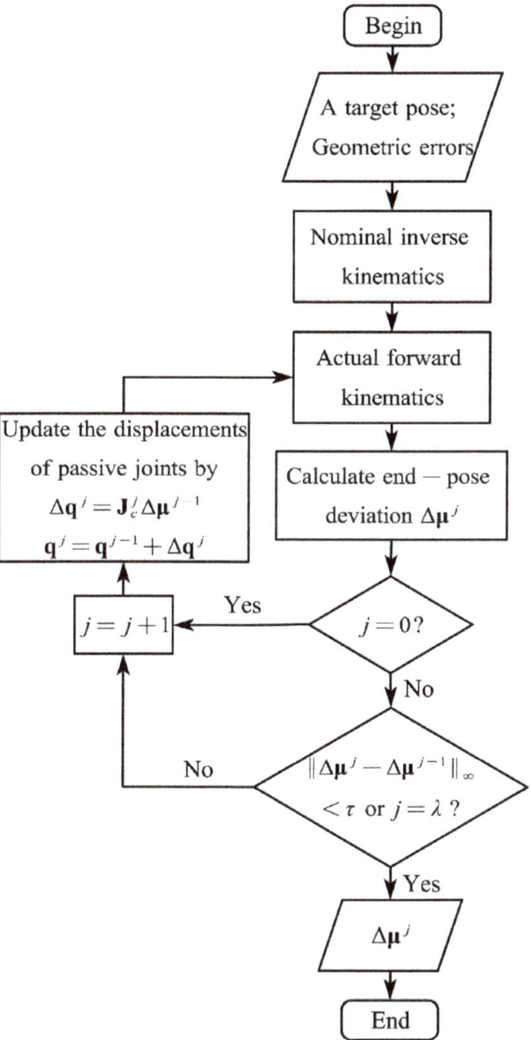

Figure 3. Scheme I: Evaluation of the limbs' comprehensive deformations caused by geometric errors of the 2UPR-RPU over-constrained manipulator.

Considering that a large amount of matrix calculation is included in the proposed evaluation model, MATLAB is used for programming in Sections 5 and 6.

5. Geometric Error Identification

Finding the internal-force-and-deformation-related geometric errors is the basis of sensitivity analysis. In this section, the reachable workspace of the 2UPR-RPU parallel manipulator is described. For geometric error identification and verification, 692 and 1738 target poses are selected in the reachable workspace. Subsequently, internal-force-and-deformation-related geometric errors in the manipulator are identified based on the proposed evaluation model and an evaluation index. Finally, simulations are conducted to verify the correctness of the identification results.

5.1. Identification Analysis

The structural parameters of the 2UPR-RPU parallel manipulator are presented in Table 4. Using the space search method [29], the reachable workspace of the manipulator can be obtained. The search results are shown in Figure 4. Because the end poses at the boundaries of the reachable workspace are more sensitive to geometric errors, the 692 target poses shown in Figure 5 are uniformly selected for geometric error identification. To identify the internal-force-and-deformation-related geometric errors, the evaluation index of the maximum comprehensive deformation of a limb can be written as

$$\Delta\mu_{\max} = \max\left(\|\Delta\mu_{2,1}^{j}\|, \|\Delta\mu_{3,2}^{j}\|, \|\Delta\mu_{1,3}^{j}\| \right) \tag{45}$$

Table 4. Structural parameters of the 2UPR-RPU parallel manipulator.

Symbols	Values	Units
l_A	0.06	m
l_B	0.15	m
c	0.025	m
d	0.115	m

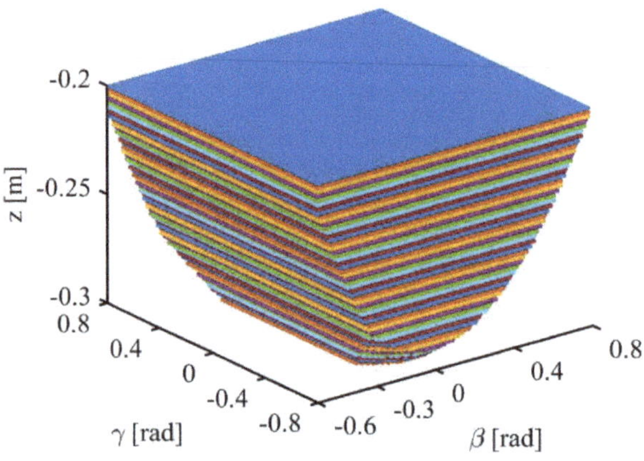

Figure 4. Reachable workspace of the 2UPR-RPU parallel manipulator.

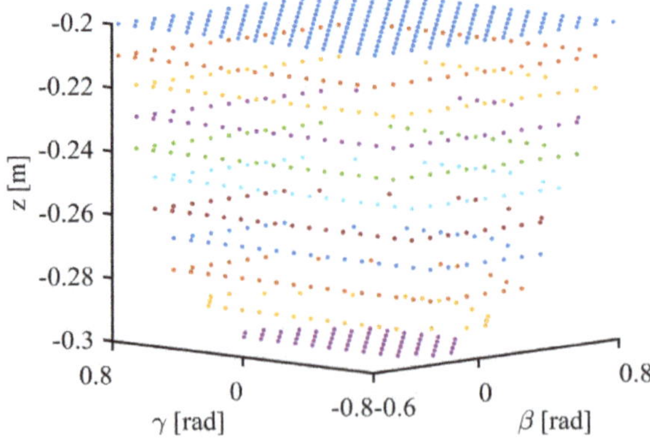

Figure 5. 692 target poses of the 2UPR-RPU parallel manipulator.

Based on Scheme I and (45), 692 $\Delta\mu_{\max}$s can be calculated for the selected target poses of the moving platform. If the 692 $\Delta\mu_{\max}$s are not close to zero, it means that there are internal-force-and-deformation-related geometric errors among the geometric errors that were assigned specified values. Without loss of generality, three groups of specified values for geometric errors are given, as listed in Table 5. The maximum iteration number λ and specified tolerance τ in Scheme I are set to 50 and 10^{-15}, respectively.

Table 5. Specified geometric errors [13,26] for geometric error identification.

Symbols	Group 1	Group 2	Group 3	Units
$\delta_{i,j}$	0.005	0.001	5×10^{-5}	m
$\varepsilon_{i,j}$	0.005	$\pi/180$	$\pi/7200$	rad

Taking group 1 as an example, the detailed processes are described as follows: (1) $\delta_{1,0}^x$ is set to 0.005 m, and the remaining geometric errors are set to 0. (2) 692 $\Delta\mu_{\max}$s are calculated according to Scheme I and (45). (3) If the number of $\Delta\mu_{\max}$s that are smaller than 10^{-15} is less than 657 ($\approx$95% of 692), then $\delta_{1,0}^x$ is referred to as an internal-force-and-deformation-related geometric error. After repeating the above steps for the 71 geometric errors, 39 internal-force-and-deformation-related geometric errors were initially identified and are listed in Table 6. In the table, "✓" denotes the internal-force-and-deformation-related geometric error; "−" denotes the error parameter that is not a geometric error.

Table 6. Initially identified internal-force-and-deformation-related geometric errors.

i	j	$\delta_{i,j}^x$	$\delta_{i,j}^y$	$\delta_{i,j}^z$	$\varepsilon_{i,j}^x$	$\varepsilon_{i,j}^y$	$\varepsilon_{i,j}^z$
1, 2	0		✓	✓	−	✓	✓
1, 2	1		✓		✓	−	
1, 2	2	−		✓	✓	✓	
1, 2	3	✓		−	−	✓	✓
1, 2	4		✓		✓	−	✓
3	0				−	✓	✓
3	1	−			✓	✓	
3	2			−	−	✓	✓
3	3				✓	−	−
3	4				✓	−	✓

Some geometric errors between any two adjacent coordinate systems in a limb may be linearly dependent. Therefore, it is necessary to analyse geometric errors simultaneously. Based on the results in Table 6, the set of the six error parameters, $\left[\delta_{i,j}^x, \delta_{i,j}^y, \delta_{i,j}^z, \varepsilon_{i,j}^x, \varepsilon_{i,j}^y, \varepsilon_{i,j}^z\right]$, are regarded as one unit. Take $\left[\delta_{1,1}^x, \delta_{1,1}^y, \delta_{1,1}^z, \varepsilon_{1,1}^x, \varepsilon_{1,1}^y, \varepsilon_{1,1}^z\right]$ as an example. $\left[\delta_{1,1}^x, \delta_{1,1}^y, \delta_{1,1}^z, \varepsilon_{1,1}^x, \varepsilon_{1,1}^y, \varepsilon_{1,1}^z\right]$ is set to [0.005 m, 0, 0.005 m, 0, 0, 0.005 rad], and the remaining units are set to [0,0,0,0,0,0]. Then, 692 $\Delta\mu_{\max}$s are calculated according to Scheme I and (45). If the number of $\Delta\mu_{\max}$s that are smaller than 10^{-15} is less than 657, the internal-force-and-deformation-related geometric errors are included in $\delta_{1,1}^x$, $\delta_{1,1}^z$, and $\varepsilon_{1,1}^z$. Then, $\delta_{1,1}^x$, $\delta_{1,1}^z$, and $\varepsilon_{1,1}^z$ are set to 0 in turn, and the remaining units are unchanged. The $\Delta\mu_{\max}$s are recalculated. If the number of $\Delta\mu_{\max}$s that decrease significantly is greater than 656, it is determined that the geometric error, which is set as 0, will cause internal forces and deformations. These steps were repeated for each error unit and the results are listed in Table 7. The identification results for groups 2 and 3 in Table 5 are the same as those shown in Table 7. The results demonstrate that there are 41 internal-force-and-deformation-related geometric errors, where the number of angular geometric errors is greater than that of linear geometric errors. In addition, the internal-force-and-deformation-related geometric errors of the first UPR limb are the same as those of the second UPR limb because of the symmetric distribution of the two limbs. For the RPU limb, the geometric errors that cause internal forces and deformations are angular geometric errors.

Table 7. Identified internal-force-and-deformation-related geometric errors.

i	j	$\delta_{i,j}^x$	$\delta_{i,j}^y$	$\delta_{i,j}^z$	$\varepsilon_{i,j}^x$	$\varepsilon_{i,j}^y$	$\varepsilon_{i,j}^z$
1, 2	0		✓	✓	–	✓	✓
1, 2	1		✓		✓	–	✓
1, 2	2	–		✓	✓	✓	
1, 2	3	✓		–	–	✓	✓
1, 2	4		✓		✓	–	✓
3	0				–	✓	✓
3	1	–			✓	✓	
3	2			–	–	✓	✓
3	3				✓	–	–
3	4				✓	–	✓

5.2. Simulation Analysis

To validate the correctness of the identified results listed in Table 7, three groups of numerical simulations were conducted using 1738 target poses of the 2UPR-RPU parallel manipulator, as shown in Figure 6. It is assumed that geometric errors are normally distributed with zero means [19,22]. Three groups of standard deviations are listed in Table 8. In the simulation, the internal-force-and-deformation-related geometric errors identified in Table 7 were set to 0, and the remaining 30 geometric errors were assigned random values generated by randn function using the standard deviations of $\delta_{i,j}$ and $\varepsilon_{i,j}$ listed in Table 8. Then, according to Scheme I and (45), 1738 $\Delta\mu_{\max}$s were calculated for each group. The simulation results are shown in Figure 7. It can be seen that $\Delta\mu_{\max}$s of Group 1, Group 2, and Group 3, are all smaller than 10^{-15}. This demonstrates that the internal-force-and-deformation-related geometric errors identified in Section 5.1 are correct.

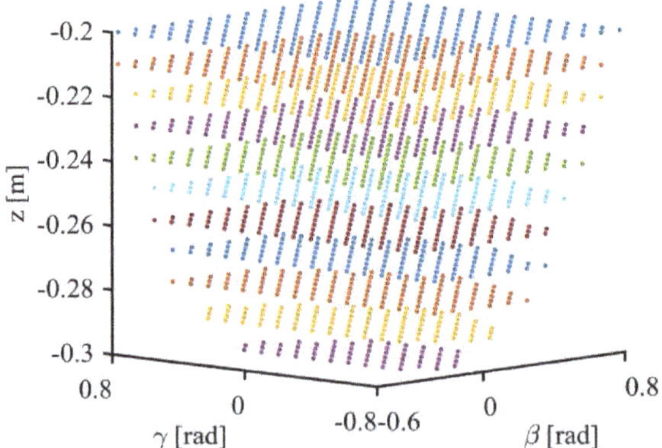

Figure 6. 1738 target poses of the 2UPR-RPU parallel manipulator.

Table 8. Standard deviations of the geometric errors for the numerical simulations.

Symbols	Group 1	Group 2	Group 3	Units
The standard deviations of $\delta_{i,j}$	1.6667×10^{-3}	3.3333×10^{-5}	1.6667×10^{-5}	m
The standard deviations of $\varepsilon_{i,j}$	1.6667×10^{-3}	$\pi/540$	$\pi/21,600$	rad

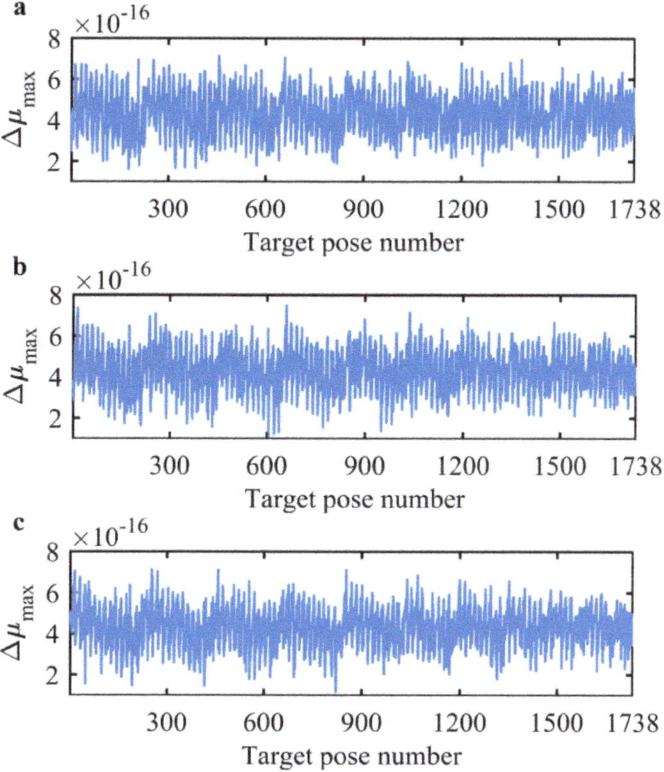

Figure 7. Simulation results using the standard deviations listed in Table 8. (**a**) Group 1; (**b**) Group 2; (**c**) Group 3.

6. Sensitivity Analysis

Sensitivity analysis can help reveal the influence of different internal-force-and-deformation-related geometric errors on the limbs' comprehensive deformations. Since $\Delta \mu^j$ is calculated iteratively in Scheme I, the Monte Carlo method [22] is utilised to conduct sensitivity analysis in this section. Two global sensitivity indices are proposed and the results of sensitivity analysis are verified through simulations.

6.1. Sensitivity Indices

According to (41), $\Delta \mu^j$ consists of $\Delta \mu^j_{2,1}$, $\Delta \mu^j_{3,2}$, and $\Delta \mu^j_{1,3}$, which can be written as

$$\Delta \mu^j_{k,i} = \begin{bmatrix} \omega^j_{k,i,1} & \omega^j_{k,i,2} & \omega^j_{k,i,3} & v^j_{k,i,1} & v^j_{k,i,2} & v^j_{k,i,3} \end{bmatrix}^{\mathrm{T}} \tag{46}$$

The end-orientation and end-position volumetric deviations between any two limbs are

$$\Delta \omega^j_{k,i} = \sqrt{\left(\omega^j_{k,i,1} \right)^2 + \left(\omega^j_{k,i,2} \right)^2 + \left(\omega^j_{k,i,3} \right)^2} \tag{47}$$

and

$$\Delta v^j_{k,i} = \sqrt{\left(v^j_{k,i,1} \right)^2 + \left(v^j_{k,i,2} \right)^2 + \left(v^j_{k,i,3} \right)^2} \tag{48}$$

Then, the evaluation indices of the average angular and linear comprehensive deformations of the three limbs can be written as

$$\Delta\omega_a^j = \frac{\Delta\omega_{2,1}^j + \Delta\omega_{3,2}^j + \Delta\omega_{1,3}^j}{3} \tag{49}$$

and

$$\Delta v_a^j = \frac{\Delta v_{2,1}^j + \Delta v_{3,2}^j + \Delta v_{1,3}^j}{3} \tag{50}$$

Under the condition that geometric errors are normally distributed with zero means, the sensitivity indices of the average angular and linear comprehensive deformations with respect to a geometric error can be written as

$$\mu_{\omega,p} = \frac{\sigma\left(\Delta\omega_{a,p}^j\right)}{\sigma\left(Ge_p\right)}, \quad p = 1, 2, \cdots, 25 \tag{51}$$

and

$$\mu_{v,p} = \frac{\sigma\left(\Delta v_{a,p}^j\right)}{\sigma\left(Ge_p\right)}, \quad p = 1, 2, \cdots, 25 \tag{52}$$

where $\sigma(\cdot)$ denotes the standard deviation and can be calculated by **std** function. Ge_ps are the internal-force-and-deformation-related geometric errors of the first and third limbs. Because of the symmetric distribution of the first and second limbs, the internal-force-and-deformation-related geometric errors of the second limb are not considered. Generally, the values of sensitivity indices vary with different target poses of the moving platform. Hence, m target poses should be chosen and the global sensitivity indices can be written as [16]

$$\mu_{\omega,p}^g = \frac{\sum\limits_{i=1}^{m} \mu_{\omega,p,i}}{m} + \sigma\left(\mu_{\omega,p}\right) \tag{53}$$

and

$$\mu_{v,p}^g = \frac{\sum\limits_{i=1}^{m} \mu_{v,p,i}}{m} + \sigma\left(\mu_{v,p}\right) \tag{54}$$

6.2. Sensitivity Analysis

Based on the equations in Section 6.1 and Scheme I, the detailed processes to calculate the two global sensitivity indices with respect to Ge_p are described as follows: (1) Set $\sigma\left(Ge_p\right)$ to 1 mm (0.001 m) or 1° ($\pi/180$ rad) for linear or angular geometric error. And the other 40 internal-force-and-deformation-related geometric errors are set to 0. In addition, the rest 30 linear or angular geometric errors are set to 1 mm or 1°. (2) Assign 1000 random values that obey the normal distribution to Ge_p and calculate 1000 $\Delta\omega_{a,p}^j$s and $\Delta v_{a,p}^j$s. (3) Calculate $\sigma\left(\Delta\omega_{a,p}^j\right)$, $\mu_{\omega,p}$, $\sigma\left(\Delta v_{a,p}^j\right)$, and $\mu_{v,p}$ for a target pose of the moving platform. (4) Repeat the above steps for m target poses and calculate the global sensitivity indices $\mu_{\omega,p}^g$ and $\mu_{v,p}^g$.

In order to improve the computational efficiency, 158 of the 1738 target poses shown in Figure 6 were selected uniformly to perform the above steps for each Ge_p. The global sensitivity indices of the average angular and linear comprehensive deformations with respect to Ge_ps are shown in Figures 8 and 9, respectively. It can be seen that the values of $\mu_{\omega,p}^g$ with respect to $Ge_5(\delta_{1,1}^y)$, $Ge_7(\varepsilon_{1,1}^z)$, $Ge_8(\delta_{1,2}^z)$, $Ge_{11}(\delta_{1,3}^x)$, and $Ge_{14}(\delta_{1,4}^y)$, are zero. This indicates that the corresponding geometric errors have no effects on the average angular comprehensive deformation. It is worth mentioning that $\delta_{2,1}^y$, $\varepsilon_{2,1}^z$, $\delta_{2,2}^z$, $\delta_{2,3}^x$, and $\delta_{2,4}^y$, have also no effects on the average angular comprehensive deformation due to the symmetric distribution of the first and second limbs. Comparing Figure 8 with Figure 9, it can also be found that the value of $\mu^g_{v,p}$ is larger than that of $\mu^g_{\omega,p}$ for each Ge_p. This demonstrates that the internal-force-and-deformation-related geometric errors have greater effects on the average linear comprehensive deformation. Thus, the distribution of the global sensitivity index $\mu^g_{v,p}$ is more useful for accuracy synthesis. According to Figure 9, Ge_ps can be sorted in descending order as follows: $Ge_{16}(\varepsilon^z_{1,4})$, $Ge_{25}(\varepsilon^z_{3,4})$, $Ge_{15}(\varepsilon^x_{1,4})$, $Ge_4(\varepsilon^z_{1,0})$, $Ge_{24}(\varepsilon^x_{3,4})$, $Ge_{12}(\varepsilon^y_{1,3})$, $Ge_6(\varepsilon^x_{1,1})$, $Ge_3(\varepsilon^y_{1,0})$, $Ge_{13}(\varepsilon^z_{1,3})$, $Ge_9(\varepsilon^x_{1,2})$, $Ge_{18}(\varepsilon^z_{3,0})$, $Ge_{23}(\varepsilon^x_{3,3})$, $Ge_{17}(\varepsilon^y_{3,0})$, $Ge_{19}(\varepsilon^x_{3,1})$, $Ge_{22}(\varepsilon^z_{3,2})$, $Ge_{10}(\varepsilon^y_{1,2})$, $Ge_{20}(\varepsilon^y_{3,1})$, $Ge_{21}(\varepsilon^y_{3,2})$, $Ge_8(\delta^z_{1,2})$, $Ge_{14}(\delta^y_{1,4})$, $Ge_5(\delta^y_{1,1})$, $Ge_{11}(\delta^x_{1,3})$, $Ge_1(\delta^y_{1,0})$, $Ge_2(\delta^z_{1,0})$, and $Ge_7(\varepsilon^z_{1,1})$. In order to lower the cost of fabrication and assembly, the allowable range of geometric errors should be larger and larger from Ge_{16} to Ge_7.

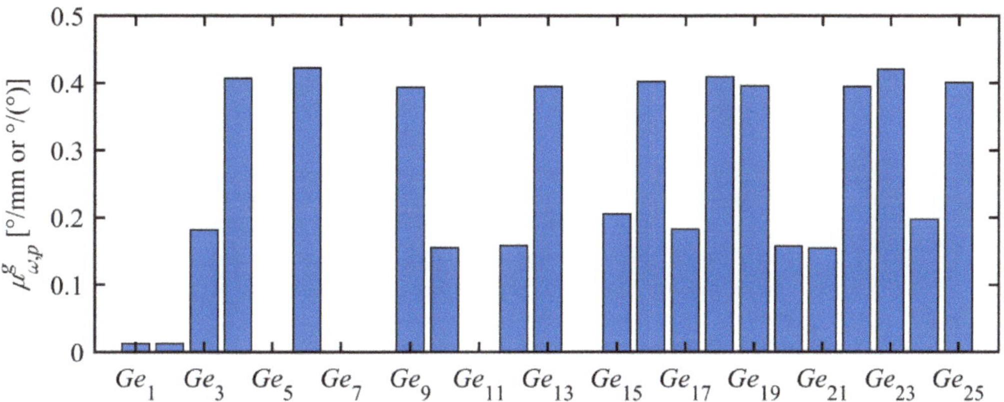

Figure 8. Global sensitivity of the average angular comprehensive deformation with respect to Ge_ps.

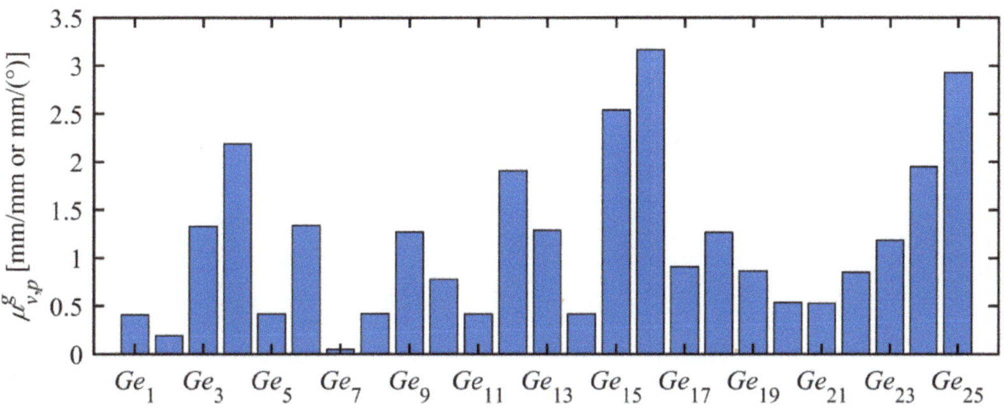

Figure 9. Global sensitivity of the average linear comprehensive deformation with respect to Ge_ps.

6.3. Verification

6.3.1. Average Angular Comprehensive Deformation

As shown in Table 9, three groups of specified values for geometric errors are given. For each group, 1738 $\Delta\omega^j_a$s were calculated according to Scheme I and using the target poses shown in Figure 6. The maximum and average values of $\Delta\omega^j_a$ are listed in Table 9. It can be seen that both the maximum and average values of $\Delta\omega^j_a$ do not change from Group 1 to Group 3. This indicates that Ge_5, Ge_7, Ge_8, Ge_{11}, Ge_{14}, $\delta^y_{2,1}$, $\varepsilon^z_{2,1}$, $\delta^z_{2,2}$, $\delta^x_{2,3}$, and $\delta^y_{2,4}$, have no effects on the average angular comprehensive deformation.

Table 9. Sensitivity analysis results of the average angular comprehensive deformation.

Group Number	Ge_5, Ge_7, Ge_8, Ge_{11}, Ge_{14}, $\delta_{2,1}^y$, $\varepsilon_{2,1}^z$, $\delta_{2,2}^z$, $\delta_{2,3}^x$, $\delta_{2,4}^y$ [mm or °]	Other Geometric Errors [mm or °]	The Maximum Value of $\Delta\omega_a^j$ [°]	The Average Value of $\Delta\omega_a^j$ [°]
Group 1	0.1	0.1	0.1430	0.0961
Group 2	0.01	0.1	0.1430	0.0961
Group 3	0.001	0.1	0.1430	0.0961

6.3.2. Average Linear Comprehensive Deformation

As shown in Table 10, $Ge_7(\varepsilon_{1,1}^z)$, which has the smallest effect on the average linear comprehensive deformation, is set to $1°$. Then, the other Ge_ps are set to $\mu_{v,7}^g / \mu_{v,p}^g$ mm or °. It is worth mentioning that the corresponding internal-force-and-deformation-related geometric errors of the second limb are assigned the same values as the first limb due to the symmetric distribution of the two UPR limbs. The remaining 30 geometric errors are set to 0.1 mm or °. According to Scheme I and using the target poses shown in Figure 6, $1738\Delta\omega_a^j$s and Δv_a^js were calculated. For comparison, the internal-force-and-deformation-related linear and angular geometric errors are set to their average values, 0.0209 mm and 0.0743 °, respectively, while the values of the remaining 30 geometric errors are unchanged. After recalculation, the maximum and average values of $\Delta\omega_a^j$ and Δv_a^j are listed in Table 11. It can be seen that both the maximum and average values of Δv_a^j are larger than that of $\Delta\omega_a^j$ for each group. This indicates that the internal-force-and-deformation-related geometric errors have greater effects on the average linear comprehensive deformation. It can also be found that from Group 2 to Group 1, the maximum and average values of $\Delta\omega_a^j$ and Δv_a^j decreased by 84%, 83%, 91%, and 89%, respectively. This demonstrates that at the same cost, restricting the values of geometric errors according to the sensitivity analysis results of the average linear comprehensive deformation can dramatically decrease the average angular and linear comprehensive deformations. Furthermore, it indirectly verifies the sensitivity analysis results of the average linear comprehensive deformation.

Table 10. Specified geometric errors for verification.

i	j	$\delta_{i,j}^x$ [mm]	$\delta_{i,j}^y$ [mm]	$\delta_{i,j}^z$ [mm]	$\varepsilon_{i,j}^x$ [°]	$\varepsilon_{i,j}^y$ [°]	$\varepsilon_{i,j}^z$ [°]
1, 2	0	0.1	0.0177	0.0381	–	0.0054	0.0033
1, 2	1	0.1	0.0174	0.1	0.0053	–	1
1, 2	2	–	0.1	0.0173	0.0056	0.0092	0.1
1, 2	3	0.0174	0.1	–	–	0.0037	0.0056
1, 2	4	0.1	0.0174	0.1	0.0028	–	0.0022
3	0	0.1	0.1	0.1	–	0.0079	0.0057
3	1	–	0.1	0.1	0.0083	0.0135	0.1
3	2	0.1	0.1	–	–	0.0137	0.0084
3	3	0.1	0.1	0.1	0.0061	–	–
3	4	0.1	0.1	0.1	0.0037	–	0.0024

Table 11. Sensitivity analysis results of the average linear comprehensive deformation.

Group Number	The Maximum Value of $\Delta\omega_a^j$ [°]	The Average Value of $\Delta\omega_a^j$ [°]	The Maximum Value of Δv_a^j [mm]	The Average Value of Δv_a^j [mm]
Group 1	0.0165	0.0118	0.0696	0.0390
Group 2	0.1061	0.0714	0.8374	0.3581

7. Conclusions

This paper deals with error modelling and sensitivity analysis of geometric errors that cause internal forces and deformations in the 2UPR-RPU over-constrained parallel manipulator. Conclusions are drawn as follows:

(1) The nominal inverse kinematics and actual forward kinematics of the over-constrained parallel manipulator are analysed according to the vector theory and the local product of the exponential formula. On this basis, an iterative model is established to indirectly evaluate the limbs' comprehensive deformations caused by geometric errors.

(2) Based on the iterative evaluation model, the maximum Euclidean norm of the end-pose deviations of limbs is defined as an evaluation index of the maximum comprehensive deformation of a limb. Programming with MATLAB, 41 internal-force-and-deformation-related geometric errors are identified. Among the 41 geometric errors, the number of angular geometric errors is greater than that of linear geometric errors; the geometric errors of the first UPR limb are the same as those of the second UPR limb; the geometric errors of the RPU limb are all angular geometric errors. The correctness of the identification results is verified through simulations under the condition that geometric errors are normally distributed with zero means.

(3) The global sensitivity indices of the average angular and linear comprehensive deformations with respect to internal-force-and-deformation-related geometric errors are proposed and calculated based on the Monte Carlo method. The results of sensitivity analysis demonstrate that $\delta_{1,1}^{y}$, $\varepsilon_{1,1}^{z}$, $\delta_{1,2}^{z}$, $\delta_{1,3}^{x}$, $\delta_{1,4}^{y}$, $\delta_{2,1}^{y}$, $\varepsilon_{2,1}^{z}$, $\delta_{2,2}^{z}$, $\delta_{2,3}^{x}$, and $\delta_{2,4}^{y}$, have no effects on the average angular comprehensive deformation. Furthermore, the internal-force-and-deformation-related geometric errors have greater effects on the average linear comprehensive deformation. Therefore, the distribution of the global sensitivity index of the average linear comprehensive deformation with respect to geometric errors is more meaningful for accuracy synthesis. Finally, the results of sensitivity analysis are verified through simulations.

Based on the work presented in this paper, we will establish a model for accuracy synthesis and determine the tolerances of the fabrication and assembly of the manipulator in the future.

Author Contributions: Conceptualization, X.D. and J.Z.; methodology, X.D.; software, B.W.; validation, X.D., B.W. and J.Z.; formal analysis, B.W.; investigation, X.D.; resources, X.D.; data curation, B.W.; writing—original draft preparation, X.D.; writing—review and editing, J.Z.; visualization, X.D.; supervision, J.Z.; project administration, J.Z.; funding acquisition, J.Z. All authors have read and agreed to the published version of the manuscript.

Funding: This research was supported by [National Natural Science Foundation of China] grant number [52275469] and [Open Fund of State Key Laboratory of Robotics and System] grant number [SKLRS-2021-KF-08].

Data Availability Statement: Not applicable.

Conflicts of Interest: The authors declare that there are no conflict of interest.

Appendix A. Lie Groups and Lie Algebras

Some equations about Lie groups and Lie algebras [25,30,31] are introduced here so that this work can be clearly understood. For a screw $\zeta = \begin{bmatrix} \omega^{T} & v^{T} \end{bmatrix}^{T}$, the $\wedge$ operation denotes

$$\hat{\zeta} = \begin{bmatrix} \hat{\omega} & v \\ 0_{1\times3} & 0 \end{bmatrix} \in se(3) \tag{A1}$$

The exponential map from the Lie algebra $se(3)$ to the special Euclidean group $SE(3)$ can be determined by

$$e^{\hat{\zeta}q} = \begin{bmatrix} \mathbf{I}_3 + \sin q\,\hat{\boldsymbol{\omega}} + (1 - \cos q)\hat{\boldsymbol{\omega}}^2 & \mathbf{V}\mathbf{v} \\ 0_{1\times 3} & 1 \end{bmatrix} \in SE(3) \tag{A2}$$

where $\mathbf{I}_3$ is an identity matrix of order three. $\hat{\boldsymbol{\omega}}$ and $\mathbf{V}$ are expressed as

$$\hat{\boldsymbol{\omega}} = \begin{bmatrix} 0 & -\omega_3 & \omega_2 \\ \omega_3 & 0 & -\omega_1 \\ -\omega_2 & \omega_1 & 0 \end{bmatrix} \in so(3) \tag{A3}$$

$$\mathbf{V} = q\mathbf{I}_3 + (1 - \cos q)\hat{\boldsymbol{\omega}} + (q - \sin q)\hat{\boldsymbol{\omega}}^2 \tag{A4}$$

The exponential map from the Lie algebra $so(3)$ to the special orthogonal group $SO(3)$ can be determined by

$$e^{\hat{\boldsymbol{\omega}}} = \mathbf{I}_3 + \frac{\sin \|\boldsymbol{\omega}\|}{\|\boldsymbol{\omega}\|}\hat{\boldsymbol{\omega}} + \frac{1 - \cos \|\boldsymbol{\omega}\|}{\|\boldsymbol{\omega}\|^2}\hat{\boldsymbol{\omega}}^2 \tag{A5}$$

For a HTM $\mathbf{g} \in SE(3)$, the Lie algebra $se(3)$ can be obtained as

$$\log(\mathbf{g}) = \frac{1}{8}\csc^3\frac{\theta}{2}\sec\frac{\theta}{2}\begin{bmatrix} \theta\cos 2\theta - \sin\theta \\ -\theta\cos\theta - 2\theta\cos 2\theta + \sin\theta + \sin 2\theta \\ 2\theta\cos\theta + \theta\cos 2\theta - \sin\theta - \sin 2\theta \\ -\theta\cos\theta + \sin\theta \end{bmatrix}^{\mathrm{T}}\begin{bmatrix} \mathbf{I}_4 \\ \mathbf{g} \\ \mathbf{g}^2 \\ \mathbf{g}^3 \end{bmatrix} \tag{A6}$$

where

$$\theta = \arccos\left(\frac{\mathrm{Tr}(\mathbf{g}) - 2}{2}\right), \theta \in (-\pi, \pi) \tag{A7}$$

The adjoint representation of $\mathbf{g}$ can be written as

$$\mathrm{Ad}(\mathbf{g}) = \mathrm{Ad}\left(\begin{bmatrix} \mathbf{R} & \mathbf{t} \\ 0_{1\times 3} & 1 \end{bmatrix}\right) = \begin{bmatrix} \mathbf{R} & 0_{3\times 3} \\ \hat{\mathbf{t}}\mathbf{R} & \mathbf{R} \end{bmatrix} \in \mathbb{R}^{6\times 6} \tag{A8}$$

References

1. Tang, T.F.; Fang, H.L.; Zhang, J. Hierarchical design, laboratory prototype fabrication and machining tests of a novel 5-axis hybrid serial-parallel kinematic machine tool. *Robot. Comput.-Integr. Manuf.* **2020**, *64*, 101944. [CrossRef]
2. Xie, F.G.; Liu, X.J.; Li, T.M. Type synthesis and typical application of 1T2R-type parallel robotic mechanisms. *Math. Probl. Eng.* **2013**, *2013*, 206181. [CrossRef]
3. Li, Q.C.; Hervé, J.M. Type synthesis of 3-DOF RPR-equivalent parallel mechanisms. *IEEE Trans. Robot.* **2014**, *30*, 1333–1343. [CrossRef]
4. Li, Q.; Xu, L.; Chen, Q.; Ye, W. New family of RPR-equivalent parallel mechanisms: Design and application. *Chin. J. Mech. Eng.* **2017**, *30*, 217–221. [CrossRef]
5. Huang, T.; Dong, C.; Liu, H.; Sun, T.; Chetwynd, D.G. A simple and visually orientated approach for type synthesis of overconstrained 1T2R parallel mechanisms. *Robotica* **2019**, *37*, 1161–1173. [CrossRef]
6. Tang, T.; Fang, H.; Luo, H.; Song, Y.; Zhang, J. Type synthesis, unified kinematic analysis and prototype validation of a family of Exechon inspired parallel mechanisms for 5-axis hybrid kinematic machine tools. *Robot. Comput.-Integr. Manuf.* **2021**, *72*, 102181. [CrossRef]
7. Zhao, Y.; Xu, Y.; Yao, J.; Jin, L. A force analysis method for overconstrained parallel mechanisms. *China Mech. Eng.* **2014**, *25*, 711–717. [CrossRef]
8. Chai, X.X.; Xiang, J.N.; Li, Q.C. Singularity analysis of a 2-UPR-RPU parallel mechanism. *J. Mech. Eng. Chin. Ed.* **2015**, *51*, 144–151. [CrossRef]
9. Jiang, Y.; Li, T.; Wang, L.; Chen, F. Kinematic error modeling and identification of the over-constrained parallel kinematic machine. *Robot. Comput.-Integr. Manuf.* **2018**, *49*, 105–119. [CrossRef]
10. He, L.; Li, Q.; Zhu, X.; Wu, C.-Y. Kinematic calibration of a three degrees-of-freedom parallel manipulator with a laser tracker. *J. Dyn. Syst. Meas. Control* **2019**, *141*, 031009. [CrossRef]

11. Song, Y.M.; Zhai, Y.P.; Sun, T. Interval analysis based accuracy design of a 3-DOF rotational parallel mechanism. *J. Beijing Univ. Technol.* **2015**, *41*, 1620–1626,1755.
12. Ni, Y.; Zhang, B.; Sun, Y.; Zhang, Y. Accuracy analysis and design of A3 parallel spindle head. *Chin. J. Mech. Eng.* **2016**, *29*, 239–249. [CrossRef]
13. Zhang, J.; Chi, C.C.; Jiang, S.J. Accuracy design of 2UPR-RPS parallel mechanism. *Trans. Chin. Soc. Agric. Eng.* **2021**, *52*, 411–420. [CrossRef]
14. Huang, T.; Whitehouse, D.J.; Chetwynd, D.G. A unified error model for tolerance design, assembly and error compensation of 3-DOF parallel kinematic machines with parallelogram struts. *CIRP Ann.* **2002**, *51*, 297–301. [CrossRef]
15. Zhang, J.T.; Lian, B.B.; Song, Y.M. Geometric error analysis of an over-constrained parallel tracking mechanism using the screw theory. *Chin. J. Aeronaut.* **2019**, *32*, 1541–1554. [CrossRef]
16. Luo, X.; Xie, F.; Liu, X.-J.; Li, J. Error modeling and sensitivity analysis of a novel 5-degree-of-freedom parallel kinematic machine tool. *Proc. Inst. Mech. Eng. Part B J. Eng. Manuf.* **2019**, *233*, 1637–1652. [CrossRef]
17. Wang, W.; Yun, C. Orthogonal experimental design to synthesize the accuracy of robotic mechanism. *J. Mech. Eng. Chin. Ed.* **2009**, *45*, 18–24. [CrossRef]
18. Yao, R.; Zhu, W.B.; Huang, P. Accuracy analysis of stewart platform based on interval analysis method. *Chin. J. Mech. Eng.* **2013**, *26*, 29–34. [CrossRef]
19. Wu, M.; Yue, X.; Chen, W.; Nie, Q.; Zhang, Y. Accuracy analysis and synthesis of asymmetric parallel mechanism based on Sobol-QMC. *Proc. Inst. Mech. Eng. Part C J. Mech. Eng. Sci.* **2020**, *234*, 4200–4214. [CrossRef]
20. Li, X.Y.; Chen, W.Y.; Han, X.G. Accuracy analysis and synthesis of 3-RPS parallel machine based on orthogonal design. *J. Beijing Univ. Aeronaut. Astronaut.* **2011**, *37*, 979–984. [CrossRef]
21. Tai-Ke, Y.; Xi, Z.; Feng, Z.; Li-Min, Z.; Yong, W. Accuracy synthesis of a 3-RPS parallel robot based on manufacturing costs. In Proceedings of the 31st Chinese Control Conference, Hefei, China, 25–27 July 2012; pp. 5168–5172.
22. Liu, H.T.; Pan, Q.; Yin, F.W.; Dong, C.L. Accuracy synthesis of the TriMule hybrid robot. *J. Tianjin Univ. Sci. Technol.* **2019**, *52*, 1245–1254.
23. Tang, T.; Luo, H.; Song, Y.; Fang, H.; Zhang, J. Chebyshev inclusion function based interval kinetostatic modeling and parameter sensitivity analysis for Exechon-like parallel kinematic machines with parameter uncertainties. *Mech. Mach. Theory* **2021**, *157*, 104209. [CrossRef]
24. Yu, D.; Zhao, Q.; Guo, J.; Chen, F.; Hong, J. Accuracy analysis of spatial overconstrained extendible support structures considering geometric errors, joint clearances and link flexibility. *Aerosp. Sci. Technol.* **2021**, *119*, 107098. [CrossRef]
25. Chen, G.; Kong, L.; Li, Q.; Wang, H.; Lin, Z. Complete, minimal and continuous error models for the kinematic calibration of parallel manipulators based on POE formula. *Mech. Mach. Theory* **2018**, *121*, 844–856. [CrossRef]
26. Kong, L.; Chen, G.; Wang, H.; Huang, G.; Zhang, D. Kinematic calibration of a 3-PRRU parallel manipulator based on the complete, minimal and continuous error model. *Robot. Comput.-Integr. Manuf.* **2021**, *71*, 102158. [CrossRef]
27. Sun, T.; Song, Y.; Li, Y.; Xu, L. Separation of comprehensive geometrical errors of a 3-DOF parallel manipulator based on Jacobian matrix and its sensitivity analysis with Monte-Carlo method. *Chin. J. Mech. Eng.* **2011**, *24*, 406–413. [CrossRef]
28. Chen, Y.; Xie, F.; Liu, X.; Zhou, Y. Error modeling and sensitivity analysis of a parallel robot with SCARA (selective compliance assembly robot arm) motions. *Chin. J. Mech. Eng.* **2014**, *27*, 693–702. [CrossRef]
29. Ma, J.G. Kinematics Analysis and Parameter Optimization of 2UPR-RPU Parallel Mechanism. Master's Thesis, Zhejiang Sci-Tech University, Hangzhou, China, 2014. [CrossRef]
30. Selig, J.M. *Geometric Fundamentals of Robotics*; Springer: New York, NY, USA, 2005. [CrossRef]
31. Dai, J.S. *Screw Algebra and Lie Groups and Lie Algebras*; Higher Education Press: Beijing, China, 2014.

Article

Consistent Solution Strategy for Static Equilibrium Workspace and Trajectory Planning of Under-Constrained Cable-Driven Parallel and Planar Hybrid Robots

Qingjuan Duan *, Quanli Zhao and Tianle Wang

School of Mechano-Electronic Engineering, Xidian University, Xi'an 710071, China
* Correspondence: qjduan@xidian.edu.cn

Abstract: This paper presents a consistent solution strategy for static equilibrium workspaces of different types of under-constrained robots. Considering the constraint conditions of cable force and taking the least squares error of the static equilibrium equation as the objective, the convex optimization solution is carried out, and the static equilibrium working space of the under-constrained system is obtained. A consistent solution strategy is applied to solve the static equilibrium workspaces of the cable-driven parallel and planar hybrid robots. The dynamic models are presented and introducing parameters that are applied to make the system stable for point-to-point movements. Based on this model, the traditional polynomial-based point-to-point trajectory planning algorithm is improved by adding unconstrained parameters to the kinematic law function. The constraints of the dynamics model are incorporated into the trajectory planning process to achieve point-to-point trajectory planning for the under-constrained cable-driven robots. Finally, under-constrained cable-driven parallel robots with three cables and planar hybrid robot with two cables are taken as examples to carry out numerical simulation. The final results show that the point-to-point trajectory planning algorithm introducing parameters is effective and feasible and can provide theoretical guidance for the design of subsequent under-constrained robots.

Keywords: static equilibrium workspace; under-constrained cable-driven robot; consistent solution strategy; trajectory planning

Citation: Duan, Q.; Zhao, Q.; Wang, T. Consistent Solution Strategy for Static Equilibrium Workspace and Trajectory Planning of Under-Constrained Cable-Driven Parallel and Planar Hybrid Robots. *Machines* **2022**, *10*, 920. https://doi.org/10.3390/machines10100920

Academic Editors: Dan Zhang, Zhufeng Shao and Stéphane Caro

Received: 18 July 2022
Accepted: 29 September 2022
Published: 10 October 2022

1. Introduction

The cable-driven robot is a mechanism that employs cables in place of rigid-body to control the end-effector pose.

The classification of cable-driven parallel robots (CDPRs) was introduced by Ming and Higuchi [1]. A cable-driven robot with n degrees-of-freedom and m cables can be classified into the following four categories: (1) $n + 1 < m$: These robots are referred to as redundantly restrained positioning mechanisms (RRPM), the static forces of the robot are generally undefined. (2) $n + 1 = m$: These robots are called completely restrained positioning mechanisms (CRPM). All degrees of freedom can be controlled through cables. (3) $m = n$: This type of robot is called incompletely restrained positioning mechanism (IRPM). When external forces such as gravity applied, the robot is fully constrained. It can withstand a limited range of wrenches. (4) $m < n$: The robot is under-constrained positioning mechanism (URPM) and in general cannot withstand arbitrary external forces and torques. Due to the under-constrained nature, these robots have one feasible solution for cable tensions and mostly works under gravity conditions. This classification method is also applicable to cable-driven planar hybrid robots.

According to the structure of cable-driven robot, it is generally divided into series mechanism and parallel mechanism.

Cable-driven robot has the characteristics of simple structure, flexibility, large workspace, low inertia, high load rate, etc. It has a wide range of applications, such as: Five-hundred-meter

Aperture Spherical radio Telescope (FAST), wind tunnel test, rehabilitation training, sports photography, etc. [2–5], it has become a hot spot in robotics research recent years.

The under-constrained cable-driven robots (UCR), with few drivers and low cost, has its special purpose, attracting more and more scholars' research interest. Carricato et al. [6–8] from Italy studied the cable-driven parallel robots with less than six cables, provided a geometrico-static model, and assessed the stability of static equilibrium within the framework of a constrained optimization problem. Several examples are provided, concerning robots with a number of cables that range from 2 to 4. Berti et al. [9] proposed a method based on interval analysis to solve the positive geometric statics problem of an under-constrained cable-driven parallel robot, and find all possible equilibrium poses of the end effector under a given cable length. Liu Xin et al. [10] proposed a consistent algorithm for solving the workspace of a cable-driven parallel robot under different constraints. Fu Ying et al. [11] conducted a dynamic analysis on the cable-driven system with four cables and six degrees of freedom. Zhao Zhigang et al. [12] proposed a comprehensive algorithm by combining the least squares method and the Monte Carlo algorithm to solve the statically balanced workspace for the cable-driven system with multi-robots. Peng Y et al. [13] analyzed the reachable workspace for spatial 3-cable under-constrained suspended cable driven parallel robots. The above-mentioned literature put forwards the solution method of static equilibrium workspace for under-constrained parallel robots. Based on this, a consistent solution strategy for the static equilibrium workspace of both cable-driven parallel & planar hybrid robots is put forward in this paper.

Ida E et al. [14] proposed a rest-to-rest trajectory planning for underactuated cable-driven parallel robots. Barbazza L et al. [15] design and optimally control an underactuated cable-driven micro–macro robot. Shang Weiwei et al. [16] proposed a geometrical approach to plan trajectories that extend beyond the static equilibrium workspace (SEW) of the mechanism. Zi B et al. [17] studied an algebraic method-based point to point trajectory planning of an under constrained cable suspended parallel robot with variable angle and height cable mast. Shi P et al. [18] studied Dimensional synthesis of a gait rehabilitation cable-suspended robot on minimum 2-norm tensions. The above-mentioned literature put forwards the planning trajectories for under-constrained parallel robots. Based on this, a consistent solution strategies of point-to-point trajectory planning introducing parameters for both cable-driven parallel & planar hybrid robots is put forward in this paper.

Firstly, the static equilibrium equations of the under constrained parallel and planar hybrid robots are established. Then, the mathematical description of the static equilibrium workspace of the under-constrained cable-driven robots (UCR) that meets the constraints of the driving motor power and cable strength is given. The characteristics of the static equilibrium equation and dynamics equation are analyzed, and the consistent solution strategy for the static equilibrium workspace and point-to-point trajectory planning introducing parameters of different types of under-constrained robots are given.

The paper is organized as follows: Section 2 presents the model for under-constrained parallel robots. Section 3 provides the model of under-constraint cable-driven planar hybrid robot. Section 4 describes the consistent solution strategy for static equilibrium workspace. Section 5 illustrates some results of static equilibrium workspace. Section 6 puts forward the consistent solution strategies for point-to-point trajectory planning. Section 7 illustrates the simulation results of trajectory planning. Finally, Section 8 draws conclusions.

2. Model of Under-Constrained Cable-Driven Parallel Robots

The schematic structure of a cable-driven robot with n degrees-of-freedom and m cables is shown in Figure 1. Here $m < n$, it's an under-constrained CDPR. The moving platform is connected to the base through m cables, the *ith* cable ($i = 1,2, \ldots ,m$) exits from the fixed base at point A_i, connected to the moving platform at point B_i. the cable length is L_i. $Oxyz$ is a Cartesian coordinate which fixed to the base, and $O'x'y'z'$ is the Cartesian coordinate fixed to the moving platform.

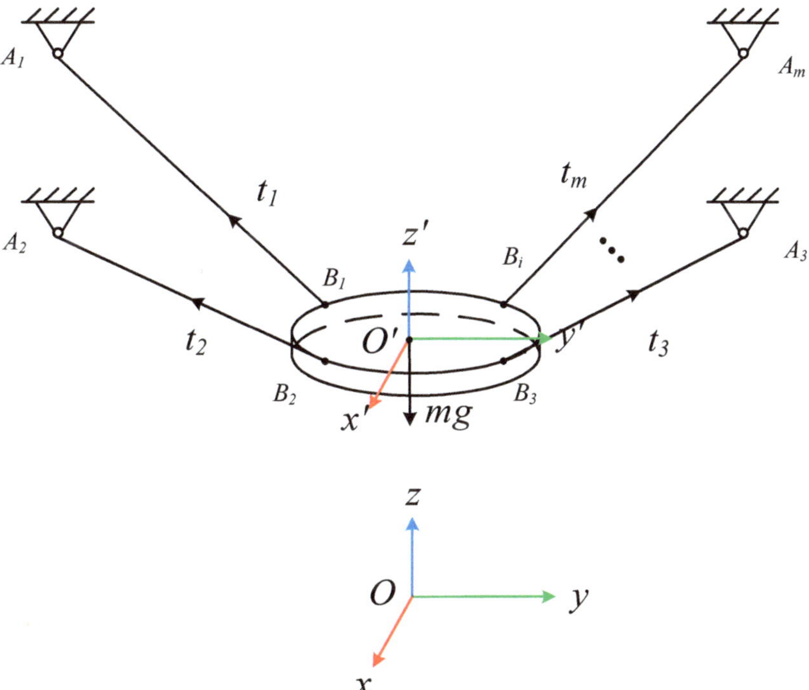

Figure 1. The schematic structure of a cable-driven parallel robot (If $m < n$, it is an under-constrained CDPR.).

$^{O}P_{O'} = [^{O}x_{O'} \ ^{O}y_{O'} \ ^{O}z_{O'}]^{\mathrm{T}}$ is the centroid of end effector in the $Oxyz$ frame. $^{O'}B_i = [^{O'}x_{Bi} \ ^{O'}y_{Bi} \ ^{O'}z_{Bi}]^{\mathrm{T}} (i = 1, 2, \cdots, m)$ is the vector connecting point O' to the point B_i in the $O'x'y'z'$ frame. $^{O}A_i = [^{O}x_{Ai} \ ^{O}y_{Ai} \ ^{O}z_{Ai}]^{\mathrm{T}} (i = 1, 2, \cdots, m)$ is the fixed base at point A_i, in the $Oxyz$ frame.

$^{O}R_{O'}$ represents the rotational matrix from frame $O'x'y'z'$ to frame $Oxyz$, in which α, β, γ are x-y-z the Euler angles.

$$^{O}R_{O'} = \mathrm{rot}(x, \alpha)\mathrm{rot}(y, \beta)\mathrm{rot}(z, \gamma) =$$
$$\begin{bmatrix} c\beta c\gamma & -c\beta s\gamma & s\beta \\ c\alpha s\gamma + s\alpha s\beta c\gamma & c\alpha c\gamma - s\alpha s\beta s\gamma & -c\beta s\alpha \\ s\alpha s\gamma - c\alpha s\beta c\gamma & s\alpha c\gamma + c\alpha s\beta s\gamma & c\alpha c\beta \end{bmatrix} \tag{1}$$

where c represents cos, s represents sin.

$^{O}B_i$ is the vector in the $Oxyz$ frame.

$$^{O}B_i = {}^{O}R_{O'} {}^{O'}B_i + {}^{O}P_{O'} \tag{2}$$

$^{O}L_i$ is the vector connecting point B_i to point A_i in the $Oxyz$ frame. e_i is the unit vector of $^{O}L_i$.

$$e_i = \frac{^{O}L_i}{\|^{O}L_i\|_2} = \frac{^{O}A_i - {}^{O}B_i}{\|^{O}A_i - {}^{O}B_i\|_2} \tag{3}$$

Thus, the dynamics equations for end effector are shown as follows:

$$JT + G = \begin{bmatrix} m\ddot{x} \\ I\dot{\omega} + \omega \times I\omega \end{bmatrix} \tag{4}$$

where $T = [t_1 \; t_2 \; \cdots \; t_m]^{\mathrm{T}}$, $t_i (i = 1, 2, \cdots, m)$ is cable tensions act on the end effector. $J = [\hat{f}_1 \; \hat{f}_2 \; \cdots \; \hat{f}_m]$ is the construction matrix, $\hat{f}_i = [e_i \quad {}^O B_i \times e_i]^{\mathrm{T}}$, and $G = [0 \; 0 - mg \; 0 \; 0]^{\mathrm{T}}$. $\ddot{x} = [{}^O \ddot{x}_{O\prime} \; {}^O \ddot{y}_{O\prime} \; {}^O \ddot{z}_{O\prime}]^{\mathrm{T}}$ is the acceleration of the end-effector centroid $O\prime$ in the world coordinate system, and $I = {}^O R_{O\prime} I_{O\prime} {}^O R_{O\prime}$, $I_{O\prime}$ is the inertia tensor of the end-effector in the local coordinate system.

ω and $\dot{\omega}$ are the angular velocity and angular acceleration of the end-effector. ω, $\dot{\omega}$ and ε, the Euler angle, satisfy the following relationship:

$$\omega = H(\varepsilon)\dot{\varepsilon} = \begin{bmatrix} 1 & 0 & s\beta \\ 0 & c\alpha & -s\alpha c\beta \\ 0 & s\alpha & c\alpha c\beta \end{bmatrix} \begin{bmatrix} \dot{\alpha} \\ \dot{\beta} \\ \dot{\gamma} \end{bmatrix} \tag{5}$$

$$\dot{\omega} = \dot{H}\dot{\varepsilon} + H\ddot{\varepsilon} \tag{6}$$

Substituting Equations (5) and (6) into Equation (4) yields a general expression for the system dynamics equation [19], i.e.,

$$M(q)\ddot{q} - s(q, \dot{q}) - J(q)T = 0$$
$$M(q) = \begin{bmatrix} mE_3 & 0 \\ 0 & IH \end{bmatrix}, s(q, \dot{q}) = G - \begin{bmatrix} 0 \\ IH\dot{\varepsilon} + \omega \times I\omega \end{bmatrix} \tag{7}$$

In Equation (7), $\ddot{q} = [\ddot{x} \; \ddot{\varepsilon}]^{\mathrm{T}}$ is the end-effector acceleration. M is the mass matrix of the end-effector. s is the force vector for a collection of Coriolis-type forces, gravitational forces, and external loads, etc. $J(q)T$ is the cable-tension to which the end-effector is subjected.

Additionally, the static equilibrium equation of the system can be expressed by

$$JT + G = 0 \tag{8}$$

3. Model of Under-Constrained Cable-Driven Planar Hybrid Robot

The schematic structure of a cable-driven planar hybrid robot with planar n -link serial robot with n degrees-of-freedom and m cables is shown in Figure 2. All links are connected by revolute joints to form a planar multi-link mechanism. This definition can also be generalized to space model. Here, $m < n$, it's an under-constrained cable-driven planar hybrid robot. $\{0\}$ is the base frame, $\{i\}$ is the link frame $x_i y_i$, $i \in \{1, 2, \cdots, n\}$. $\theta_i (i = 1, 2, \cdots, n)$ is the angle between link frame $\{i - 1\}$ and $\{i\}$. $[\theta_1 \; \theta_2 \; \cdots \; \theta_n]^{\mathrm{T}}$ is the Joint coordinates of the system.

${}^0 C_i$ is the position vector of the ith link centroid in the fixed frame $\{0\}$. It can be expressed by

$$ {}^0 C_i = [{}^0 x_{C_i} \; {}^0 y_{C_i}]^{\mathrm{T}} = ({}^0 R_1 {}^1 R_2 \cdots {}^{i-1} R_i) {}^i C_i \tag{9}$$

where ${}^0 R_1 = \begin{bmatrix} c\theta_1 & -s\theta_1 & 0 \\ s\theta_1 & c\theta_1 & 0 \\ 0 & 0 & 1 \end{bmatrix}$, ${}^i R_{i+1} = \begin{bmatrix} c\theta_{i+1} & -s\theta_{i+1} & 0 \\ s\theta_{i+1} & c\theta_{i+1} & l_i \\ 0 & 0 & 1 \end{bmatrix}$, ${}^i R_{i+1}$ represents the rotational matrix from frame $i + 1$ to frame i, $i = 1, 2, \cdots, n - 1$.

${}^0 B_j^i$ is the jth cable position vector of the ith link in the fixed frame. It can be expressed by

$$ {}^0 B_j^i = [{}^0 x_{B_j^i} \; {}^0 y_{B_j^i}]^{\mathrm{T}} = ({}^0 R_1 {}^1 R_2 \cdots {}^{i-1} R_i) {}^i B_j^i \tag{10}$$

where ${}^i B_j^i = [{}^i x_{B_j^i} \; {}^i y_{B_j^i}]^{\mathrm{T}}$ is the cable position vector of the ith link in the link frame $\{i\}$.

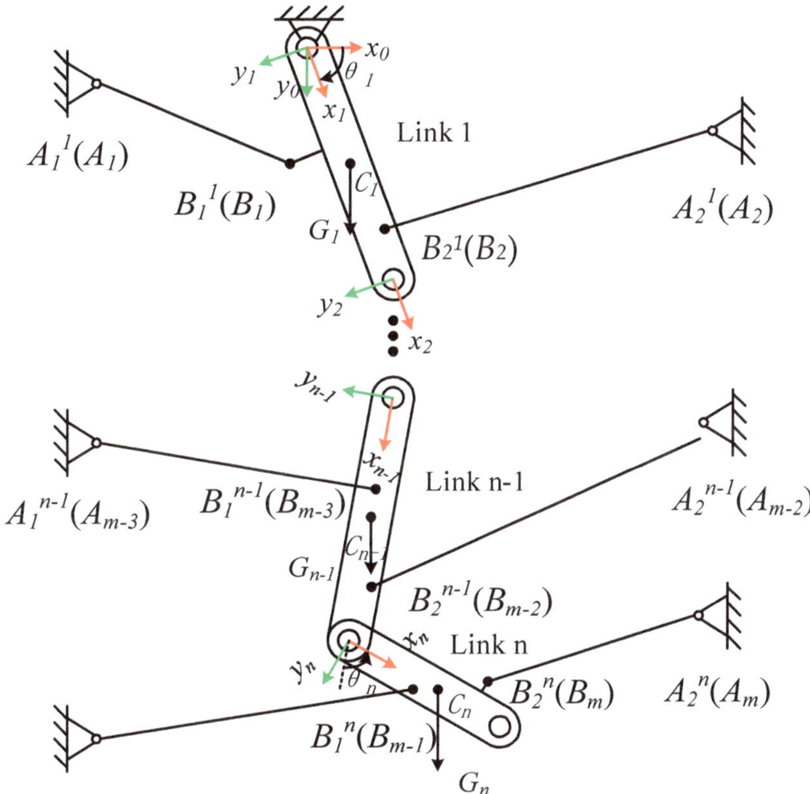

Figure 2. The schematic structure of a cable−driven planar hybrid robot with planar serial $n-$link with n degrees-of-freedom and m cables (If $m < n$, it's an under-constrained cable-driven planar hybrid robot.).

The generalized force is solved by Lagrange equation,

$$\tau = \frac{d}{d_t}\left(\frac{\partial L}{\partial \dot{\theta}}\right) - \frac{\partial L}{\partial \theta}$$

$$L = E_k - E_p \tag{11}$$

$$\tau = \frac{\partial\left(\frac{\partial L}{\partial \dot{\theta}}\right)}{\partial \dot{\theta}}\ddot{\theta} + \frac{\partial\left(\frac{\partial L}{\partial \dot{\theta}}\right)}{\partial \theta}\dot{\theta} - \frac{\partial L}{\partial \theta} \tag{12}$$

where τ is the generalized force, L is the Lagrange function. $\theta = \begin{bmatrix} \theta_1 & \theta_2 & \cdots & \theta_n \end{bmatrix}^{\mathrm{T}}$ is the generalized coordinate. E_k and E_p are the kinetic energy and potential energy of the system, respectively.

Write Equation (12) as a generic expression,

$$M(\theta)\ddot{\theta} - s(\theta, \dot{\theta}) = \tau$$

$$M(\theta) = \frac{\partial\left(\frac{\partial L}{\partial \dot{\theta}}\right)}{\partial \dot{\theta}}, s(\theta, \dot{\theta}) = -\frac{\partial\left(\frac{\partial L}{\partial \dot{\theta}}\right)}{\partial \theta}\dot{\theta} + \frac{\partial L}{\partial \theta} \tag{13}$$

where $M(\theta)$ is the inertia matrix, and $s(\theta, \dot{\theta})$ is the gravity, centrifugal force, Coriolis force, etc.

The expression (14) can be obtained according to the principle of virtual work

$$\tau^{\mathrm{T}} \delta\theta = \sum_{i=1}^{m} T_i e_i^{\mathrm{T}} \delta(^0 B_i) = \sum_{i=1}^{m} T_i e_i^{\mathrm{T}} \left(\frac{\partial^0 B_i}{\partial\theta} \delta\theta \right) \tag{14}$$

where m represents the number of cables, T is the cable tension, e is the unit vector of the cable tension, and $^0 B_i$ is the position of the *ith* pulling point in the coordinate system.

From Equation (14), the relationship between generalized force and cable tension can be obtained

$$\tau = J_v^{\mathrm{T}} T \tag{15}$$

where

$$J_v^{\mathrm{T}} = \begin{bmatrix} e_1^{\mathrm{T}} \frac{\partial^0 B_1}{\partial\theta_1} & e_1^{\mathrm{T}} \frac{\partial^0 B_1}{\partial\theta_2} & \cdots & e_1^{\mathrm{T}} \frac{\partial^0 B_1}{\partial\theta_n} \\ e_2^{\mathrm{T}} \frac{\partial^0 B_2}{\partial\theta_1} & e_2^{\mathrm{T}} \frac{\partial^0 B_2}{\partial\theta_2} & \cdots & e_2^{\mathrm{T}} \frac{\partial^0 B_2}{\partial\theta_n} \\ \vdots & \vdots & \ddots & \vdots \\ e_m^{\mathrm{T}} \frac{\partial^0 B_m}{\partial\theta_1} & e_m^{\mathrm{T}} \frac{\partial^0 B_m}{\partial\theta_2} & \cdots & e_m^{\mathrm{T}} \frac{\partial^0 B_m}{\partial\theta_n} \end{bmatrix}^{\mathrm{T}}$$

where J_v is the pseudo-Jacobian of the constraint equations, J_v^{T} is the $n \times m$ matrix, T is the $m \times 1$ cable tension matrix.

Combining Equations (13) and (15), the system dynamics equation is obtained as follows.

$$M(\theta)\ddot{\theta} - s(\theta, \dot{\theta}) - J_v^{\mathrm{T}}(\theta)T = 0 \tag{16}$$

Since the system is in static equilibrium, and the generalized velocity $\dot{\theta} = 0$, Formula (11) can be simplified as

$$\tau = \frac{\partial E_p}{\partial\theta} \tag{17}$$

where $E_p = \sum_{i=1}^{n} (-n_i g)^0 y_{C_i}$. Taking the origin of the coordinate system as the reference point of potential energy.

Combine Equations (15) and (17), yields

$$J_v^{\mathrm{T}} T = \frac{\partial E_p}{\partial\theta} \tag{18}$$

4. Consistent Solution Strategy for Static Equilibrium Workspace

$\begin{bmatrix} ^O x_O, & ^O y_O, & ^O z_O, & \alpha & \beta & \gamma \end{bmatrix}^{\mathrm{T}}$ is the pose of the parallel robot, $\begin{bmatrix} \theta_1 & \theta_2 & \cdots & \theta_n \end{bmatrix}^{\mathrm{T}}$ is the Joint coordinates of the planar hybrid robot. They are uniformly written into generalized coordinates $X = \begin{bmatrix} x_1 & x_2 & \cdots & x_n \end{bmatrix}^{\mathrm{T}}$.

Combine Equations (8) and (18), yields

$$J(X)T = W(X) \tag{19}$$

where W is the external force.

4.1. The Definition of Static Equilibrium Workspace

Due to the limitation of motor torque and cable strength, the tension of the cable must be within a certain range. Therefore, for a robot with n degrees of freedom pulled by m cables, the mathematical description of its static equilibrium workspace is as follows:

$$X = \begin{bmatrix} x_1 & x_2 & \cdots & x_n \end{bmatrix}^{\mathrm{T}}$$

$$\exists t_{\min} \leq t_i \leq t_{\max}(t_{\min} > 0 \cap t_{\max} > 0 \cap t_{\max} > t_{\min}, i = 1, 2, \cdots, m) \tag{20}$$

$$J(X)T = W(X)$$

where X represents the generalized coordinate. $t_{\min}, t_{\max}$ are the minimum and the maximum allowable tension of the cable, respectively. $T = \begin{bmatrix} t_1 & t_2 & \cdots & t_m \end{bmatrix}^{\mathrm{T}}$ is the cable tension, and X belongs to the Static equilibrium workspace.

4.2. Consistent Solution Strategy

For a cable-driven robot with n degrees-of-freedom and m cables, here $m < n$, it's an Under-constrained Cable-driven Robot (UCR). The Static equilibrium workspace of the under-constrained cable-driven system is analyzed as follows:

In static equilibrium Equation (19), J is $n \times m$ matrix, T is $m \times 1$ matrix. For the under-constrained cable-driven system, the solution of the equation is in the form of the least squares solution, and it can be expressed by $T = J^\dagger W$. When this solution is within the limit of cable force, it is considered that the pose (generalized coordinate) is the static equilibrium point satisfying Equation (20).

Then, the following inequality holds

$$\|J(X)T - W(X)\|_2 < \sigma* \\ T = J^\dagger(X)W(X), t_{\min} \leq t_i \leq t_{\max} \tag{21}$$

where $\sigma*$ is the error of the least square solution.

To sum up, the consistent solution strategy for the Static equilibrium workspace of the UCRs is as follows:

1. For the under-constrained robot, given the lower limit and the upper limit of cable tension $t_{\min}$ and $t_{\max}$ separately, selected the controllable degrees of freedom and uncontrollable degrees of freedom $X_a = [x_1, \ldots, x_k]$ and $X_b = [x_1, \ldots, x_{n-k}]$ separately.
2. Judge the search range k of the system dimension, set the search step $\delta_1, \ldots, \delta_k$, and generate the pose set $\mathbf{Q}$ to be searched.
3. Take a pose $X_a{}^i$ to be searched from the set and bring it into Formula (19) to solve the Jacobian matrix J.
4. Set convex optimization solution goal: $\sigma = \|J(X^i)T - W(X^i)\|_2$, nonlinear constraint: $t_{\min} \leq t_i \leq t_{\max}$. Use the interior point method to solve it.
5. Check whether the results of step 4 meets $\sigma < \sigma^*$. If the condition is true, there is a Static equilibrium point $X^i = [X_a{}^i, X_b{}^i]$ in the controllable generalized coordinate $X_a{}^i$ that meets the cable force condition, turn to step 6, otherwise, there is no Static equilibrium point in the controllable posture.
6. Calculate the next pose $X_a{}^{i+1}$ to be searched.

5. Simulation Results of Static Equilibrium Workspace

5.1. 3—6 Under-Constrained Parallel Robot

Solve the Static equilibrium workspace of 3—6 (3 cables pulling 6 degrees of freedom) under-constrained parallel robot. The mass of end effector is $m = 0.935$ kg. See Table 1 for specific parameters.

Table 1. 3−6 Configuration of cable-driven parallel robot.

Coordinates of Anchor Point (m)	Coordinates of the Hinge Point between the Cable and the End Actuator (m)
$^{O}A_1 = \begin{bmatrix} 0.9 & 0 & 1.8 \end{bmatrix}^T$	$^{O'}B_1 = \begin{bmatrix} 0.2265 & -0.2275 & 0 \end{bmatrix}^T$
$^{O}A_2 = \begin{bmatrix} 0 & 1.8 & 1.8 \end{bmatrix}^T$	$^{O'}B_2 = \begin{bmatrix} 0 & 0.2275 & 0 \end{bmatrix}^T$
$^{O}A_3 = \begin{bmatrix} -0.9 & 0 & 1.8 \end{bmatrix}^T$	$^{O'}B_3 = \begin{bmatrix} -0.2265 & -0.2275 & 0 \end{bmatrix}^T$

$t_{\min} = 0.1$ N is the lower limit of cable tension, and $t_{\max} = 10$ N the upper limit of cable tension. $\sigma* = 1 \times 10^{-8}$ is the error. The controllable degree of freedom is $X_a = \begin{bmatrix} ^{O}x_{O\prime}, ^{O}y_{O\prime}, ^{O}z_{O\prime} \end{bmatrix}$, and the uncontrollable degree of freedom is $X_b = [\alpha, \beta, \gamma]$, the rotation angle of x-axis, y-axis, and z-axis, respectively. Considering the volume of end effector and cable, set the search interval is

$$\mathbf{S}_x \in (-0.6, 0.6), \mathbf{S}_y \in (0.3, 1.6), \mathbf{S}_z \in (0.27, 1.52)$$

The search step of $\delta_x, \delta_y, \delta_z$ is 0.05m, and the pose set $\mathbf{Q}$ to be searched is generated.

Matlab is used to solve the static equilibrium workspace of 3−6 cable-driven parallel robot.

It can be seen from Figure 3 the 3D view, *xoy* planar view and *xoz* planar view of the Static equilibrium workspace that the Static equilibrium workspace of the 3−6 under-constrained parallel robot is approximately a triangular prism, whose cross section is symmetrical along the x-axis, and there are two narrow workspace cracks in the lower half ($z < 0.75$ m) of the triangular prism. Compared with the system schematic diagram 1, this situation occurs because when the center of mass of the end actuator is low, the tension force of the three cables cannot be balanced with gravity, so the pose of the end actuator cannot be kept stable. The anchor points $^{O}A_1$ and $^{O}A_3$ are symmetrical along the y-axis, and the anchor point $^{O}A_2$ is on the axis, so that the cross section of its Static equilibrium workspace is symmetrical about the y-axis.

5.2. 0−0−2 Under-Constrained Planar Hybrid Robot

Solve the Static equilibrium workspace of the 0−0−2 under-constrained planar hybrid robot (no cable attached on the first and second links, and two cables attach on the third link). The Tables 2 and 3 show the specific parameters of 0−0−2 cable-driven planar hybrid robot system.

Table 2. 0−0−2 configuration of cable-driven planar hybrid robot.

Coordinates of Anchor Point (m)	Coordinates of the Hinge Point on the End Effector (m)
$^{O}A_1 = \begin{bmatrix} -0.9 & 0.6 \end{bmatrix}^T$	$^{3}B_1 = \begin{bmatrix} 0.09 & 0.08 \end{bmatrix}^T$
$^{O}A_2 = \begin{bmatrix} 0.9 & 0.5 \end{bmatrix}^T$	$^{3}B_2 = \begin{bmatrix} 0.16 & -0.03 \end{bmatrix}^T$

Table 3. Link parameters.

The Quality of the Link(kg)	The Length of the Link(m)
$m_1 = 11.8$	$l_1 = 0.45$
$m_2 = 4.5$	$l_2 = 0.4$
$m_3 = 1.1$	$l_3 = 0.25$

The lower limit of the cable tension is $t_{\min} = 5$ N, the upper limit of the cable tension is $t_{\max} = 200$ N, $\sigma* = 1 \times 10^{-6}$ is the error.

$X_a = [\theta_1, \theta_2]$ is the controllable degrees of freedom, and $X_b = [\theta_3]$ is the uncontrollable degrees of freedom.

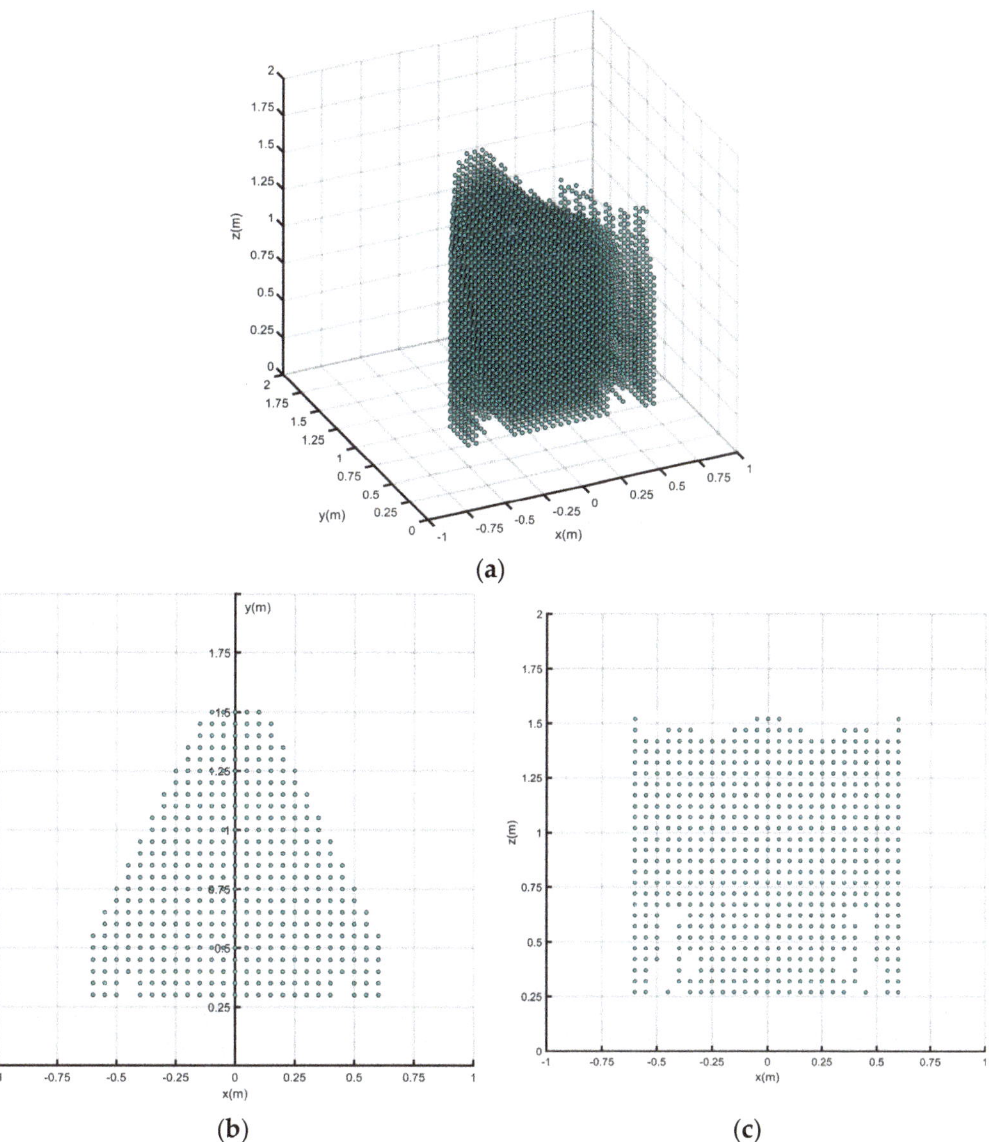

Figure 3. 3−6 Static equilibrium workspace of cable-driven parallel robot. (**a**) 3D Static Equilibrium Workspace; (**b**) *xoy* planar view of Static Equilibrium Workspace; (**c**) *xoz* planar view of Static Equilibrium Workspace.

Set the search interval is $\mathbf{S}_{\theta_1} \in (65°, 95°) \& \mathbf{S}_{\theta_2} \in (-40°, 0°)$.

The search step $\delta_{\theta_1}, \delta_{\theta_2}$ is $1°$, and the pose set to be searched is $\mathbf{Q}$.

The static equilibrium working space of the $0-0-2$ cable-driven robot is solved by MATLAB programming.

According to Figure 4, the static equilibrium workspace of the $0-0-2$ under-constrained cable-driven planar hybrid robot is a curved surface. From Figure 4b, $\theta_1\theta_2$ planar view of the Static equilibrium workspace, the Static equilibrium points in the pose set are mainly distributed in the lower right corner area. The uncontrollable degrees of freedom of the static equilibrium point belongs to $\theta_3 \in (-80°, -20°)$ from Figure 4c,d.

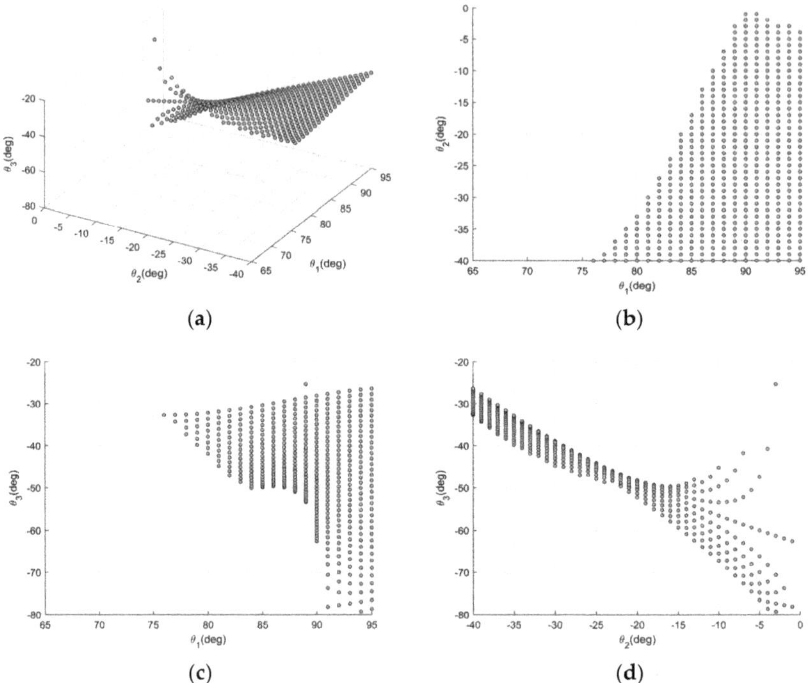

(a)

(b)

(c)

(d)

Figure 4. Static Equilibrium Workspace of $0-0-2$ unconstrained cable-driven planar hybrid robot. (a) 3D Static Equilibrium Workspace; (b) $\theta_1\theta_2$ planar view of Static Equilibrium Workspace; (c) $\theta_1\theta_3$ planar view of Static Equilibrium Workspace; (d) $\theta_2\theta_3$ planar view of Static Equilibrium Workspace.

6. Consistent Solution Strategies for Point-to-Point Trajectory Planning

6.1. Analysis of the Dynamics of Under-Constrained Systems

The end poses $\begin{bmatrix} {}^O x_O, & {}^O y_O, & {}^O z_O, & \alpha & \beta & \gamma \end{bmatrix}^{\mathrm{T}}$ of the parallel robot and Joint coordinates $\begin{bmatrix} \theta_1 & \theta_2 & \cdots & \theta_n \end{bmatrix}^{\mathrm{T}}$ of the planar hybrid robot are written in generalized coordinates $X = \begin{bmatrix} x_1 & x_2 & \cdots & x_n \end{bmatrix}^{\mathrm{T}}$.

Write Equations (7) and (16) uniformly as:

$$M(X)\ddot{X} - s(X, \dot{X}) - J(X)T = 0 \tag{22}$$

where $\ddot{X}$ is the acceleration in generalized coordinates, $M(X)$ is the mass matrix, and $J(X)T$ is the vector of the cable tensions.

For the under-constrained system, the controllable degrees of freedom, X_a, are m dimensional vectors. The uncontrollable degrees of freedom, X_u, are $n - m$ dimensional vectors. As a result, Equation (22) can be written in the form of

$$\begin{bmatrix} M_{aa} & M_{au} \\ M_{ua} & M_{uu} \end{bmatrix} \begin{bmatrix} \ddot{X}_a \\ \ddot{X}_u \end{bmatrix} - \begin{bmatrix} s_a \\ s_u \end{bmatrix} - \begin{bmatrix} J_a \\ J_u \end{bmatrix} T = 0 \tag{23}$$

For Equation (23), when $X, \dot{X}, \ddot{X}_a$ is known, the acceleration $\ddot{X}_u$ and the cable force T for the system at this moment can be obtained.

By presenting the acceleration terms for the (UDFS) uncontrollable degrees of freedom, we obtain the expressions for $\ddot{X}_u$

$$\ddot{X}_u = M_{uu}^{-1}\left(s_u + J_u T - M_{ua}\ddot{X}_a\right) \tag{24}$$

Substituting the expression for $\ddot{X}_u$ into Equation (23), we obtain the expression for the cable extension T:

$$T = \left(J_a - M_{au}M_{uu}^{-1}J_u\right)^{-1}\left[\left(M_{aa} - M_{au}M_{uu}^{-1}M_{ua}\right)\ddot{X}_a + M_{au}M_{uu}^{-1}s_u - s_a\right] \tag{25}$$

6.2. Point-to-Point Trajectory Planning Introducing Parameters

In the process of point-to-point trajectory planning for under-constrained systems, the start and end points of the motion should be chosen as static equilibrium points and the speed at the start and end points should be zero.

For an under-constrained system with n degrees-of-freedom and m cables, there are m controllable degrees of freedom and $n - m$ uncontrollable degrees of freedom. The dynamics Equations (22) and (23) shows that during trajectory planning, only controllable degrees of freedom, X_a, can be trajectory planned, while uncontrollable degrees of freedom X_u are influenced by dynamics [20]. When the controllable degree of freedom reaches the end, an uncontrollable degree of freedom cannot be guaranteed to reach the end with zero velocity. Therefore, it is necessary to choose a suitable trajectory plan for the controllable degrees of freedom to ensure that the uncontrollable degrees of freedom reach the end point with zero velocity.

Assume that the trajectory planned for X_a is $x_a(t), t \in [0, T]$, the trajectory of X_u to be found is $x_u(t)$, and the set of differential equations is established by Equation (24) as follows.

$$y = \begin{bmatrix} x_u(t) \\ \dot{x}_u(t) \end{bmatrix}$$

$$\dot{y} = \begin{bmatrix} \dot{x}_u(t) \\ M_{uu}^{-1}\left(s_u + J_u T - M_{ua}\ddot{x}_a\right) \end{bmatrix} = f(y, x_a, \dot{x}_a, \ddot{x}_a) \tag{26}$$

$$y(0) = \begin{bmatrix} x_u(0) \\ 0 \end{bmatrix} := y_0, \ y(T) = \begin{bmatrix} x_u(T) \\ 0 \end{bmatrix} := y_T$$

where y_0, y_T are the position and velocity of the non-controllable degrees of freedom at the start and end points, respectively, which are both static equilibrium points with zero velocity.

The optimal trajectory for controllable degrees of freedom is achieved when the trajectory $x_a(t)$ is such that Equation (26) holds.

Since Equation (26) has $2 \times (n - m)$ variables and its boundary conditions have $4 \times (n - m)$ constraint. In order to make Equation (26) solvable, $2 \times (n - m)$ parameter $\kappa_1, \cdots, \kappa_{2\lambda}$ needs to be added to the planned trajectory $x_a(t)$.

As an example of a conventional polynomial, a trajectory is planned for controllable degrees of freedom X_a and parameters are introduced for the planned trajectory.

A conventional polynomial point-to-point trajectory is planned as a straight-line path [21] with a trajectory equation of the form:

$$x_a(t) = x_a(0) + (x_a(T) - x_a(0))u(t) \tag{27}$$

where $x_a(0), x_a(T)$ is the position of the controllable degrees of freedom at the start and end points, respectively, $0 \leq u \leq 1$.

For a smoothly derivable point-to-point trajectory $x_a(t)$ of order r, the law of motion $u(t)$ is designed as follows:

$$u(t) = \sum_{i=r+1}^{2r+1} a_i \left(\frac{t}{T}\right)^i, \quad t \in [0, T] \tag{28}$$

where

$$a_i = \frac{(-1)^{i-r-1}(2r+1)!}{i \cdot r!(i-r-1)!(2r+1-i)!} \tag{29}$$

Introducing parameters to the trajectory of a traditional polynomial programming, the equation of the trajectory is

$$x_a(\kappa, t) = x_a(0) + (x_a(T) - x_a(0))u(\kappa, t) \tag{30}$$

The law-of-motion function $u(\kappa, t)$ can be set to

$$u(\kappa, t) = u(\gamma(\kappa, t)) = \sum_{i=r+1}^{2r+1} a_i \gamma^i(\kappa, t) \tag{31}$$

where $a_i = \frac{(-1)^{i-r-1}(2r+1)!}{i \cdot r!(i-r-1)!(2r+1-i)!}$. In order to maintain $u(0) = 0, u(T) = 1, \gamma(\kappa, t)$. is expressed as follows.

$$\gamma(\kappa, t) = \alpha t + \sum_{i=2}^{2\lambda+1} \kappa_{i-1} t^i \tag{32}$$

where $\alpha = \dfrac{1 - \sum\limits_{i=2}^{2\lambda+1} \kappa_{i-1} T^i}{T}$.

Equation (26) is a multivariate marginal differential equation containing parameters that are converted to a multivariate initial differential equation for solution, and the expression of the converted multivariate initial differential equation is as follows.

$$\begin{cases} \dot{y} = f(y, x_a(\kappa, t), \dot{x}_a(\kappa, t), \ddot{x}_a(\kappa, t)) \\ \quad\quad y(0) = y_0 \end{cases} \tag{33}$$

Since the solution of the multivariate initial differential equation removes the constraint $y(T) = y_T$ at the end point, the result can be solved again by setting up the Newton iteration equation $F(\kappa) = y(\kappa, T) - y_T$ to ensure that the end point of the solved trajectory is the same as the planned trajectory.

The trajectory parameters κ_i for controlled degree of freedom planning is solved as follows.

1. Set the initial equations of the Newton iterative method: $F(\kappa) = y(\kappa, T) - y_T$, the iterative convergence value: ζ, and the initial values of the parameters to be solved: κ_0.
2. Solve the multivariate initial differential Equation (33) by taking $t = T$ into the solution $y(\kappa, t)$, and calculating equation $F(\kappa_i)$.
3. If the condition $\|F(\kappa_i)\| \leq \zeta$ is satisfied, substitute the coefficient into the trajectory equation $x_a(\kappa, t)$ to obtain the best planning trajectory for the controllable degrees of freedom, and the solution is finished; otherwise, let $\kappa_{i+1} = \kappa_i + J_F^{-1}(\kappa_i)F(\kappa_i)$, and substitute κ_{i+1} as the initial value into step 2 to continue the solution.

7. Simulation Analysis of Trajectory Planning

7.1. 3−6 Under-Constrained Parallel Robot

Point-to-point trajectory planning was performed for the 3−6 (6 degrees of freedom parallel robots with 3 cables) under-constrained parallel robot with the articulation shown in Table 1, and Table 4 shows the end-effector and trajectory planning parameters.

Table 4. End-effector and trajectory planning parameters.

$m(\mathrm{kg})$	$I_{o\prime}(\mathrm{kg}\cdot\mathrm{m}^2)$	X_0	$T(\mathrm{s})$	X_T
0.935	$\begin{bmatrix} 0.01613 & 0 & 0 \\ 0 & 0.01599 & 0 \\ 0 & 0 & 0.03211 \end{bmatrix}$	$\begin{bmatrix} 0 \\ 0.59 \\ 1 \\ -0.43 \\ 0 \\ 0 \end{bmatrix}$	1.5	$\begin{bmatrix} -0.15 \\ 0.8 \\ 1.17 \\ -0.05 \\ 0.26 \\ 0 \end{bmatrix}$

In Table 4, m is the mass of the end-effector and $I_{o\prime}$ is the inertia tensor of the end-effector in a local coordinate system where the origin of the local coordinate system coincides with the center of mass. X_0, X_T is the initial and end pose of the trajectory planning and both are static equilibrium points. Figure 5 shows the initial state and end state for unconstrained parallel robot trajectory planning.

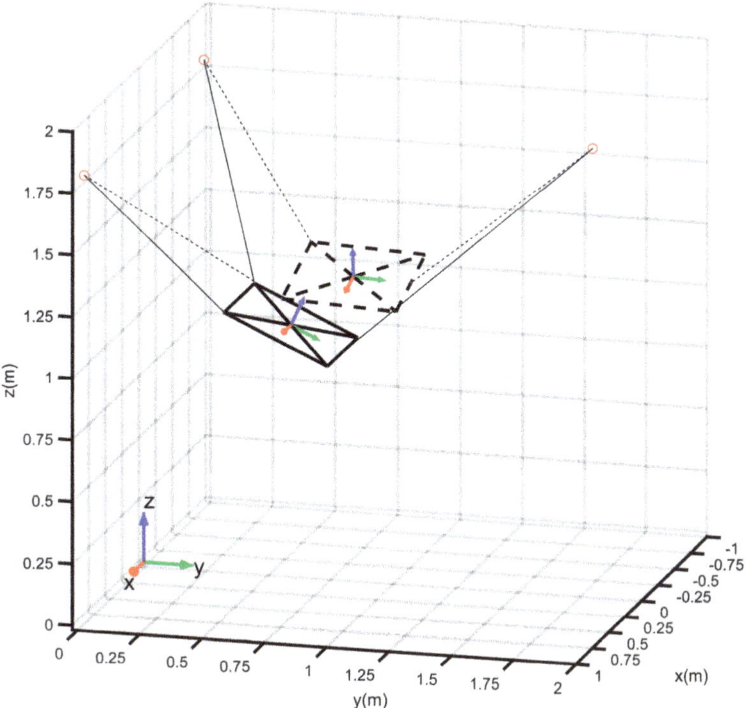

Figure 5. Initial state (solid line) and end state (dashed line) for unconstrained parallel robot trajectory planning.

The controllable degrees of freedom are $X_a = \begin{bmatrix} {}^{O}x_{O\prime}, {}^{O}y_{O\prime}, {}^{O}z_{O\prime} \end{bmatrix}$ and the uncontrollable degrees of freedom are $X_b = [\alpha, \beta, \gamma]$. These are the angles of rotation around the x-axis, y-axis and z-axis respectively.

For the traditional polynomial trajectory planning method, $r = 3$ is taken to ensure that the trajectory is smoothly derivable to the third order, and a_i is calculated via Equation (29) with the following results.

$$a_i = \begin{bmatrix} a_4 \\ a_5 \\ a_6 \\ a_7 \end{bmatrix} = \begin{bmatrix} 35 \\ -84 \\ 70 \\ -20 \end{bmatrix}$$

Substitute a_i into Equation (27) to find the conventional polynomial trajectory $x_a(t)$, and solve Equation (33) for the multivariate initial value differential equation.

For the polynomial trajectory planning method with the introduction of parameters, take $r = 3$, whose coefficients a_i are the same as those of a conventional polynomial trajectory. Let the initial value of the parameter vector κ be the $6-$dimensional zero vector $\zeta = 1 \times 10^{-8}$, and solve for the parameters using Newton's iterative method to obtain the following results.

$$\kappa = \begin{bmatrix} \kappa_1 \\ \kappa_2 \\ \kappa_3 \\ \kappa_4 \\ \kappa_5 \\ \kappa_6 \end{bmatrix} = \begin{bmatrix} -5.52007049 \\ 15.73380808 \\ -24.49999798 \\ 21.38293401 \\ -9.84359244 \\ 1.85786785 \end{bmatrix}$$

Substituting the parameter κ into Equation (30), the traditional polynomial trajectory $x_a(\kappa, t)$ is obtained and solved for the multivariate initial differential Equation (33).

The solution results of the traditional polynomial trajectory planning method are compared with those of the polynomial trajectory planning method with parameters, and the comparison results are shown below.

Figure 6 shows the pose of the end-effector in the trajectory. In this case, the direction drops from 0 m to -0.15 m, the direction increases from 0.59 to 0.8 and the direction increases from 1 m to 1.17 m. The orientation angle also changes under the influence of the system dynamics, where α rises from -0.43 rad to -0.05 rad and β from 0 rad to 0.26 rad (rotated about z-axis) and γ remains largely stable during the process. The trajectory start and end points of both planning algorithms are consistent with X_0, X_T in Table 4 in terms of the pose curves.

Figure 7 shows the velocity curves of the trajectories solved by the two planning algorithms. It can be seen that the difference in velocity between the trajectories planned by the two algorithms is more obvious. Both planning algorithms achieve the boundary condition of zeroing the velocity at the start and end point in the direction of x, y, z. However, for the angular velocity profile of α, β, γ, the traditional trajectory planning algorithm cannot guarantee stationary at the end point, and the trajectory does not zero at α, β, γ of 2.5 s, which cannot guarantee the trajectory boundary condition.

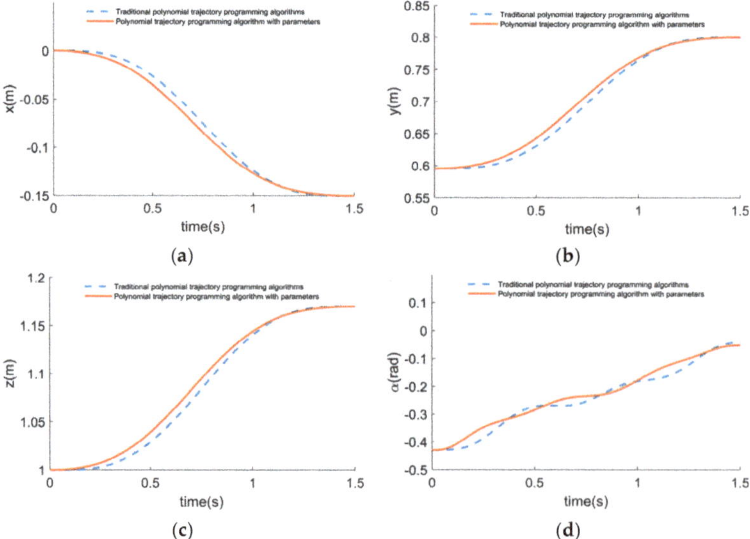

Figure 6. *Cont.*

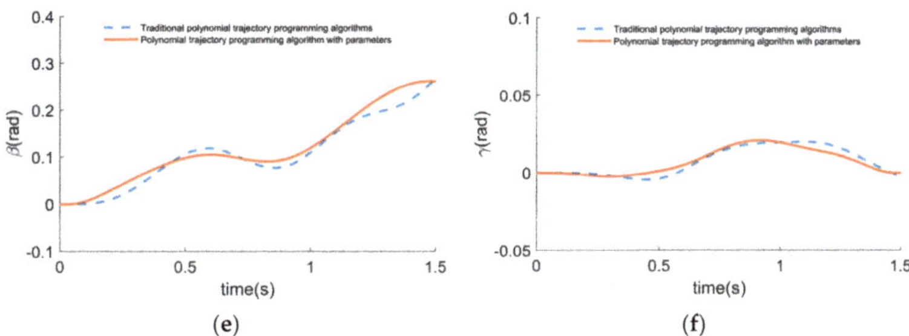

Figure 6. Trajectory poses solved under two planning algorithms. (**a**) Comparison of x-direction displacement; (**b**) Comparison of y-direction displacement; (**c**) Comparison of z-direction displacement; (**d**) Comparison of α angular; (**e**) Comparison of β angular; (**f**) Comparison of γ angular.

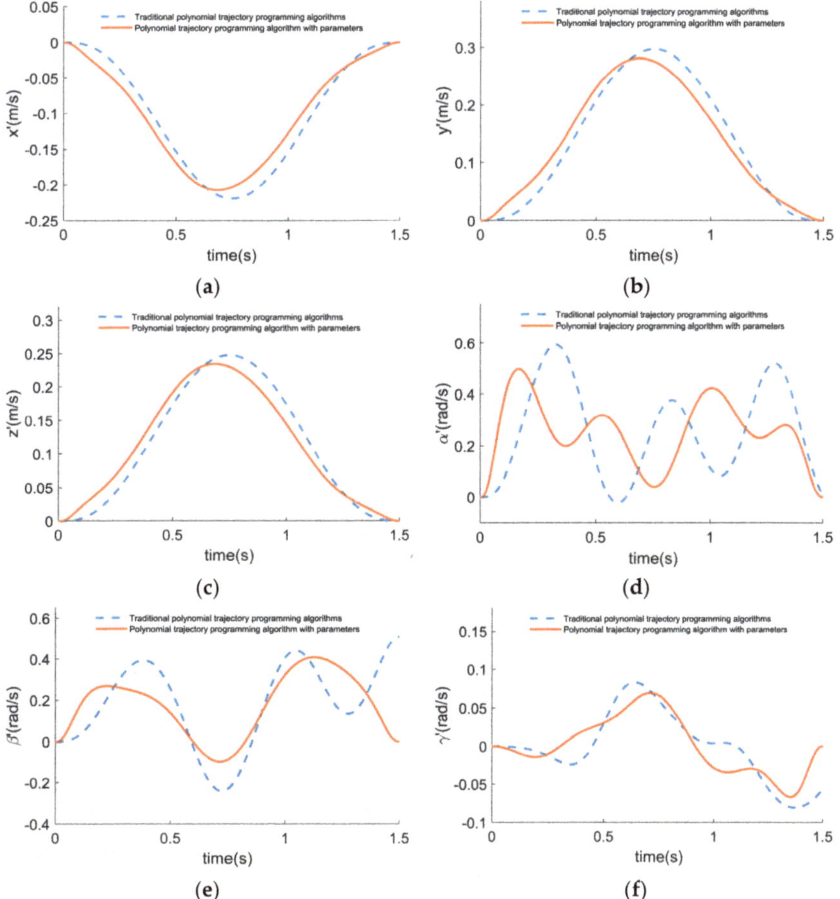

Figure 7. Velocity solved under two planning algorithms. (**a**) Comparison of $x-$direction velocity; (**b**) Comparison of $y-$direction velocity; (**c**) Comparison of $z-$direction velocity; (**d**) Comparison of $\alpha-$angle velocity; (**e**) Comparison of $\beta-$angle velocity; (**f**) Comparison of $\gamma-$angle velocity.

7.2. 0−0−2 Under-Constrained Planar Hybrid Robot

Point-to-point trajectory planning for an under-constrained cable-driven robot with 0-2-2 (Figure 8). The center of mass of each link is at its geometric center. The coordinates of the articulation points and the link parameters for the cable-driven robot are shown in Tables 2 and 3, while the following Table 5 shows the link inertia and trajectory planning related parameters.

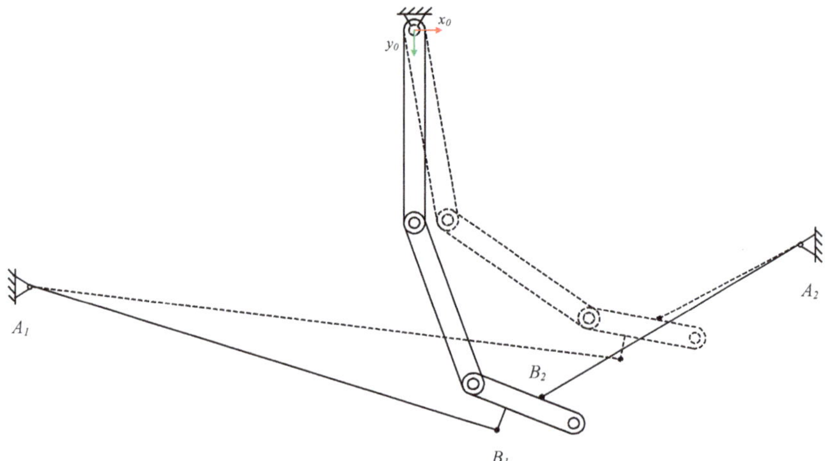

Figure 8. Initial state (solid line) and end state (dashed) for under-constrained cable-driven robot in trajectory planning.

Table 5. Parameters about Figure 8 to link rotational inertia and trajectory planning.

$I_{zi}(\mathrm{kg}\cdot\mathrm{m}^2)$	$X_0(\mathrm{rad})$	$T(\mathrm{s})$	$X_T(\mathrm{rad})$
$I_{z1} = 0.2$ $I_{z2} = 0.06$ $I_{z3} = 0.006$	$\begin{bmatrix} 90 \\ -20 \\ -48.61 \end{bmatrix} \times \frac{\pi}{180}$	1	$\begin{bmatrix} 80 \\ -45 \\ -24.44 \end{bmatrix} \times \frac{\pi}{180}$

In Table 5, I_{zi} is the mass product of inertia of the i link around the z-axis. X_0, X_T are the starting and ending points of the trajectory plan, and both are static equilibrium points.

The controllable degrees of freedom are $X_a = [\theta_1, \theta_2]$, and the uncontrollable degrees of freedom are $X_b = [\theta_3]$.

For the traditional polynomial trajectory planning method, $r = 3$ is taken to ensure that the trajectory is smoothly derivable to the third order, and a_i is calculated via Equation (29) with the following results.

$$a_i = \begin{bmatrix} a_4 \\ a_5 \\ a_6 \\ a_7 \end{bmatrix} = \begin{bmatrix} 35 \\ -84 \\ 70 \\ -20 \end{bmatrix}$$

Substitute a_i into Equation (27) to obtain the traditional polynomial trajectory $x_a(t)$, and solve Equation (33) for the multivariate initial differential equation.

For the polynomial trajectory planning method with the introduction of parameters, take $r = 3$, whose coefficients a_i are the same as those of a conventional polynomial trajectory. Let the initial value of the parameter vector κ be a 2−dimensional zero vector, $\zeta = 1 \times 10^{-8}$, and solve for the parameters using Newton's iterative method to obtain the following results.

$$\kappa = \begin{bmatrix} \kappa_1 \\ \kappa_2 \end{bmatrix} = \begin{bmatrix} 0.58865332 \\ -0.29981383 \end{bmatrix}$$

Substituting the parameter κ into Equation (30), the conventional polynomial trajectory $x_a(\kappa, t)$ is obtained and solved for the multivariate initial differential Equation (33).

A comparison of the solution results of the traditional polynomial trajectory planning method with those of the polynomial trajectory planning method with the introduction of parameters is shown below.

The joint angle curves for the two planning algorithms are shown in Figure 9. The joint angle θ_1, θ_2 starts and ends at the same point in both planning algorithms, while angle θ_1 decreases from $90°$ to $80°$ and angle θ_2 changes from $-25°$ to $-45°$. As can be seen from Figure 7c, in the polynomial trajectory planning method with the introduction of parameters, the start and end points of the θ_3 curve are the same as X_0, X_T in Table 5, whereas the traditional polynomial trajectory planning method does not guarantee this.

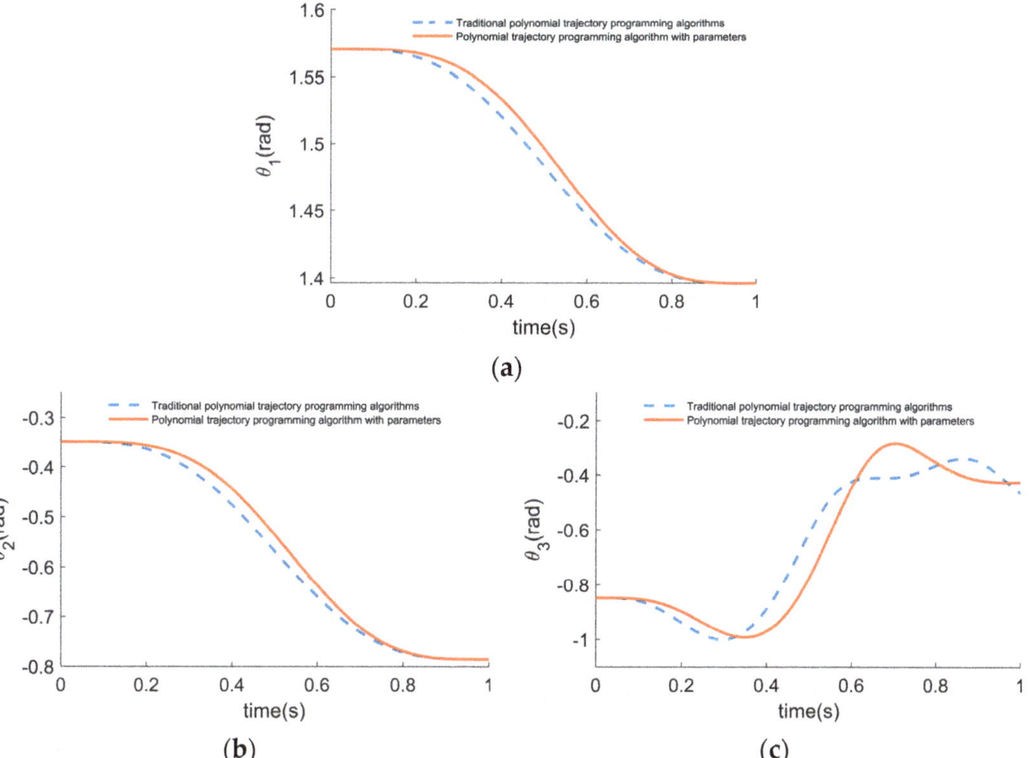

Figure 9. Trajectory joint angles solved under the two planning algorithms. (**a**) Comparison of θ_1−angle; (**b**) Comparison of θ_2−angle; (**c**) Comparison of θ_3−angle.

The angular velocity of the two planning algorithms is shown in Figure 10. Both planning algorithms reach the boundary condition of zero at the start and end points for the angular velocity of θ_1, θ_2. As can be seen in Figure 10c, for the angular velocity profile of θ_3, stationarity is not guaranteed at the end point using the conventional trajectory planning algorithm, whereas the polynomial trajectory planning method with the introduction of parameters maintains stationarity at the end point.

In summary, for under-constrained parallel or planar hybrid systems, the traditional polynomial trajectory planning algorithm cannot guarantee zero velocity at the end of the trajectory. In contrast, the polynomial trajectory planning algorithm with the introduction of parameters at the start and end points keeps the under-constrained system stable without oscillation and satisfies the trajectory planning requirements.

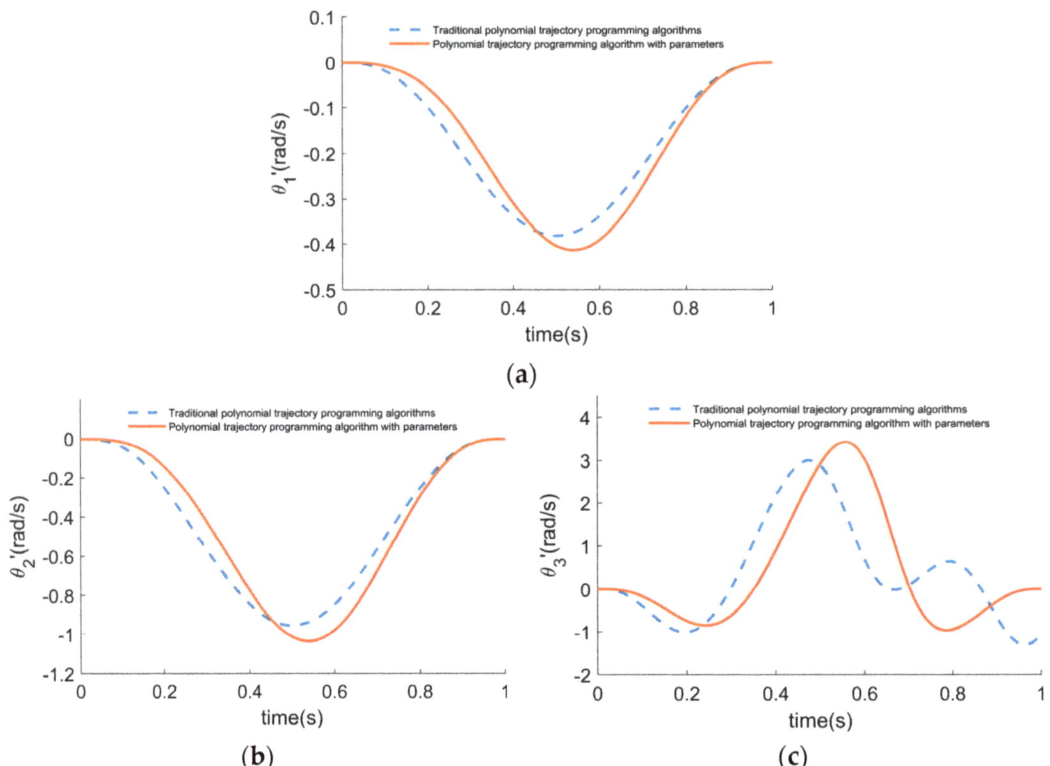

Figure 10. Angular velocity of the solution under the two planning algorithms. (**a**) Comparison of θ_1−angular velocity; (**b**) Comparison of θ_2−angle velocity; (**c**) Comparison of θ_3−angle velocity.

8. Conclusions

In this paper, the force closure equations and geometric closure equations for incompletely constrained cable traction systems (parallel and planar hybrid) are developed, the equations are solved jointly using a convex optimization solution method with boundary conditions, a static equilibrium inverse kinematic model of the incompletely constrained cable traction system is developed, and the static equilibrium workspace is solved.

Point-to-point trajectory planning algorithms for incompletely constrained cable traction systems are investigated. Traditional purely algebraic methods such as polynomial functions or trigonometric functions for planning the trajectory of the end-effector do not take into account the dynamical model of the system, which can cause oscillations of the system at the start and end points due to insufficient controllable degrees of freedom on the UCR. In this paper, a dynamical model of the UCR is developed, which consider the cable force, external force, controllable position and orientation and uncontrollable position and orientation of the end-effector, etc. Based on this model, the traditional polynomial-based point-to-point trajectory planning algorithm is improved by adding $2 \times (n - m)$ parameters to the kinematic law function $u(t)$. The constraints of the dynamics model are incorporated into the trajectory planning process to achieve point-to-point trajectory planning for the UCR. The results show that the trajectory of the improved algorithm is smooth and derivable, and the end-effector is stationary and stable without oscillation at the start and end points, which proves the effectiveness of the optimized algorithm.

Author Contributions: Conceptualization, Q.D., Q.Z. and T.W.; methodology, Q.D.; software, Q.Z.; validation, Q.D. and Q.Z.; writing—original draft preparation, Q.Z.; writing—review and editing, Q.D. and Q.Z. All authors have read and agreed to the published version of the manuscript.

Funding: This research was funded by Key Research and Development Program of Shaanxi (Program No. 2022GY-314).

Institutional Review Board Statement: Not applicable.

Informed Consent Statement: Not applicable.

Data Availability Statement: Not applicable.

Conflicts of Interest: The authors declare no conflict of interest. The funders had no role in the design of the study; in the collection, analyses, or interpretation of data; in the writing of the manuscript; or in the decision to publish the results.

References

1. Ming, A.; Higuchi, T. Study on multiple degree of freedom positioning mechanisms using wires (Part 1): Concept, design and control. *Int. J. Jpn. Soc. Precis. Eng.* **1994**, *28*, 131–138.
2. Colombo, G.; Joerg, M.; Schreier, R.; Dietz, V. Treadmill training of paraplegic patients using a robotic orthosis. *J. Rehabil. Res. Dev.* **2000**, *37*, 693. [PubMed]
3. Surdilovic, D.; Radojičić, J.; Bremer, N. Efficient calibration of cable-driven parallel robots with variable structure. In *Cable-Driven Parallel Robots*; Mechanisms and Machine Science; Springer: Cham, Switzerland, 2015.
4. Reed, W.H.; Abbott, F.T. A New "Free-Flight" Mount System for High-Speed Wind-Tunnel Flutter Models. *Proc. Symp. Aeroelastic Dyn. Model. Technol.* **1963**, *1*, 169–206.
5. Bennett, R.M.; Farmer, M.G.; Mohrt, R.L.; Hall, W.E. Wind-tunnel technique for determining stability derivatives from cable-mounted models. *J. Aircr.* **1978**, *15*, 304–310. [CrossRef]
6. Carricato, M.; Merlet, J.P. Stability analysis of underconstrained cable-driven parallel robots. *IEEE Trans. Robot.* **2013**, *29*, 288–296. [CrossRef]
7. Carricato, M.; Merlet, J.P. Geometrico-static analysis of under-constrained cable-driven par-allel robots. In *Advances in Robot Kinematics: Motion in Man and Machine*; Lenarcic, J., Stanisic, M.M., Eds.; Springer: Dordrecht, The Netherlands, 2010; pp. 309–319. [CrossRef]
8. Carricato, M. Inverse Geometrico-Static Problem of Underconstrained Cable-Driven Parallel Robots with Three Cables. *J. Mech. Robot.* **2013**, *5*, 031002. [CrossRef]
9. Berti, A.; Merlet, J.P.; Carricato, M. Solving the direct geometrico-static problem of underconstrained cable-driven parallel robots by interval analysis. *Int. J. Robot. Res.* **2016**, *35*, 723–739. [CrossRef]
10. Liu, X.; Qiu, Y.; Sheng, Y. Proof of the existence condition of the workspace of the rope traction parallel robot and the consistent solution strategy. *J. Mech. Eng.* **2010**, *46*, 27–34. [CrossRef]
11. Fu, Y. *Dynamic Modeling and Control of a Four Rope Traction Parallel Robot System*; Huaqiao Chinese University: Quanzhou, China, 2015.
12. Zhao, Z.; Li, W.; Li, Y.; Wang, Y.; Ma, Y. Static equilibrium workspace analysis of multi robot under constrained rope traction system. *J. Syst. Simul.* **2017**, *29*, 708–713.
13. Peng, Y.; Bu, W. Analysis of reachable workspace for spatial 3-cable under-constrained suspended cable driven parallel robots. *J. Mech. Robot.* **2021**, *13*, 061002. [CrossRef]
14. Ida, E.; Bruckmann, T.; Carricato, M. Rest-to-Rest Trajectory Planning for Underactuated Cable-Driven Parallel Robots. *IEEE Trans. Robot.* **2019**, *99*, 1–14. [CrossRef]
15. Barbazza, L.; Zanotto, D.; Rosati, G.; Agrawal, S.K. Design and Optimal Control of an Underactuated Cable-Driven Micro–Macro Robot. *IEEE Robot. Autom. Lett.* **2017**, *2*, 896–903. [CrossRef]
16. Shang, W.; Zhang, N. Dynamic trajectory planning of a 3-DOF under-constrained cable-driven parallel robot. *Mech. Mach. Theory* **2016**, *98*, 21–35.
17. Zhao, T.; Zi, B.; Qian, S.; Zhao, J. Algebraic Method Based Pointto Point Trajectory Planning of an Under Constrained Cable Suspended Parallel Robot with Variable Angle and Height Cable Mast. *Chin. J. Mech. Eng. Engl. Ed.* **2020**, *33*, 18.
18. Shi, P.; Tang, L. Dimensional synthesis of a gait rehabilitation cable-suspended robot on minimum 2-norm tensions. *J. Mech. Med. Biol.* **2021**, *21*, 2150046. [CrossRef]
19. Oriolo, G. Control of mechanical systems with second-order nonholonomic constraints: Under-actuated manipulators. In Proceedings of the IEEE Conference on Decision & Control, Brighton, UK, 11–13 December 1991.
20. Seifried, R. *Dynamics of Underactuated Multibody Systems*; Springer: Cham, Switzerland, 2014.
21. Piazzi, A.; Visioli, A. Optimal noncausal set-point regulation of scalar systems. *Automatica* **2001**, *37*, 121–127. [CrossRef]

Article

Singularity Analysis and Geometric Optimization of a 6-DOF Parallel Robot for SILS

Doina Pisla [1], Iosif Birlescu [1,*], Nicolae Crisan [2], Alexandru Pusca [1], Iulia Andras [2], Paul Tucan [1], Corina Radu [3], Bogdan Gherman [1] and Calin Vaida [1]

[1] Research Center for Industrial Robots Simulation and Testing CESTER, Technical University of Cluj-Napoca, 400641 Cluj-Napoca, Romania
[2] Department of Urology, Iuliu Hatieganu University of Medicine and Pharmacy, 400000 Cluj-Napoca, Romania
[3] Department of Internal Medicine, Iuliu Hatieganu University of Medicine and Pharmacy, 400000 Cluj-Napoca, Romania
* Correspondence: iosif.birlescu@mep.utcluj.ro

Abstract: The paper presents the singularity analysis and the geometric optimization of a 6-DOF (Degrees of Freedom) parallel robot for SILS (Single-Incision Laparoscopic Surgery). Based on a defined set of input/output constraint equations, the singularities of the parallel robotic system are determined and geometrically interpreted. Then, the geometric parameters (e.g., the lengths of the mechanism links) for the 6-DOF parallel robot for SILS are optimized such that the robotic system complies with an operational workspace defined in correlation with the SILS task. A numerical analysis of the singularities showed that the operational workspace is singularity free. Furthermore, numerical simulations validate the parallel robot for the next developing stages (e.g., designing and prototyping stages).

Keywords: parallel robot; singularity analysis; geometric optimization; single-incision laparoscopic surgery

Citation: Pisla, D.; Birlescu, I.; Crisan, N.; Pusca, A.; Andras, I.; Tucan, P.; Radu, C.; Gherman, B.; Vaida, C. Singularity Analysis and Geometric Optimization of a 6-DOF Parallel Robot for SILS. *Machines* **2022**, *10*, 764. https://doi.org/10.3390/machines10090764

Academic Editor: Zhufeng Shao

Received: 29 July 2022
Accepted: 29 August 2022
Published: 2 September 2022

Publisher's Note: MDPI stays neutral with regard to jurisdictional claims in published maps and institutional affiliations.

1. Introduction

Single-incision laparoscopic surgery (SILS) is a type of minimally invasive surgery (MIS) where the surgeon uses a special multi-lumen trocar (a trocar with multiple insertion points) inserted through a single incision in the patient body for all instruments required for surgery. SILS was described in the gynecological literature as early as 1969, when Wheesess reported the first 4.000 cases of tubal ligation [1]. As the literature suggests, the main advantages of SILS are: (i) the short recovery time for the patient, (ii) lesser degree of pain (when compared to classical MIS), (iii) and better cosmetics (since it uses a single access port) [2]. However, the challenges of SILS (performed manually) are caused by its poor ergonomics [3]. As history shows, robotic systems for surgery were developed to circumvent various limitations of the classical, human-performed interventions. MIS is one example in which robots had a significant impact [4,5]. Multiple approaches have been proposed and developed for robotic systems for MIS, namely: multi-arm robotic systems [4], flexible robotic systems [4], laparoscope holders [6], etc. The analysis of the intraoperative data from robotic surgery has enabled the addition of new technologies, one of them referring to automated gesture recognition, which assists the surgeon during delicate procedures such as suturing. For this task, multiple intelligent algorithms were developed using supervised, unsupervised, or semi-supervised learning paradigms [7]. This is an important step toward autonomous robotic-assisted surgery, where the robot uses a confidence-based shared control strategy to perform certain tasks under supervision but without the involvement of the human operator [8]. The advancements in computer science, which are now providing more reliable intelligent algorithms and the increased availability of large quantities of information (big data)—pre-operative digital information, instruments motion for repetitive tasks, recorded surgeries, patient evolution in the short and long term—are indicators and supporters of the next generation of devices, intelligent surgical robots [9].

The year 2018 marks the starting point for surgical robots dedicated to SILS due to the first use of the robotic da Vinci SP system in SILS surgery [4,10,11]. The da Vinci SP robot received FDA approval in 2018, becoming a benchmark in SILS robotic surgery, with a serial architecture capable of manipulating and orienting active instruments and the laparoscopic camera by using a single port through which these instruments are inserted. The system has an articulated laparoscopic camera and instruments, thus offering much better visibility and dexterity in the operating field compared to classical instruments [12]. The drawbacks are the limited forces (for tissue manipulation) due to the multiple bends of the instruments. Another robot dedicated to SILS surgery that has received FDA approval is the Senhance robot, which in contrast to da Vinci SP is a multi-arm system with three independent serial manipulators capable of orienting and manipulating surgical instruments [13]. Furthermore, there are various robotic systems for SILS which have not yet received FDA approval, such as SORT, SPIDER, MASTER, Virtual Incision, and so on [4].

Since a robotic system for SILS must comply with complex design specifications (e.g., footprint in the operating room, occupied volume, quick removal in case of an emergency), various optimizations are achieved in the early design stages. In a general interpretation, optimization aims to find the best solution for a given problem based on the imposed restrictions. In robotics, the optimization process often faces conflicting criteria such as speed versus energy efficiency, accuracy versus stability, or geometric dimensions versus working volume. When dealing with multi-objective optimization problems, the results are often given as a set of Pareto optimal solutions which must be further interpreted. In [14], the authors present an optimal dimensional synthesis of a parallel mechanism using two objective functions: finding the smallest dimensions for the geometric links of the robot for a given workspace, and the second aims to ensure the best overall dexterity within this workspace. As the two functions have conflicting criteria, it was shown that favoring the first function can lead to a robot with low dexterity while the second leads to a bulky robot. A numerical optimization methodology is presented in [15] to achieve an optimal design synthesis for a planar parallel manipulator for a prescribed dexterous workspace, using the condition number (an index that describes the dexterity of a robot). Merlet emphasizes in [16] the importance of the proper use of the condition number and global conditioning index (GCI) in the optimal design of a robot, illustrating their limitations (namely, when they are applied to robots that have both translation and orientation motions). GCI was first introduced by Gosselin and Angeles in [17], and it was one of the first approaches to describe the global performance of a manipulator, also demonstrating that in some cases, GCI can provide conflicting results in workspace optimization problems. In [18], the authors successfully applied the GCI and condition number to a 2-DOF parallel mechanism enabling them to obtain a generalized characterization of the manipulator with respect to its operational workspace. In [19], the authors used the condition number to optimize Orthoglide, a 3-DOF parallel robot with only translational motions used in milling applications. The dimensional synthesis of the 3-DOF Delta parallel robot for a prescribed workspace is presented in [20], where the authors define several optimization objectives solved using the Lagrange multipliers method, demonstrating that existing industrial robots could have almost half their size for the same given workspace. Furthermore, for a Delta-like parallel robot, in [21], the authors achieved an optimum design using multi-objective optimization algorithms, i.e., Pareto-optimization. A methodology to achieve an optimal design for a 6-DOF parallel manipulator having as objective its accuracy is presented in [22], where the authors are using the distribution of the condition number to determine the best solutions. The Stewart–Gough platform was analyzed in detail in [23], where the authors used a multi-objective optimization algorithm (NSGA-II) to obtain the optimal geometric parameters and leg stroke lengths, demonstrating that this approach is more efficient than the classical numerical methods.

The workspace maximization of the Delta robot using a geometrical technique implemented in CATIA is shown in [24], where the use of the Simulated Annealing Algorithm provided promising results. The process of optimization is also found in the scientific literature concerning the development of medical robots. In [25], the authors optimized a 3-DOF serial robot considering the robot workspace and the surgical instrument insertion points. Other authors [26] used optimization processes for the static balancing of a surgical robot.

A novel approach for robotic-assisted SILS was proposed in [27], which is based on hybrid robotic systems with three major components: (1) a 6-DOF parallel robot that simultaneously guides (2) two interconnected orientation platforms (mounted on the mobile platform of the 6-DOF parallel robot) together with a laparoscopic camera, and (3) two active surgical instruments (mounted in the orientation platforms) to perform the SILS task. The kinematics of the surgical instruments was studied in [28], and in [29], the input–output equations for the orientation platform were provided. Furthermore, a recent work [27] achieved the kinematic modelling of the 6-DOF parallel robot.

The present paper illustrates the results regarding the further development of the SILS robotic system, namely: (i) a singularity analysis for the 6-DOF parallel robot presented in [27] (correlated with the operational workspace of the robot) and (ii) a geometric optimization of the 6-DOF parallel robot, to reduce its operating room footprint while maintaining the capabilities of performing the SILS task in a defined operational workspace. The singularity analysis was achieved using the vanishing conditions of the Jacobian matrices A and B. The input singularities are straightforward to interpret for the 6-DOF parallel robot, and for the output singularities, a geometric interpretation was achieved based on the characteristic tetrahedron [30]. For the geometric optimization, the Multi-Objective Genetic Algorithm implemented in Matlab gamultiobj function [31] is used to minimize the dimensions of the robot for a defined operational workspace. The GCI index is used in the optimization process only as guidance and not as a decision parameter due to its drawbacks when used on robots that have both translation and orientation motions [16,17].

The paper is structured as follows: Section 2 defines the parallel robotic system with all its modules and presents the inverse geometric model for the 6-DOF parallel robot. Section 3 defines the robotic system's operational workspace to create a clear description of the robotic system task. Section 4 presents a singularity analysis for the general 6-DOF parallel robot. Section 5 presents the proposed optimization algorithm for the 6-DOF parallel robot and the general optimization criteria. Section 6 presents the optimized version of the 6-DOF parallel robot, showing that the operational workspace is singularity free. Furthermore, numerical simulations are provided to validate the optimized robotic system. The conclusions are presented in Section 7.

2. An Innovative Parallel Robotic System for SILS

SILS requires the simultaneous independent manipulation of two surgical instruments while the operating field is viewed through a laparoscopic camera. The position of the laparoscope is not fixed, and it can change depending on the target tissue location. Furthermore, both the surgical instruments and the laparoscope must be manipulated with respect to Remote Center of Motion (RCM) points, i.e., the insertion points of the instrument within the trocar. Lastly, the position of the insertion points (or the trocar) on the patient's abdomen may vary based on various medical factors (e.g., the position of the resected tissue, areas that must be avoided, etc.) [3].

To comply with the required design parameters, an innovative parallel robotic system for SILS was proposed with the following components:

1. **One 6-DOF parallel robot** (patent pending [32])—guides a triangular mobile platform (Figure 1a) that contains a 1-DOF insertion mechanism for the laparoscopic camera;
2. **Two 3-DOF orientation platforms** (described in [29])—both mounted on the mobile platform (Figure 1b) on the sides of the laparoscopic camera insertion mechanism. The orientation platform can orient the surgical instruments using RCM with 2-DOF. The third DOF is for the linear insertion/retraction of the surgical instrument;

3. **Two surgical instruments** (described in [28])—each mounted in one orientation platform. Each surgical instrument has a serial architecture with 4-DOF: 1-DOF for the rotation about its longitudinal axis (the third rotation for the surgical instrument), 1-DOF for the (articulated) bending, and 2-DOF for the gripper (each jaw of the gripper is actuated separately enabling grabbing and gripper turning) (Figure 1c).

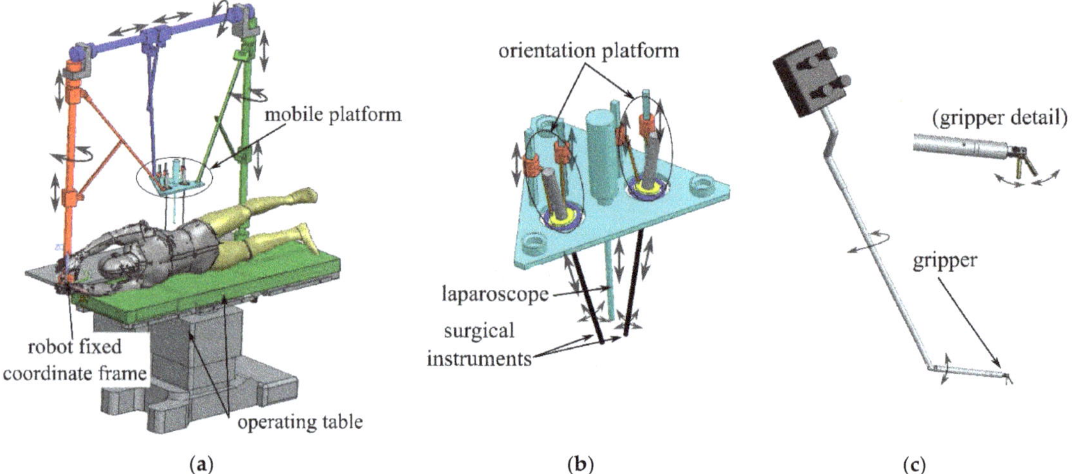

Figure 1. The initial modules for the SILS robotic system: (**a**) the 6-DOF parallel robot; (**b**) mobile platform containing the laparoscope and two orientation platforms; (**c**) surgical instrument.

With respect to a defined robotic-assisted medical protocol [3], the proposed robotic system must comply with the following stepwise procedure:

1. A preplanning procedure is performed to establish the adequate therapeutic conduit, also with respect to the robotic-assisted medical task. Among other medical parameters (patient history, etc.), the patient position is established, insertion points for the instruments are defined, and the relative position of the robot with respect to the insertion points is defined to ensure the required ranges of motions for the surgical instruments;

2. The 6-DOF parallel robot positions the mobile platform at the trocar (insertion points) and inserts the laparoscope and the surgical instruments on a linear trajectory. Due to the mechanically constrained RCM position (near the mobile platform), the 6-DOF parallel robot guides the mobile platform in close proximity to the patient;

3. The combined motion of the 3-DOF orientation platforms and the 4-DOF surgical instruments allows the surgeon to perform the task from the control console;

4. The laparoscope position can be adjusted by reorienting the mobile platform. Consequently, the laparoscope RCM must be maintained through the robot control, and simultaneously, the instrument position must be corrected by the 3-DOF orientation platforms (e.g., to maintain the instrument gripper-tissue relative position while changing the laparoscope position).

The kinematic scheme of the 6-DOF parallel robot is presented in Figure 2. Although the mechanism topology is the same as in [27], in this work, the kinematic chains are not considered identical (removing this design constraint may improve the final solution for the robot architecture). In [27], the links l_2, l_3, and l_4 had the same values.

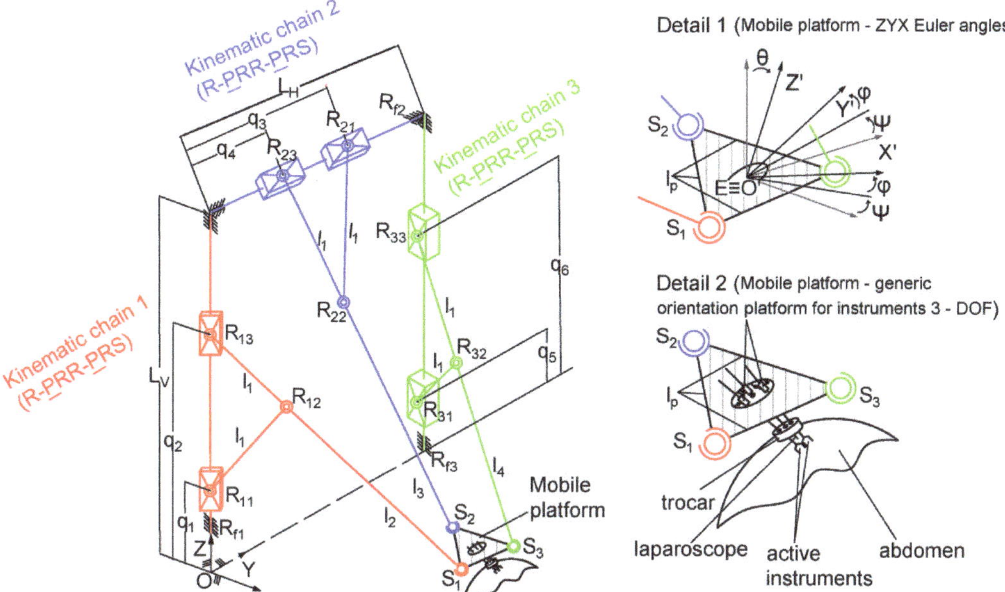

Figure 2. Kinematic scheme of the 6-DOF parallel robot for SILS.

The 6-DOF parallel robot mechanism topology is as follows [27]:

- The parallel robot consists of three (R-$\underline{P}$RR-$\underline{P}$RS) kinematic chains actuated by the prismatic joints pairs q_1 and q_2 for the first kinematic chain, q_3 and q_4 for the second kinematic chain, and q_5 and q_6 for the third kinematic chain. Furthermore, each kinematic chain has a free rotation motion around the actuation axis of the prismatic joints (R_{f1}, R_{f2}, and R_{f3}, respectively), and three passive revolute joints (R_{11}, R_{12}, and R_{13} for the first kinematic chain, R_{21}, R_{22}, and R_{23} for the second kinematic chain, R_{31}, R_{32}, and R_{33} for the third kinematic chain);
- The three kinematic chains guide the mobile platform via three passive spherical joints, S_1, S_2, and S_3, respectively. The mobile platform contains two orientation platforms which are illustrated in Figure 2 (Detail 2) as generic mechanisms (to point out the functionality). The kinematic scheme of these orientation platforms is presented in [29];
- The fixed coordinate frame $OXYZ$ is attached to the robot base, and the mobile coordinate frame $O'X'Y'Z'$ is attached to the mobile platform with its origin at the geometric center of the equilateral triangle formed by the centers of the three passive spherical joints (S_1, S_2, S_3). Furthermore, point E [X_E, Y_E, Z_E] is defined as the origin of the mobile coordinate frame $O'X'Y'Z'$.

Inverse Geometric Modelling for the 6-DOF Parallel Robot

The inverse geometric model will provide constraint equations which, in turn, will be used as objective functions for the optimization algorithm.

For the inverse geometric model, the inputs are the Cartesian coordinates of point E [X_E, Y_E, Z_E] and the orientation angles ψ, θ, φ, whereas the outputs are the active joint generalized coordinates q_i ($i = 1 \ldots 6$). With respect to the ZYX Euler convention, the coordinates of the passive spherical joints are:

$$S_1 : \begin{cases} X_{S1} = X_E + \frac{\sqrt{3}}{6}l_p c_\psi c_\theta - \frac{1}{2}l_p c_\psi s_\theta s_\varphi + \frac{1}{2}l_p s_\psi c_\theta \\ Y_{S1} = Y_E + \frac{\sqrt{3}}{6}l_p s_\psi c_\theta - \frac{1}{2}l_p s_\psi s_\theta s_\varphi - \frac{1}{2}l_p c_\psi c_\theta \\ Z_{S1} = Z_E - \frac{\sqrt{3}}{6}l_p s_\theta - \frac{1}{2}l_p s_\varphi c_\theta \end{cases},$$

$$S_2 : \begin{cases} X_{S2} = X_E - \frac{\sqrt{3}}{3}l_p c_\psi c_\theta \\ Y_{S2} = Y_E - \frac{\sqrt{3}}{3}l_p s_\psi c_\theta \\ Z_{S2} = Z_E + \frac{\sqrt{3}}{3}l_p s_\theta \end{cases}, \tag{1}$$

$$S_3 : \begin{cases} X_{S3} = X_E + \frac{\sqrt{3}}{6}l_p c_\psi c_\theta + \frac{1}{2}l_p c_\psi s_\theta s_\varphi - \frac{1}{2}l_p s_\psi c_\theta \\ Y_{S3} = Y_E + \frac{\sqrt{3}}{6}l_p s_\psi c_\theta + \frac{1}{2}l_p s_\psi s_\theta s_\varphi + \frac{1}{2}l_p c_\psi c_\theta \\ Z_{S3} = Z_E - \frac{\sqrt{3}}{6}l_p s_\theta + \frac{1}{2}l_p s_\varphi c_\theta \end{cases}$$

where c_ψ, c_θ, c_φ represent the cosines of the ψ, θ, φ Euler angles, and s_ψ, s_θ, s_φ represent the sines of the ψ, θ, φ, Euler angles, respectively. Furthermore, the distances between S_1, S_2, S_3 and the actuation axes of the three kinematic chains (Figure 2) are given by [27]:

$$\begin{aligned} R_1 &= \tfrac{1}{2l_1}(l_1 + l_2)\sqrt{4l_1^2 - (q_2 - q_1)^2}, \\ R_2 &= \tfrac{1}{2l_1}(l_1 + l_3)\sqrt{4l_1^2 - (q_4 - q_3)^2}, \\ R_3 &= \tfrac{1}{2l_1}(l_1 + l_4)\sqrt{4l_1^2 - (q_6 - q_5)^2} \end{aligned} \tag{2}$$

In addition, the following circle equation must be fulfilled [27]:

$$\begin{cases} X_{S1}^2 + Y_{S1}^2 - R_1^2 = 0 \\ X_{S2}^2 + (Z_{S2} - L_V)^2 - R_2^2 = 0 \\ X_{S3}^2 + (Y_{S3} - L_H)^2 - R_3^2 = 0 \end{cases} \tag{3}$$

And the following equations derived for each kinematic chain of the 6-DOF parallel robot (Figure 2) must be fulfilled:

$$\begin{cases} q_2 - \tfrac{1}{2l_1}(l_1 + l_2)(q_2 - q_1) - Z_{S1} = 0 \\ q_4 + \tfrac{1}{2l_1}(l_1 + l_3)(q_3 - q_4) - Y_{S2} = 0 \\ q_6 - \tfrac{1}{2l_1}(l_1 + l_4)(q_6 - q_5) - Z_{S3} = 0 \end{cases} \tag{4}$$

Using Equations (1)–(3) yields the solution for the inverse geometric model:

$$\begin{cases} q_1 = \frac{1}{l_1+l_2}\left[(l_1 + l_2)q_2 - 2l_1\sqrt{-X_{S1}^2 - Y_{S1}^2 + (l_1 + l_2)^2}\right] \\ q_2 = \sqrt{(l_1 + l_2)^2 - X_{S1}^2 - Y_{S1}^2} + Z_{S1} \\ q_3 = \frac{1}{l_1+l_3}\left[(l_1 + l_3)q_4 + 2l_1\sqrt{-X_{S2}^2 + (l_1 + l_3)^2 - (L_V - Z_{S2})^2}\right] \\ q_4 = Y_{S2} - \sqrt{(l_1 + l_3)^2 - X_{S2}^2 - (L_V - Z_{S2})^2} \\ q_5 = \frac{1}{l_1+l_4}\left[(l_1 + l_4)q_6 - 2l_1\sqrt{-X_{S3}^2 + (l_1 + l_4)^2 - (L_H - Y_{S3})^2}\right] \\ q_6 = \sqrt{(l_1 + l_4)^2 - X_{S3}^2 - (L_H - Y_{S3})^2} + Z_{S3} \end{cases} \tag{5}$$

3. The Proposed Operational Workspace for the Parallel Robotic System for SILS

For the SILS task, the discussion of the operational workspace must be split into two components: (1) the intraoperative operational workspace (inside the patient body with respect to the insertion points), defined by the orientation platforms, the surgical instruments, and the mobile platform of the

6-DOF parallel robot (for the laparoscopic camera), and (2) the external operating workspace (outside the patient body with respect to the insertion points), defined only by the 6-DOF parallel robot.

Considering the intraoperative workspace, in previous work [29], the maximum values for the orientation angles of the surgical instruments were proposed as follows:

- The maximum laparoscope angles in all directions with respect to a vertical axis passing through the insertion point to be 20 [°], achieved with the 6-DOF parallel robot;
- The maximum angle for the active instruments in all directions with respect to an axis orthogonal to the mobile platform (of the 6-DOF parallel robot) passing through the insertion point to be 30 [°], achieved with the orientation platforms described in [29].

Larger angles may affect patient safety. The intraoperative workspace was defined based on a sphere of 240 mm in diameter. Figure 3 illustrates the proposed intraoperative workspace, showing possible insertion points for the surgical instruments and their maximum orientation angles

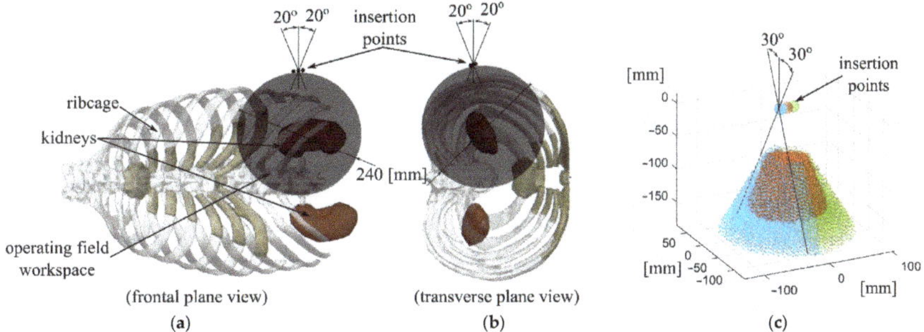

Figure 3. The proposed intraoperative workspace for the SILS robotic system: (**a**) frontal plane view showing the desired workspace with respect to the ribcage; (**b**) transverse plane view of the desired workspace; (**c**) the workspace of the surgical instruments with respect to the insertion points (brown—laparoscope, blue and green—active instruments).

A single set of insertion points for the RCM motion is not sufficient since adjustments may be required for the relative position between the patient and the robot. Consequently, for the external operational workspace, a cylindrical volume was proposed that should contain the sets of insertion points. The proposed cylinder was defined with radius $R = 75$ [mm] and height $h = 75$ [mm], and its position (approximately) relative to the patient is shown in Figure 4. This cylinder represents the operational workspace of the 6-DOF parallel robot.

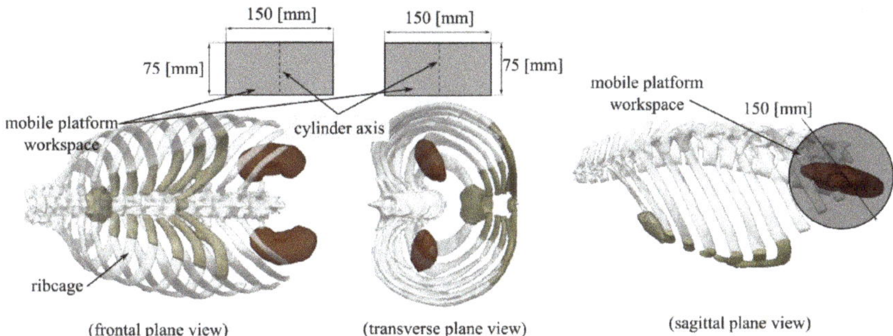

Figure 4. The proposed external workspace for the 6-DOF SILS robotic system (cylinder with radius 75 [mm] and height 75 [mm] and its position with respect to the ribcage).

4. Singularities of the 3-R-<u>P</u>RR-<u>P</u>RS Parallel Robot

The singularity analysis is achieved for the 3-R-<u>P</u>RR-<u>P</u>RS parallel robot without the loss of generality (e.g., no numerical values will be substituted for the geometric parameters). Later in Section 5, the design solution for the 6-DOF parallel robot for SILS is selected (based on the 3-R-

PRR-PRS parallel robot), and the singularities are correlated with the operational workspace for the SILS task.

The singularities are studied using the vanishing conditions of the determinants of the Jacobian matrices $\mathbf{A}$ and $\mathbf{B}$ from the matrix relation [33–35]:

$$\mathbf{A} \cdot \dot{\mathbf{X}} + \mathbf{B} \cdot \dot{\mathbf{Q}} = 0 \tag{6}$$

where $\dot{\mathbf{Q}}$ and $\dot{\mathbf{X}}$ represent the velocity vectors for the active joints $\mathbf{Q} = [q_1, q_2, q_3, q_4, q_5, q_6]^{\mathrm{T}}$ and for the mobile platform coordinates $\mathbf{X} = [X_E, Y_E, Z_E, \psi, \theta, \varphi]^{\mathrm{T}}$, respectively. With respect to the Jacobi matrices $\mathbf{A}$ and $\mathbf{B}$, three types of singularities can be defined, namely: type I singularities (input singularities) when $\det(\mathbf{B}) = 0$, type II singularities (output singularities) when $\det(\mathbf{A}) = 0$, and type III singularities when both $\det(\mathbf{B}) = 0$ and $\det(\mathbf{A}) = 0$.

The implicit equations used in the singularity analysis were defined using Equations (3) and (4) with Equations (1) and (2) substituted, yielding:

$$
\begin{cases}
f_1 : \frac{1}{6l_1}\left[l_p l_1 \left(\sqrt{3}s_\theta + 3c_\theta s_\varphi \right) + 3(q_1 + q_2 - 2Z_E) + 3l_2(q_1 - q_2) \right] = 0 \\[4pt]
f_2 : \frac{1}{12l_1}\left[4\sqrt{3}l_1 l_p \left(X_E c_\psi + Y_E s_\psi - \tfrac{1}{2}l_p s_\varphi s_\theta \right) c_\theta + l_1 l_p^2 \left(3c_\varphi^2 - 2 \right) c_\theta^2 - 12 l_1 l_p \left((X_E c_\psi + Y_E s_\psi)s_\varphi s_\theta - (X_E s_\psi - Y_E c_\psi)c_\varphi \right) - \right. \\[4pt]
\qquad \left. -12 l_1^2 (l_1 + 2l_2) + 12 l_1 \left(X_E^2 + Y_E^2 - l_2^2 \right) + 3l_1 \left(l_p^2 - (q_1 + q_2)^2 \right) + 6l_2(q_1 - q_2)^2 + 3\tfrac{l_2^2}{l_1}(q_1 - q_2)^2 \right] = 0 \\[4pt]
f_3 : \frac{1}{6l_1}\left[2\sqrt{3}l_p l_1 s_\psi c_\theta + 3(q_3 + q_4 - 2Y_E) + 3l_3(q_3 - q_4) \right] = 0 \\[4pt]
f_4 : \frac{1}{12l_1}\left[8\sqrt{3}l_1 l_p \left(X_E c_\psi c_\theta + (L_V - Z_E)s_\theta \right) + 4 l_1 l_p^2 \left(3c_\psi^2 - 1 \right) c_\theta^2 - 12 l_1^2 (l_1 + 2l_2) + 12 l_1 \left((L_V - Z_E)^2 + X_E^2 - l_3^2 \right) + \right. \\[4pt]
\qquad \left. + l_1 \left(4l_p^2 - 3(q_3 - q_4)^2 \right) + 3\left(2l_3 + \tfrac{l_3^2}{l_1} \right)(q_3 - q_4)^2 \right] = 0 \\[4pt]
f_5 : \frac{1}{6l_1}\left[l_p l_1 \left(\sqrt{3}s_\theta - 3s_\varphi c_\theta \right) + 3(q_5 + q_6 - 2Z_E) + 3l_4(q_5 - q_6) \right] = 0 \\[4pt]
f_6 : \frac{1}{12l_1}\left[4\sqrt{3}l_1 l_p \left((L_H - Y_E)s_\psi - \tfrac{1}{2}l_p s_\varphi s_\theta - X_E c_\psi \right) c_\theta + l_1 l_p^2 \left(3c_\varphi^2 - 2 \right) c_\theta^2 - 12 l_1 l_p \left((L_H - Y_E)s_\varphi s_\theta + X_E c_\varphi \right) + \right. \\[4pt]
\qquad \left. + 12 l_1 l_p \left(X_E s_\varphi c_\psi s_\theta - (L_H - Y_E)c_\psi c_\theta \right) - 12 l_1^2 (l_1 + 2l_4) + 12 l_1 \left((L_H - Y_E)^2 + X_E^2 - l_2^2 \right) + 3l_1 \left(l_p^2 + (q_5 - q_6)^2 \right) + \right. \\[4pt]
\qquad \left. + 3\left(2l_4 + \tfrac{l_4^2}{l_1} \right)(q_5 - q_6)^2 \right] = 0
\end{cases}
\tag{7}
$$

4.1. Type I Singularities

Computing the determinant of the Jacobian Matrix $\mathbf{B}$ yields a factored result:

$$\det(\mathbf{B}) = -\frac{1}{8l_1^6}(l_1 + l_2)^2(l_1 + l_3)^2(l_1 + l_4)^2(q_1 - q_2)(q_3 - q_4)(q_5 - q_6) \tag{8}$$

The singularity analysis is as follows:

1. $l_1 = 0$; this condition causes $\det(\mathbf{B})$ to be undefined; however, this condition is disregarded since it will be avoided in the mechanism design (link lengths cannot be 0);
2. $l_1 + l_2 = 0$ or $l_1 + l_3 = 0$ or $l_1 + l_4 = 0$; these conditions imply negative values for the link lengths, and are not regarded as possible singularities due to the mechanism design;
3. $q_1 - q_2 = 0$ or $q_3 - q_4 = 0$ or $q_5 - q_6 = 0$; these conditions require that the active joints of a kinematic chain are equal, meaning the active joints are overlapping. This condition is avoidable in the robot design or in the robot control.

4.2. Type II Singularities

Computing the determinant of the Jacobian matrix $\mathbf{A}$ yields a factored result:

$$\det(\mathbf{A}) = l_p^3 \cos(\theta) F\left(X_E, Y_E, Z_E, \psi, \theta, \varphi, l_p, L_H, L_V \right) \tag{9}$$

which describes singularities when:

1. $l_p = 0$, which is impossible in the robot design (link lengths cannot be zero);
2. $\cos(\theta) = 0$, which describes a parametric singularity of the ZYX Euler angles; since the parameter θ describes the second rotation in the ZYX Euler angles, for $\theta = \pi/2$ there is a gimbal lock, i.e., the parameters ψ and φ describe rotations about the same axis [36].
3. $F\left(X_E, Y_E, Z_E, \psi, \theta, \varphi, l_p, L_H, L_V \right) = 0$. The factor F (shown in Appendix A) could not be further factorized in this work. However, the following relation can be written:

$$\det(\mathbf{A}) = -8H_0\cos(\theta) \tag{10}$$

where the factor H_0 (shown in Equation (11)) is written in a compact manner using Equation (1). Note that the l_p parameter is not present in H_0 (but is still encoded in Equation (1)).

$$
\begin{aligned}
H_0 \;=\; & L_H L_V\big[(X_{S2}Y_{S2} - X_{S2}Y_{S3} - X_{S3}Y_{S2} + X_{S3}Y_{S3})X_{S1}^2 - (X_{S1}Y_{S1} - X_{S1}Y_{S3})X_{S2}^2 - (X_{S1}Y_{S1} - X_{S1}Y_{S2})X_{S3}^2 + \\
& + (2Y_{S1} - Y_{S2} - Y_{S3})X_{S1}X_{S2}X_{S3}\big] - L_H\big[(X_{S2}Y_{S2}Z_{S3} - X_{S2}Y_{S3}Z_{S2} - X_{S3}Y_{S2}Z_{S2} + X_{S3}Y_{S3}Z_{S2})X_{S1}^2 - (X_{S1}Y_{S1}Z_{S3} - \\
& - X_{S1}Y_{S3}Z_{S1} - X_{S3}Y_{S1}Z_{S1} + X_{S3}Y_{S1}Z_{S3})X_{S2}^2 - (X_{S1}Y_{S1}Z_{S2} - X_{S1}Y_{S2}Z_{S2} - X_{S2}Y_{S1}Z_{S1} + X_{S2}Y_{S1}Z_{S2})X_{S3}^2 + \\
& + (2Y_{S1}Z_{S2} - Y_{S2}Z_{S3} - Y_{S3}Z_{S1})X_{S1}X_{S2}X_{S3}\big] - L_V\big[(X_{S2}Y_{S2}Y_{S3} - X_{S2}Y_{S3}^2)X_{S1}^2 - (X_{S1}Y_{S1}Y_{S3} + X_{S3}Y_{S1}Y_{S3} - \\
& - X_{S1}Y_{S3}^2 - X_{S3}Y_{S1}^2)X_{S2}^2 + (X_{S2}Y_{S1}Y_{S2} - X_{S2}Y_{S1}^2)X_{S3}^2 + (2Y_{S1}Y_{S3} - Y_{S1}Y_{S2} - Y_{S2}Y_{S3})X_{S1}X_{S2}X_{S3}\big] + \\
& + (Y_{S2}Y_{S3}Z_{S3} - Y_{S3}^2Z_{S2})X_{S1}^2X_{S2} - (Y_{S1}Y_{S3}Z_{S3} - Y_{S3}^2Z_{S1})X_{S2}^2X_{S1} - (Y_{S1}Y_{S2}Z_{S3} - 2Y_{S1}Y_{S3}Z_{S2} + Y_{S2}Y_{S3}Z_{S1})X_{S1}X_{S2}X_{S3} - \\
& - (Y_{S1}Y_{S3}Z_{S1} - Y_{S1}^2Z_{S3})X_{S2}^2X_{S3} + (Y_{S1}Y_{S2}Z_{S1} - Y_{S1}^2Z_{S2})X_{S3}^2X_{S2}
\end{aligned}
\tag{11}
$$

To study the output singularities, the characteristic tetrahedron was used [30], which states that a singularity occurs when the geometry of the tetrahedron is degenerate (e.g., faces are coplanar, etc.). The tetrahedron is composed of three faces defined by characteristic planes spanned by reciprocal wrenches at each spherical joint of the three kinematic chains and a base defined by the plane of the mobile platform. To define the characteristic planes associated with the three kinematic chains (for the 6-DOF parallel robot for SILS), first, the actuators are considered fixed, and then two reactive forces for the remaining passive motion are defined (at the level of the spherical joint). Figure 5 illustrates this concept on the kinematic chain 1, with the actuators q_1 and q_2, the spherical joint S_1, and the reactive forces R_1 and R_2, respectively (which span the characteristic plane P_1). Figure 6 illustrates how these planes intersect to form the characteristic tetrahedron (in a nonsingular pose).

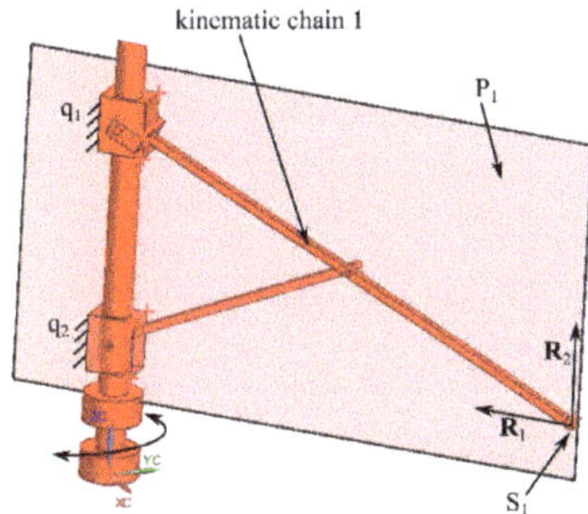

Figure 5. Characteristic plane defined for the kinematic chain 1.

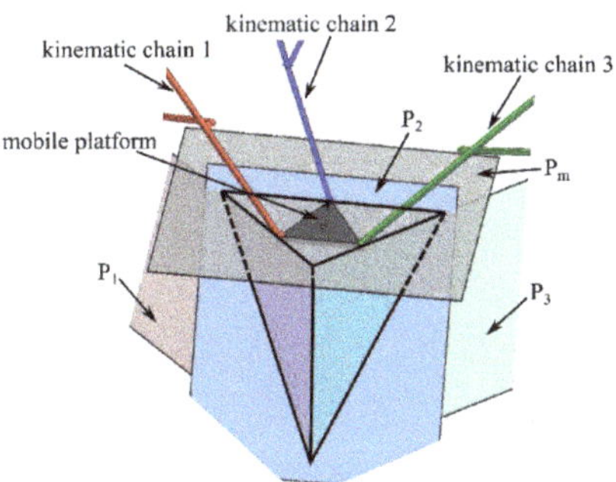

Figure 6. Characteristic tetrahedron for the 6-DOF parallel robot for SILS (nonsingular pose).

The first characteristic plane, P_1 (describing the first kinematic chain), contains the OZ axis (the actuation axis of q_1 and q_2) of the fixed coordinate system and the center of the passive spherical joint S_1 and has the following equation:

$$P_1 : (-Y_{S1})x + (X_{S1})y = 0 \tag{12}$$

Furthermore, the characteristic plane P_2 (for the second kinematic chain) contains the actuation axis of q_3 and q_4 and the center of the passive spherical joint S_2 and has the following equation:

$$P_2 : (Z_{S2} - L_V)x - (X_{S2})z + L_V X_{S2} = 0 \tag{13}$$

The characteristic plane P_3 (for the third kinematic chain) contains the actuation axis of q_5 and q_6 and the center of the passive spherical joint S_3 and has the equation:

$$P_3 : (-Y_{S3} + L_H)x + (X_{S3})y - L_H X_{S3} = 0 \tag{14}$$

Lastly, the characteristic plane P_m (for the mobile platform) contains the centers of all passive spherical joints and has the equation:

$$\begin{aligned}
&P_m : (Coef_x)x + (Coef_y)y + (Coef_z)z + Coef_d = 0 \\
&Coef_x = (Z_{S2} - Z_{S3})Y_{S1} + (Z_{S3} - Z_{S1})Y_{S2} + (Z_{S1} - Z_{S2})Y_{S3} \\
&Coef_y = (Z_{S3} - Z_{S2})X_{S1} + (Z_{S1} - Z_{S3})X_{S2} + (Z_{S2} - Z_{S1})X_{S3} \\
&Coef_z = (Y_{S2} - Y_{S3})X_{S1} + (Y_{S3} - Y_{S1})X_{S2} + (Y_{S1} - Y_{S2})X_{S3} \\
&Coef_d = (X_{S2}Y_{S1} - X_{S1}Y_{S2})Z_{S3} + (Y_{S2}Z_{S1} - Y_{S1}Z_{S2})X_{S3} + (X_{S1}Z_{S2} - X_{S2}Z_{S1})Y_{S3}
\end{aligned} \tag{15}$$

There are eight type II (output) singularities (described in a general form in [30]) that are geometrically interpreted by the degeneracy of the geometry of the characteristic tetrahedron. Each singularity case for the 3-R-$\underline{P}$RR-$\underline{P}$RS parallel robot may be checked independently using the planes defined in Equations (12)–(15) (the planes contain the faces and the base of the characteristic tetrahedron) and Equation (11), which generally describes the degeneracy conditions of the characteristic tetrahedron:

In Case 1, all faces of the characteristic tetrahedron and the base intersect in a point [30]; consequently, P_1, P_2, P_3, and P_m intersect in a point. The proof that this case represents a singularity for the 3-R-$\underline{P}$RR-$\underline{P}$RS parallel robot is straightforward. Assuming that the characteristic planes P_1, P_2, P_3, intersect at point $I(x, y, z)$. Equations (12)–(14) can be solved simultaneously to compute the intersection point's Cartesian coordinates:

$$I : \begin{cases}
x = \dfrac{L_H X_{S1} X_{S3}}{(L_H - Y_{S3})X_{S1} + X_{S3}Y_{S1}} \\[2mm]
y = \dfrac{L_H X_{S3} Y_{S1}}{(L_H - Y_{S3})X_{S1} + X_{S3}Y_{S1}} \\[2mm]
z = \dfrac{L_H L_V (X_{S1}X_{S2} - X_{S1}X_{S3}) + L_V(X_{S2}X_{S3}Y_{S1} - X_{S1}X_{S2}Y_{S2}) + L_H X_{S1}X_{S3}Z_{S2}}{((L_H - Y_{S3})X_{S1} + X_{S3}Y_{S1})X_{S2}}
\end{cases} \tag{16}$$

If point I is also contained in the characteristic plane P_m, then Equation (15) must be fulfilled for the x, y, and z Cartesian components shown in Equation (16). Substituting Equation (16) into Equation (15) leads to an equation that can be linearly solved for one Cartesian component of the three passive spherical joints. The computed solution for Z_{S1} is:

$$\begin{aligned}
Z_{S1} &= \frac{H_1}{((L_H - Y_{S3})X_{S2} - (L_H - Y_{S2})X_{S3})(X_{S1}Y_{S3} - X_{S3}Y_{S1})X_{S2}} \\
H_1 &= L_H L_V [(X_{S2} - X_{S3})((Y_{S3} - Y_{S1})X_{S1} + (Y_{S2} - Y_{S3})X_{S2} + (Y_{S1} - Y_{S2})X_{S3})X_{S1}] - \\
&\quad - L_H [(Z_{S2}(Y_{S3} - Y_{S2})X_{S2} + (Y_{S2}Z_{S3} - Y_{S3}Z_{S2})X_{S3})X_{S1}^2 + (Z_{S2}(Y_{S2} - Y_{S1})X_{S3}^2 + \\
&\quad + X_{S2}(2Y_{S1}Z_{S2} - Y_{S2}Z_{S3})X_{S3} - X_{S2}^2 Y_{S1}Z_{S3})X_{S1} + (X_{S2}Z_{S3} - X_{S3}Z_{S2})X_{S2}X_{S3}Y_{S1}] - \\
&\quad - L_V[((Y_{S2} - Y_{S3})X_{S1} + (Y_{S3} - Y_{S1})X_{S3} + (Y_{S1} - Y_{S2})X_{S3})(X_{S1}Y_{S3} - X_{S3}Y_{S1})X_{S2}] + \\
&\quad + ((Y_{S2}Z_{S3} - Y_{S3}Z_{S2})X_{S1} - (X_{S2}Z_{S3} - X_{S3}Z_{S2})Y_{S1})(X_{S1}Y_{S3} - X_{S3}Y_{S1})X_{S2}
\end{aligned} \tag{17}$$

Equation (17) describes the constraint of the Z_{S1} parameter when all four planes intersect at a point. Substituting Equation (17) into Equation (11) causes H_0 to vanish, proving the singularity for the 3-R-$\underline{P}$RR-$\underline{P}$RS parallel robot.

To illustrate an example of this singularity, the geometric parameters $\{L_H = 1420, L_V = 1000, l_p = 260\}$ [mm] and the mobile platform coordinates $\{X_E = 400$ mm, $Y_E = 700$ mm, $Z_E = 500$ mm, $\psi = 0$ rad, $\varphi = \pi/10$ rad$\}$ (arbitrary chosen) are considered (only five output parameters are considered, the last one, θ, is computed). Substituting the numerical values into Equation (A1) (see Appendix A) and solving the equation for θ (using the solve function in Maple), yields four real solutions $\{\theta = 0.93$ rad, $\theta = -1.755$ rad, $\theta = 1.82$ rad, $\theta = -2.40$ rad$\}$. Figure 7a shows the parallel robot in the singular pose (for $\theta = 0.93$), whereas Figure 7b shows how the characteristic planes P_1, P_2, P_3, and P_m intersect at a point.

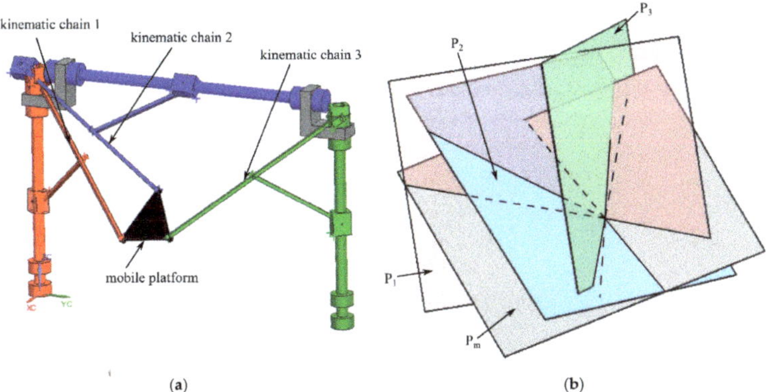

Figure 7. Type II singularity of the 6-DOF parallel robot for SILS (case 1—the characteristic tetrahedron faces and base intersect in a point): (**a**) parallel robot pose; (**b**) characteristic planes intersect in a point.

In Case 2, the faces of the characteristic tetrahedron (i.e., P_1, P_2, P_3) intersect in a line, but no faces are coplanar [30]. For the 6-DOF parallel robot for SILS, this case is impossible since P_1 and P_3 are always vertical, whereas P_2 has only two configurations when it is vertical (when the kinematic chain 2 points upwards or downwards). However, when all of the characteristic planes, P_1, P_2, and P_3, are vertical, they are all coplanar (due to the mechanism topology); therefore, P_1, P_2, and P_3 cannot intersect in a line. This is straightforward to prove using Equations (12)–(14). Assuming that the planes P_1, P_2, and P_3 intersect in a line, then the rank of the coefficient's matrix (matrix containing the x, y, and z coefficients from the plane equations) $\mathbf{M}_1$ and the rank of the augmented matrix (matrix containing the x, y, and z coefficients and the free terms) $\mathbf{M}_2$ must be 2. The two matrices are defined using Equations (12)–(14):

$$\mathbf{M}_1 : \begin{bmatrix} -Y_{S1} & X_{S1} & 0 \\ -L_V + Z_{S2} & 0 & -X_{S2} \\ L_H - Y_{S3} & X_{S3} & 0 \end{bmatrix}, \mathbf{M}_2 : \begin{bmatrix} -Y_{S1} & X_{S1} & 0 & 0 \\ -L_V + Z_{S2} & 0 & -X_{S2} & L_V X_{S2} \\ L_H - Y_{S3} & X_{S3} & 0 & -L_H X_{S3} \end{bmatrix} \tag{18}$$

It can be checked that the rank of $\mathbf{M}_1$ is 2 for:

$$X_{S2} = 0, \text{ and/or } X_{S1} = \frac{Y_{S1} X_{S3}}{Y_{S3} - L_H} \tag{18a}$$

but for the solutions shown in Equation (18a) the rank of $\mathbf{M}_2$ is 3 (the planes intersect in two or three lines). The rank of $\mathbf{M}_2$ is 2 for:

$$X_{S2} = 0, \text{ and } X_{S1} = 0 \tag{18b}$$

in which case the rank of $\mathbf{M}_1$ is also 2 (describing two coincident planes and the third one intersecting it). Furthermore, for:

$$X_{S2} = 0, \text{ and } X_{S1} = 0, \text{ and } X_{S3} = 0 \tag{18c}$$

the rank of both $\mathbf{M}_1$ and $\mathbf{M}_2$ is 1 (describing three coincident planes).

In Case 3, two of the tetrahedron faces and its base intersect in a line [30]. Considering the characteristic planes P_1, P_3, and P_m, the coefficient and augmented matrices are:

$$\mathbf{M}_3 : \begin{bmatrix} -Y_{S1} & X_{S1} & 0 \\ L_H - Y_{S3} & X_{S3} & 0 \\ Coef_x & Coef_y & Coef_z \end{bmatrix}, \mathbf{M}_4 : \begin{bmatrix} -Y_{S1} & X_{S1} & 0 & 0 \\ L_H - Y_{S3} & X_{S3} & 0 & -L_H X_{S3} \\ Coef_x & Coef_y & Coef_z & Coef_d \end{bmatrix} \tag{19}$$

It can be checked that the rank of both matrices $\mathbf{M}_3$ and $\mathbf{M}_4$ is 2 for:

$$X_{S1} = \frac{X_{S3} Y_{S1}}{Y_{S3}}, \text{ and } X_{S2} = \frac{X_{S3} Y_{S2}}{Y_{S3}} \tag{19a}$$

Furthermore, substituting the values of Equation (20) into Equation (19) shows no proportionality between the matrix's rows; therefore, the intersection describes a line. In addition, substituting Equation (20) into Equation (11) causes the factor H_0 to vanish.

Another geometric configuration for Case 3 is represented by the characteristic planes P_1, P_2, and P_m intersecting in a line (the case when P_3, P_2, and P_m intersect in a line is redundant with this one due to the configuration symmetry of the parallel robot). The coefficient and augmented matrices are:

$$\mathbf{M_5}:\begin{bmatrix} -Y_{S1} & X_{S1} & 0 \\ -L_V + Z_{S2} & 0 & -X_{S2} \\ Coef_x & Coef_y & Coef_z \end{bmatrix},\ \mathbf{M_6}:\begin{bmatrix} -Y_{S1} & X_{S1} & 0 & 0 \\ -L_V + Z_{S2} & 0 & -X_{S2} & L_V X_{S2} \\ Coef_x & Coef_y & Coef_z & Coef_d \end{bmatrix} \tag{20}$$

The rank of both matrices $\mathbf{M_5}$ and $\mathbf{M_6}$ is 2 for:

$$X_{S1} = \frac{X_{S3}(L_V - Z_{S1})}{L_V - Z_{S3}},\ \text{and}\, X_{S2} = \frac{X_{S3}(L_V - Z_{S2})}{L_V - Z_{S3}} \tag{20a}$$

Substituting the values of Equation (20a) into Equation (20) shows no proportionality between the matrix's rows (the intersection describes a line), and substituting Equation (20a) into Equation (11) causes factor H_0 to vanish.

As an example, for the 3-R-P̲R̲R-P̲RS parallel robot, for $\theta = 0$ rad, $\varphi = \pi/2$ rad, the factor F (Equation (A1)—Appendix A) vanishes. In this configuration, the mobile platform is in a vertical pose. Figure 8 shows an example of this singularity, (a) the parallel robot poses and (b) the characteristic planes (P_1, P_3, P_m) intersecting in a line. It can be shown (at least for this example) that the characteristic plane P_2 also intersects that line; therefore, this singularity is the special case of Case 1 discussed previously (since all the characteristic planes intersect at a point).

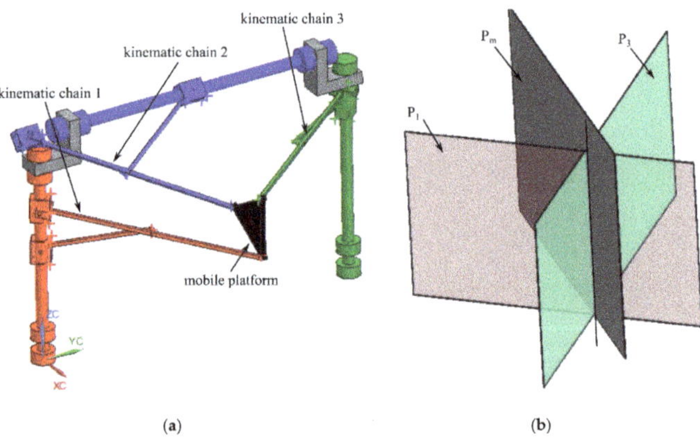

Figure 8. Type II singularity of the 6-DOF parallel robot for SILS (case 3—two faces of the characteristic tetrahedron and its base intersect in a line): (**a**) parallel robot pose; (**b**) characteristic planes intersect in a line.

In Case 4, two of the characteristic tetrahedron faces are coplanar [30]. Due to the arguments presented in Case 2 for the 6-DOF parallel robot for SILS, the only possible tetrahedron faces that can become coplanar are the ones defined by P_1 and P_3. The coefficient and augmented matrices in this case are:

$$\mathbf{M_7}:\begin{bmatrix} -Y_{S1} & X_{S1} & 0 \\ L_H - Y_{S3} & X_{S3} & 0 \end{bmatrix},\ \mathbf{M_8}:\begin{bmatrix} -Y_{S1} & X_{S1} & 0 & 0 \\ L_H - Y_{S3} & X_{S3} & 0 & -L_H X_{S3} \end{bmatrix} \tag{21}$$

If P_1 and P_3 are coplanar then the ranks of $\mathbf{M_7}$ and $\mathbf{M_8}$ must both be 1. It can be checked that this is the case for:

$$X_{S1} = 0,\ \text{and}\, X_{S3} = 0 \tag{21a}$$

Substituting Equation (21a) into Equation (11) causes factor H_0 to vanish, proving the singularity. As an example, for the numerical values $X_E = -\frac{\sqrt{3}}{6} l_p$ mm, $\psi = 0$ rad, $\theta = 0$ rad, the factor F vanishes. Figure 9a shows this singular configuration for the parallel robot, and Figure 9b shows P_1 and P_3 being coplanar.

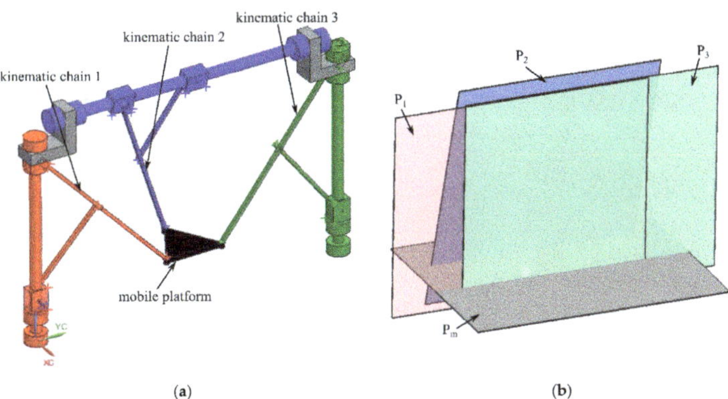

Figure 9. Type II singularity of the 6-DOF parallel robot for SILS (Case 4—two faces of the characteristic tetrahedron are coplanar): (**a**) parallel robot pose; (**b**) characteristic planes.

In Case 5, one side and the base of the characteristic tetrahedron are coplanar [30]. There are three general configurations for the 3-R-$\underline{P}$RR-$\underline{P}$RS parallel robot that correspond to this case. The first of these general configurations is represented by P_2 and P_m being coplanar, whereas the second and third general configurations for this singularity are achieved when either P_1 or P_2 are coplanar with P_m. Considering the first case, the coefficient and augmented matrices are:

$$\mathbf{M}_9 : \begin{bmatrix} -L_V + Z_{S2} & 0 & -X_{S2} \\ Coef_x & Coef_y & Coef_z \end{bmatrix}, \mathbf{M}_{10} : \begin{bmatrix} -L_V + Z_{S2} & 0 & -X_{S2} & L_V X_{S2} \\ Coef_x & Coef_y & Coef_z & Coef_d \end{bmatrix} \tag{22}$$

It can be shown that both matrices $\mathbf{M}_9$ and $\mathbf{M}_{10}$ have rank 1 for:

$$\begin{aligned} X_{S2} &= \frac{(Y_{S2} - Y_{S3})X_{S1} + (Y_{S1} - Y_{S2})X_{S3}}{Y_{S1} - Y_{S3}}, \text{ and} \\ Z_{S1} &= \frac{(Y_{S1} - Y_{S3})Z_{S2} + (Y_{S2} - Y_{S1})Z_{S3}}{Y_{S2} - Y_{S3}} \end{aligned} \tag{22a}$$

Substituting Equation (22a) into Equation (11) causes factor H_0 to vanish. It is easy to find relationships that describe this singularity case when P_1 or P_2 are coplanar with P_m.

An example is given for the geometric parameters $\{L_H = 1420$ mm, $L_V = 1000$ mm, $l_p = 260$ mm$\}$, and for the mobile platform coordinates $\{X_E = 350$ mm, $Y_E = 750$ mm, $Z_E = 200$ mm, $\psi = 0$ rad, $\varphi = 0$ rad$\}$ (arbitrary chosen for the example purpose). Solving the factor F for the numerical values yields two real solutions $\{\theta = 1.158$ rad, $\theta = 1.9$ rad$\}$. Figure 10 illustrates this singularity for $\theta = 1.158$ rad. Another example is shown for the mobile platform coordinates $\{X_E = 350$ mm, $Y_E = 750$ mm, $Z_E = 200$ mm, $\psi = 0$ rad, $\theta = \pi/2$ rad$\}$ and the computed angle for $\varphi = 0.44$ rad. In this example (Figure 11) P_1 and P_m are coplanar.

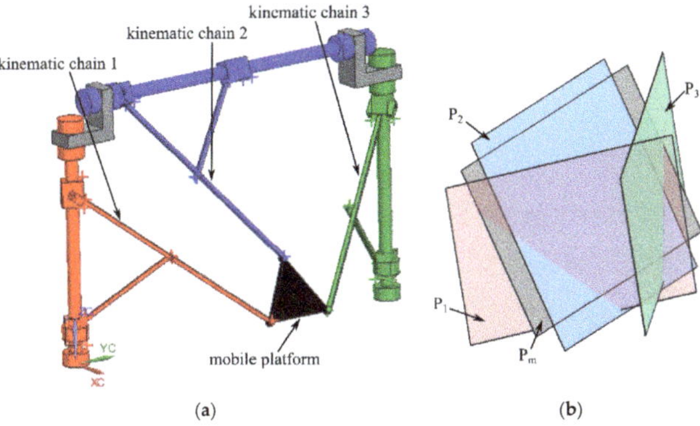

Figure 10. Type II singularity of the 6-DOF parallel robot for SILS (case 5—one face and the base of the characteristic tetrahedron are coplanar): (**a**) parallel robot pose; (**b**) P_2 and P_m are coplanar.

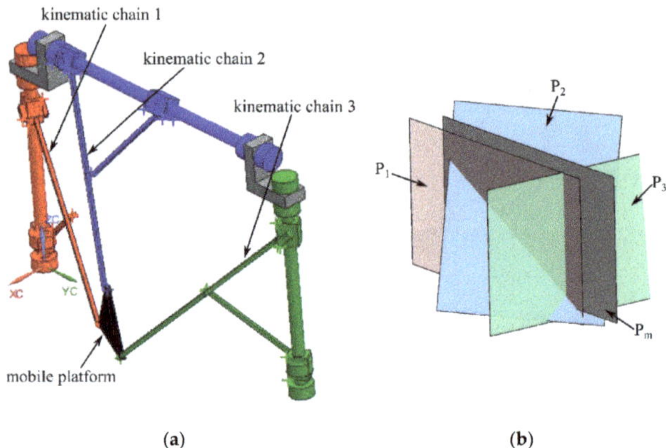

Figure 11. Type II singularity of the 6-DOF parallel robot for SILS (case 5—one face and the base of the characteristic tetrahedron are coplanar): (**a**) parallel robot pose; (**b**) P_1 and P_m are coplanar.

In Case 6, two sides and the base of the characteristic tetrahedron are coplanar [30]. For the 6-DOF parallel robot for SILS this case is impossible (due to the mechanism topology). Any two faces of the characteristic tetrahedron are coplanar if and only if the faces are also coplanar with the YOZ plane (Figure 2). This condition can be proven in two steps. In the first step, P_1, P_3, and P_m are assumed to be all coplanar. Consequently, the ranks of $\mathbf{M}_3$ and $\mathbf{M}_4$ (Equation (19)) must both be 1, which can be achieved if:

$$X_{S1} = 0, \text{ and } X_{S2} = 0, \text{ and } X_{S3} = 0 \tag{23}$$

It can be checked that Equation (23) is the only solution for which P_1, P_3, and P_m are all coplanar. This is also easy to see geometrically. The actuation axes of q_1 and q_2 (line contained in P_1) and q_5 and q_6 (line contained in P_3) are both included in the YOZ plane of the fixed coordinate system. If P_1 and P_3 are coplanar, then the actuation axis of q_1 and q_2 must be contained in P_3, and reciprocally, the actuation axis of q_5 and q_6 must be contained in P_1. Therefore, the planes P_1 and P_3 (and any other plane, e.g., P_m) are coplanar if they are coplanar with YOZ, the result is shown by Equation (23). In the second step, P_1, P_2, and P_m are assumed to be all coplanar. Consequently, the ranks of $\mathbf{M}_5$ and $\mathbf{M}_6$ (Equation (20)) must both be 1, which can be achieved if (again) the conditions from Equation (23) are met. It can be shown, by the same argument as above, that if the planes P_1, P_2, and P_m are coplanar, then they must be coplanar with YOZ, the result is shown by Equation (23). Since Equation (23) represents a (unique) solution for both cases (P_1, P_3, and P_m being coplanar and P_1, P_2, and P_m being coplanar), the conclusion is that if P_1, P_3, and P_m are coplanar, they must also be coplanar with P_2 (condition discussed in Case 8).

In Case 7, one side and the base of the tetrahedron are coplanar and the other two sides of the tetrahedron are also coplanar [30]. Considering all four characteristic planes, the coefficient and augmented matrices are:

$$\mathbf{M}_{11}: \begin{bmatrix} -Y_{S1} & X_{S1} & 0 \\ -L_V + Z_{S2} & 0 & -X_{S2} \\ L_H - Y_{S3} & X_{S3} & 0 \\ Coef_x & Coef_y & Coef_z \end{bmatrix}, \ \mathbf{M}_{12}: \begin{bmatrix} -Y_{S1} & X_{S1} & 0 & 0 \\ -L_V + Z_{S2} & 0 & -X_{S2} & L_V X_{S2} \\ L_H - Y_{S3} & X_{S3} & 0 & -L_H X_{S3} \\ Coef_x & Coef_y & Coef_z & Coef_d \end{bmatrix} \tag{24}$$

The ranks of the matrices $\mathbf{M}_{11}$ and $\mathbf{M}_{12}$ is 2 (describing pairs of coplanar planes that intersect in a line), for:

$$X_{S1} = 0, \text{ and } X_{S3} = 0, \text{ and } Z_{S1} = L_V, \text{ and } Z_{S3} = L_V \tag{24a}$$

Substituting Equation (26) in Equation (11) causes H_0 to vanish.

For the 3-R-$\underline{P}$RR-$\underline{P}$RS parallel robot, this case can only be achieved if P_1 is coplanar with P_3 and P_2 is coplanar with P_m. The factor F vanishes for the geometric values parameters $\{L_H = 1420 \text{ mm}, L_V = 1000 \text{ mm}, l_p = 260 \text{ mm}\}$, and for the mobile platform coordinates $\{X_E = -75.05 \text{ mm}, Y_E = 750 \text{ mm}, Z_E = 1000 \text{ mm}, \psi = 0 \text{ rad}, \theta = 0 \text{ rad}, \varphi = 0 \text{ rad}\}$. This configuration is illustrated in Figure 12.

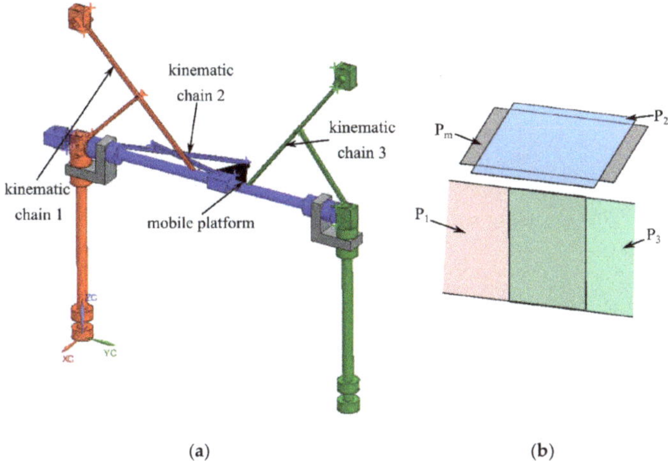

(a) **(b)**

Figure 12. Type II singularity of the 6-DOF parallel robot for SILS (case 7—one face and the base of the characteristic tetrahedron are coplanar and the other two faces of the tetrahedron are also coplanar): (**a**) parallel robot pose; (**b**) characteristic planes.

In Case 8, all of the faces and the base of the characteristic tetrahedron are coplanar [30]. The coefficient and augmented matrices for this case are already shown in Equation (24). All four planes are coplanar if the ranks of the matrices $\mathbf{M}_{11}$ and $\mathbf{M}_{12}$ are both 1. This is achieved by setting:

$$X_{S1} = 0, \text{ and } X_{S2} = 0, \text{ and } X_{S3} = 0 \tag{25}$$

Substituting Equation (25) into Equation (11) causes factor H_0 to vanish. For the 3-R-$\underline{P}$RR-$\underline{P}$RS parallel robot, this singularity is achieved if $\{X_E = 0 \text{ mm}, \theta = \pm\pi/2 \text{ rad}\}$. This configuration is illustrated in Figure 13.

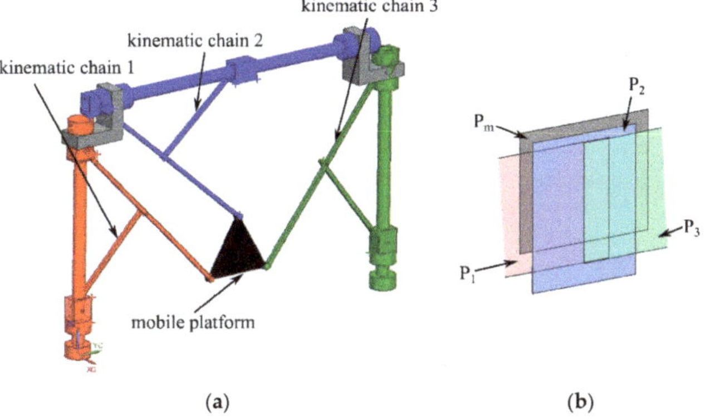

(a) **(b)**

Figure 13. Type II singularity of the 6-DOF parallel robot for SILS (case 8—all the faces and the base of the characteristic tetrahedron are coplanar): (**a**) parallel robot pose; (**b**) characteristic planes.

4.3. Type III Singularities

Since the determinant of the Jacobian matrix A is free of input parameters (i.e., it only depends on the geometric parameters and the outputs $\{X_E, Y_E, Z_E, \psi, \theta, \varphi\}$), whereas the determinant of the Jacobian matrix B is free of output parameters (it depends only on the geometric parameters and the inputs $\{q_1, q_2, q_3, q_4, q_5, q_6\}$), it is feasible to assume that type III singularities may occur. However, as stated before, the type I singularities are easily avoidable in the design stage; hence it can be stated that the 6-DOF parallel robot for SILS will have no type III singularities.

5. Geometric Optimization Algorithm

The 6-DOF parallel robot for SILS was geometrically optimized with respect to an operational workspace defined to comply with the medical task. Several optimization criteria were defined, and based on these criteria, appropriate input data were defined for the optimization algorithm.

5.1. Optimization Criteria

The following criteria (with the following importance order *Criterion 1* is more important than *Criterion 2*, which is equally important as *Criterion 3*) were defined with respect to the SILS task:

- *Criterion 1—Operational workspace.* The proposed operational workspace is a cylinder shape (see Section 3). Furthermore, the intervals for the orientation angles of the endoscopic camera must be (for the entire operational workspace):

$$\psi, \theta, \varphi \in [-20 \ 20][°] \tag{26}$$

- *Criterion 2—Footprint.* To minimize the robot footprint, the following conditions were imposed:

$$L_H < 600 \ [\text{mm}], \ L_V < 600 \ [\text{mm}] \tag{27}$$

- *Criterion 3—Dexterity.* The robot should have adequate performance with respect to dexterity. This work uses an approximation of the Global Conditioning Index (GCI_a) to assess the 6-DOF parallel robot dexterity which is computed as [16]:

$$\text{GCI}_a = \frac{\sum\limits_{i=1}^{n} \frac{1}{k_i}}{n} \tag{28}$$

where k_i represents the condition number [16] of the ith point from the (discrete generated) operational workspace, which is computed using Equation (29), and n is the number of points within the operational workspace. For a given point, the condition number is:

$$k = \| \mathbf{J} \| \ \| \mathbf{J}^{-1} \| \tag{29}$$

where $\| \cdot \|$ represents the norm of the Jacobian matrix $\mathbf{J}$, which is computed in this work with:

$$\| \mathbf{J} \| = \sqrt{\text{trace}(\mathbf{J}\mathbf{J}^{\mathrm{T}})} \tag{30}$$

where $\mathbf{J}$ is computed using the input and output Jacobian matrices (Equation (6)), as follows:

$$\mathbf{J} = -\mathbf{B}^{-1} \cdot \mathbf{A} \tag{31}$$

An important note is that condition number k is a measure of dexterity at one specific robot configuration, and the lower its value, the better (a minimum value of 1 represents isotropy [16]). For the approximation of GCI, the inverse of k is used, which is bounded by 0 and 1 (in this case the higher the better). GCI_a is a measure of the average of the inverse of k computed from all points within the operational workspace. However, due to the drawbacks of the condition number and the GCI (when the robot has translation and rotational motions), the values of GCI_a (i.e., Criterion 3) will be used as a guiding value, not as a definite decision parameter.

5.2. Geometric Optimization Algorithm Description

The optimization algorithm is presented next in pseudocode (Algorithm 1).

The test operational workspace WS_DATA is defined (1) and is used in future steps to determine the validity of the possible optimized solutions. Note that the proposed algorithm yields solutions (stacked in SOLS(M)) for a single parallel robot pose (Cartesian coordinates and orientation) within the desired cylindrical operational workspace. These solutions must be subsequently validated for the entire WS_DATA. The objectives OBJ of optimization (2) are to minimize the seven geometric parameters of the parallel robot subject to the kinematic constraint C, which are evaluated with values for the mobile platform coordinates within the operational workspace. The numerical intervals for the optimization process are defined (3); L defines intervals for the robot geometric parameters (for the optimization objectives), whereas CYL ensures that the Cartesian coordinates (for the optimum solutions) are within the desired cylinder, and ANG ensures that the mobile platform orientations are within the proper ranges. Note, CYL and ANG are different from WS_DATA; CYL and ANG are used by the gamultiobj function (from Matlab [31]) to ensure that the objectives are always minimized with respect to robot poses within the operational workspace, whereas (as pointed out before) WS_DATA is used to evaluate (in a discrete manner) if the generated solutions are valid for the entire operational workspace. The optimization is performed (4) iteratively until the solutions set SOLS(M) fills as best as possible the operational workspace (considering, of course, the Cartesian coordinates and the orientations of the mobile platform). Then, a solution subset SOL(K) (for the minimized geometric parameters) is defined for the optimized solutions that are valid for the entire WS_DATA (i.e., geometric parameters that yield real solutions for all WS_DATA). If no such subset exists, (3) is repeated. Then, the solution subset is analyzed to determine the ranges of motion Q_RANGE(K) for the active joints q_i ($i = 1, \ldots ,6$) and if there exist ranges that yield feasible mechanisms (without crossing the actuation axes of the active joints, etc.), the solutions (yielding the adequate ranges) are subsequently saved in FINAL(H). If there are no adequate ranges for the active joints (to yield feasible mechanisms), (3) is repeated. The GCI is computed for all of the remaining solutions in FINAL(H), and the design solution is defined (not automatically but by the robot designers).

Algorithm 1. The optimization algorithm.

---

0. **BEGIN OPTIMIZATION**

1. **Define the test operational workspace**

    1.1. Input the cartesian coordinates $E(N)$ [X_{Ei}, Y_{Ei}, Z_{Ei}] ($i = 1, \ldots ,n$) for the operating workspace

    1.2. Input the required orientation angles ψ, θ, φ

    1.3. Compute **WS_DATA** by assigning all angles ψ, θ, φ for all $E(i) \in E(N)$

    1.4. *Goto* **2**

2. **Define objective functions and constraints**

    2.1. Define objective functions **OBJ** [$o_1, o_2, o_3, o_4, o_5, o_6, o_7$]

    2.2. Define constraints: C [$C_1, C_2, C_3, h_1, h_2, h_3, h_4, h_5, h_6$]

    2.3. *Goto* **3**

3. **Input the dimension intervals for the optimization process**

    3.1. Define and input **L** [$l_{p\_min}, l_{p\_max}, L_{H\_min}, L_{H\_max}, L_{V\_min}, L_{V\_max}, l_{1\_min}, l_{1\_max},$ $l_{2\_min}, l_{2\_max}, l_{3\_min}, l_{3\_max}, l_{4\_min}, l_{4\_max}$]

    3.2. Define and input **CYL** [$R_{\_min}, R_{\_max}, H_{\_min}, H_{\_max}, \alpha_{\_min}, \alpha_{\_max}$]

    3.3. Define and input **ANG** [$\psi_{\_min}, \psi_{\_max}, \theta_{\_min}, \theta_{\_max}, \phi_{\_min}, \phi_{\_max}$]

    3.4. *Goto* **4**

4. **Minimize robot dimensions**

    4.1. Compute the Pareto front (iteratively) denoted **SOLS(M)** by minimizing **OBJ** subject to the constraints C and the intervals **L, CYL, ANG**.

    4.2. Define a smaller solution set **SOL(K)** $\in$ **SOLS(M)** such that for every **SOL(i)** ($i = 1, \ldots ,K$), q_i ($i = 1, \ldots ,6$) $\in \mathbb{R}$ for every point in **WS_DATA**

    4.3. *If* **SOL(K)** exists, *then Goto* 4.4, *else Goto* **3**

    4.4. Compute the active joints ranges **Q_RANGE(K)** [q_{i_min}, q_{i_max}] ($i = 1, \ldots ,6$) for every **SOL(i)** ($i = 1, \ldots ,K$)

    4.5. *If* **Q_RANGE(j)** ($j = 1, \ldots ,K$) is acceptable (mechanism is feasible), *then* save the solution **SOL(j)** into **FINAL(H)**

    4.6. *If* **FINAL(H)** is empty, *then Goto* **3**, *else Goto* 4.7

    4.7. Compute GCI for every **FINAL(h)** ($h = 1, \ldots ,H$)

    4.8. Define the most optimal solution

    4.9. *Goto* **5**

5. **END OPTIMIZATION**

---

6. The Optimized 6-DOF Parallel Robot for SILS

6.1. The Geometric Optimization

The parallel robot was optimized using the algorithm presented in Section 5.2 based on the following inputs:

- The test workspace data (WS_DATA) was defined based on a cylinder with:

$$C(X_C = 290, Y_C = 415, Z_C = -75) \ [\text{mm}]$$
$$R = 75 \ [\text{mm}], h = 75 \ [\text{mm}]$$

$$(32)$$

where C is the center of the base circle (with respect to the fixed coordinate system of the robot), R is the circle's radius, and h is the cylinder height. WS_DATA was generated by discretizing the cylinder and adding the orientations for the mobile platform, based on the following:

$$E_i = \begin{cases} X_{Ei} = X_C + R_C \cos(\alpha) \\ Y_{Ei} = Y_C + R_C \sin(\alpha) \\ Z_{Ei} = H_C \\ \psi_i = \psi \\ \theta_i = \theta \\ \varphi_i = \varphi \end{cases}, \quad \begin{aligned} & R_C \in [0, 75]\,[\text{mm}], \text{increment} = 7.5\,[\text{mm}] \\ & H_C \in [-75, 0]\,[\text{mm}], \text{increment} = 15\,[\text{mm}] \\ & \alpha \in [0, 360]\,[°], \text{increment} = 10\,[°] \\ & \psi, \theta, \varphi \in [-20, 20]\,[°], \text{increment} = 2.5\,[°] \end{aligned} \tag{33}$$

where E_i represents the ith mobile platform configuration within the operational workspace. WS_DATA contained approximately 9.78 million unique sets of mobile platform coordinates (a better resolution was not achievable due to computation power limitations). Figure 14 illustrates the point cloud defining the discrete operational workspace (only in Cartesian coordinates).

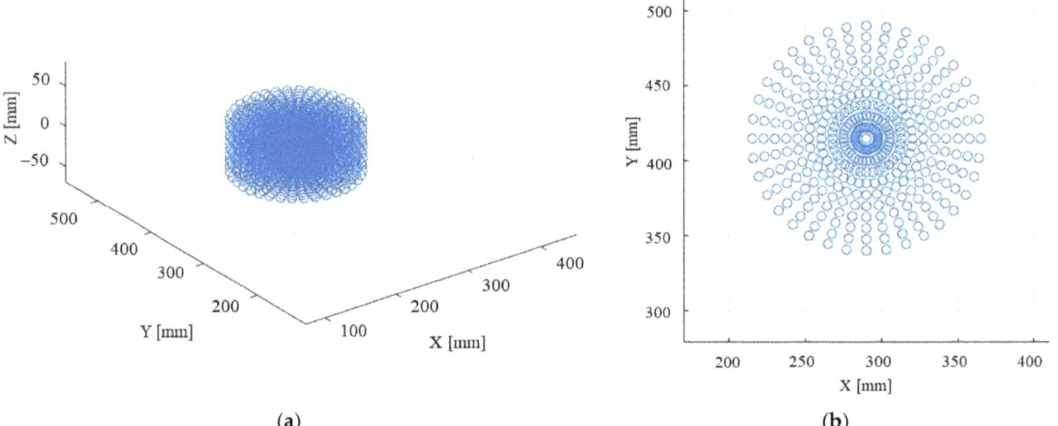

(a)   (b)

Figure 14. The generated discrete workspace (WS_DATA) for the SILS robotic systems (only the Cartesian coordinates): (**a**) isometric view; (**b**) Z axis view.

- The input intervals L for the geometric parameters for the 6-DOF SILS robot where:

$$l_1 \in [140, 200]\,[\text{mm}], l_2 \in [350, 600]\,[\text{mm}], l_3 \in [200, 350]\,[\text{mm}], l_4 \in [200, 350]\,[\text{mm}],$$
$$l_p \in [200, 250]\,[\text{mm}], L_H \in [500, 600]\,[\text{mm}], L_V \in [200, 600]\,[\text{mm}] \tag{34}$$

  Furthermore, the CYL intervals were:

$$R \in [0, 75]\,[\text{mm}], H \in [-75, 0]\,[\text{mm}], \alpha \in [0, 360]\,[°] \tag{35}$$

  and the ANG intervals:

$$\psi \in [-20, 20]\,[°], \theta \in [-20, 20]\,[°], \varphi \in [-20, 20]\,[°] \tag{36}$$

- The objective functions OBJ to be minimized (to reduce the size of the robot for a given operational workspace) and the constraints C were:

$$\text{OBJ} = [l_p, L_H, L_V, l_1, l_2, l_3, l_4],$$
$$C = [X_E, Y_E, Z_E, h_1, h_2, h_3, h_4, h_5, h_6] \tag{37}$$

where h_i $(i = 1, \ldots, 6)$ are due to Equation (5):

$$\begin{cases} h_1 : \frac{1}{l_1 + l_2} \left[(l_1 + l_2)q_2 - 2l_1 \sqrt{-X_{S1}^2 - Y_{S1}^2 + (l_1 + l_2)^2} \right] - q_1 = 0 \\ h_2 : \sqrt{(l_1 + l_2)^2 - X_{S1}^2 - Y_{S1}^2} + Z_{S1} - q_2 = 0 \\ h_3 : \frac{1}{l_1 + l_3} \left[(l_1 + l_3)q_4 + 2l_1 \sqrt{-X_{S2}^2 + (l_1 + l_3)^2 - (L_V - Z_{S2})^2} \right] - q_3 = 0 \\ h_4 : Y_{S2} - \sqrt{(l_1 + l_3)^2 - X_{S2}^2 - (L_V - Z_{S2})^2} - q_4 = 0 \\ h_5 : \frac{1}{l_1 + l_4} \left[(l_1 + l_4)q_6 - 2l_1 \sqrt{-X_{S3}^2 + (l_1 + l_4)^2 - (L_H - Y_{S3})^2} \right] - q_5 = 0 \\ h_6 : \sqrt{(l_1 + l_4)^2 - X_{S3}^2 - (L_H - Y_{S3})^2} + Z_{S3} - q_6 = 0 \end{cases} \tag{38}$$

and:

$$\begin{aligned} X_E &= X_C + R\cos(\beta), \\ Y_E &= Y_C + R\sin(\beta), \\ Z_E &= H \end{aligned} \tag{39}$$

- The gamultiobj function was iterated 200 times with the defined parameters, and with each iteration, the data were saved within the solution set SOLS. The gamultiobj function uses random number generators, and multiple iterations ensured that the optimized solution set spanned the "entire" operational workspace. The hypothesis is that the large number of solutions that spanned the entire operational workspace led to a better probability of finding feasible solutions (that are tested with the WS_DATA) in the next step. The set SOLS(m) (m = 1, … ,28,568) includes numerical values for the mobile platform coordinates, the geometric parameters, and the active joints. Figure 15 illustrates a point cloud based on the Cartesian coordinates within SOLS(m) (m = 1, … ,28,568), whereas Figure 16 illustrates the distribution of SOLS(m) (m = 1, … ,28,568) with respect to the mobile platform coordinates.
- A subset SOL(k) (k = 1, … ,931) was selected from SOLS(m) (m = 1, … ,28,568) which yield real values for q_i $(i = 1, \ldots, 6)$ for all mobile platform coordinates in WS_DATA. Furthermore, the ranges Q_RANGE(k) for q_i $(i = 1, \ldots, 6)$ of each solution the subset SOL(k) were evaluated to determine which solutions yield feasible mechanisms (where, e.g., the actuation axes do not cross). For the viable solutions FINAL(h) (h = 1, … ,9), the GCI was computed for the WS_DATA. Table 1 shows the resulting feasible solutions (FINAL(h)) for the geometric parameters of the 6-DOF parallel robot for SILS.
- The design solution was Sol. no. 1 from Table 1, not because it has the best value for GCI_a, but because it shows a "good" compromise between the footprint (e.g., L_H) and the computed GCI_a index (it ranks third with respect to L_H and first with respect to GCI_a). Table 2 shows the ranges of the active joins for the selected design solution.

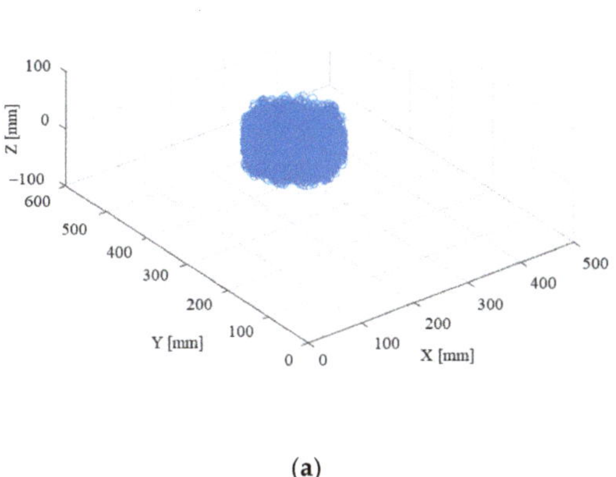

(a)

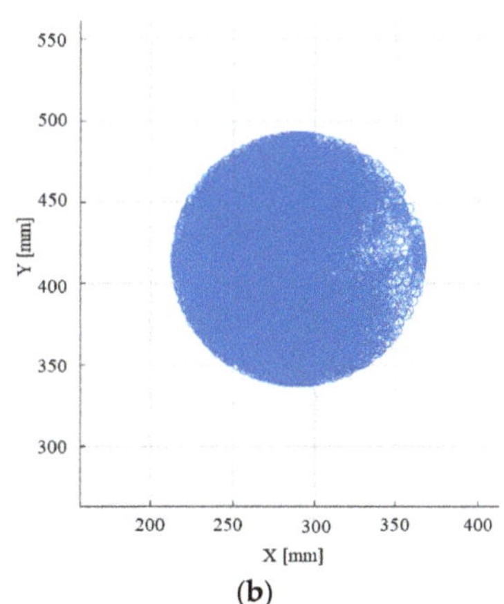

(b)

Figure 15. Point cloud spanned by the solution set SOLS(m) (m = 1, … ,28,568): (**a**) isometric view; (**b**) Z axis view.

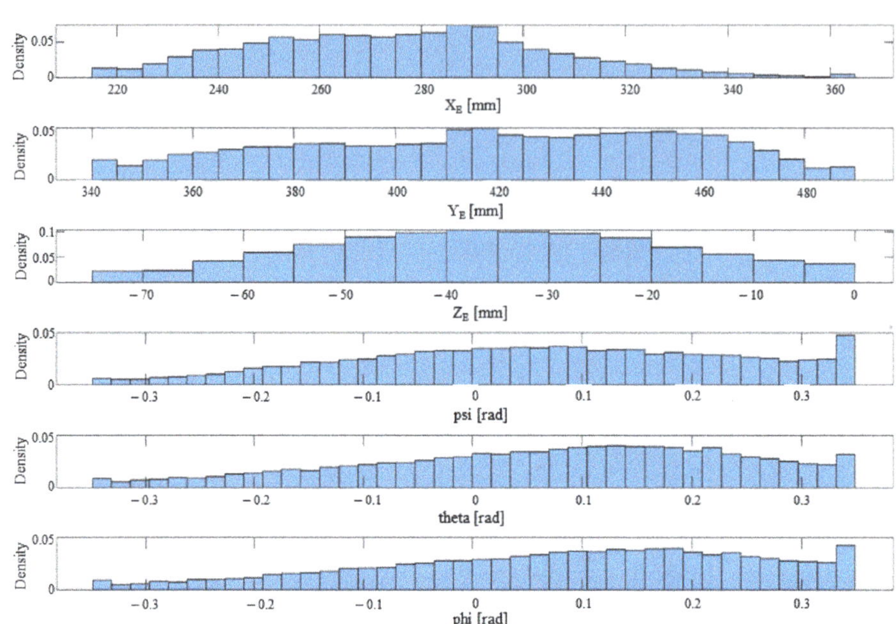

Figure 16. Mobile platform coordinates distribution density for the solution set SOLS(m) (m = 1, . . . , 28,568).

Table 1. Results from the geometric optimization algorithm.

Sol. No.	l_p [mm]	L_H [mm]	L_V [mm]	l_1 [mm]	l_2 [mm]	l_3 [mm]	l_4 [mm]	GCI_a
1	215.25	558.86	237.03	158.57	596.12	283.74	342.68	0.192
2	215.51	599.99	206.68	156.04	552.89	269.82	328.02	0.188
3	211.56	575.65	237.92	178.22	480.05	349.99	326.12	0.175
4	222.53	580.55	238.12	173.55	576.46	283.69	314.10	0.180
5	232.36	546.54	228.56	155.83	536.28	291.79	328.11	0.157
6	214.9	506.28	231.16	155.45	501.26	282.31	312.53	0.156

Table 2. Active joints ranges for the design solution.

	q_1 [mm]	q_2 [mm]	q_3 [mm]	q_4 [mm]	q_5 [mm]	q_6 [mm]
min	38.2	179.6	197.8	6.8	−60.8	80.4
max	347.5	592.1	476.6	398.8	193.2	465.2

Other authors used the NSGA-II algorithm [37] for multi-objective optimization problems (see, e.g., [38]) due to its computational efficiency and algorithm stability. Furthermore, there is no guarantee that the optimization algorithm used in this work yields a global optimum solution. However, based on multiple runs of the optimization algorithm (which yielded very similar results), the conclusion that the resulting design solutions are at least stable local ones is not implausible.

6.2. The Optimized Model of the 6-DOF Parallel Robot

Figure 17 shows the CAD model of the design solution of the 6-DOF parallel robot for SILS. Figure 17a shows the proposed relative position between the robotic system and the patient. Figure 17b shows the CAD model of the 6-DOF parallel robot with its actuators. The actuator positions are an initial concept that are subject to change in the later design stages based on the technical requirements and constraints.

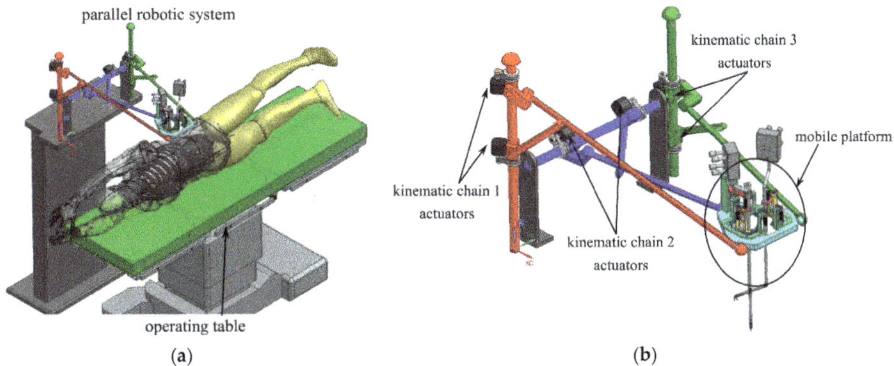

Figure 17. CAD model of the optimized version of the parallel robotic system for SILS: (**a**) relative position between the robotic system and the patient; (**b**) the 6-DOF parallel robotic system for SILS.

6.3. The Singularities of the Optimized Model of the 6-DOF Parallel Robot

The output singularities for the optimized model were studied by slicing the 6-dimensional (singular) configuration hyperspace, i.e., the numerical values for the angles ψ, θ, and φ were substituted into the singularity factor F (Equation (A1)—Appendix A), and the implicit surfaces (for X_E, Y_E, and Z_E) were plotted in Cartesian space.

Figure 18a illustrates the implicit surfaces for the given values of angles ψ, θ, and φ within the interval [−20, 20] [°] using an increment of 2°. Note that not all orientations defined by ψ, θ, and φ were associated with a color, but rather only the ψ angle was used in the surface color definition to avoid using a large number for surface colors (9261 colors were needed if each surface was assigned with a color). One important note is that no singularity surface intersects the operational workspace cylinder. Furthermore, these surfaces describe the output singularities without considering the inverse kinematic model. A second computation was made by considering the inverse kinematic model, i.e., the points on the surface were checked to yield real solutions for $\{q_1, q_2, q_3, q_4, q_5, q_6\}$, and are also within the intervals defined in Table 2. Figure 18b illustrates the results of this computation as implicit singularity surfaces. These surfaces show roughly where a singularity may occur for the 6-DOF parallel robot for SILS.

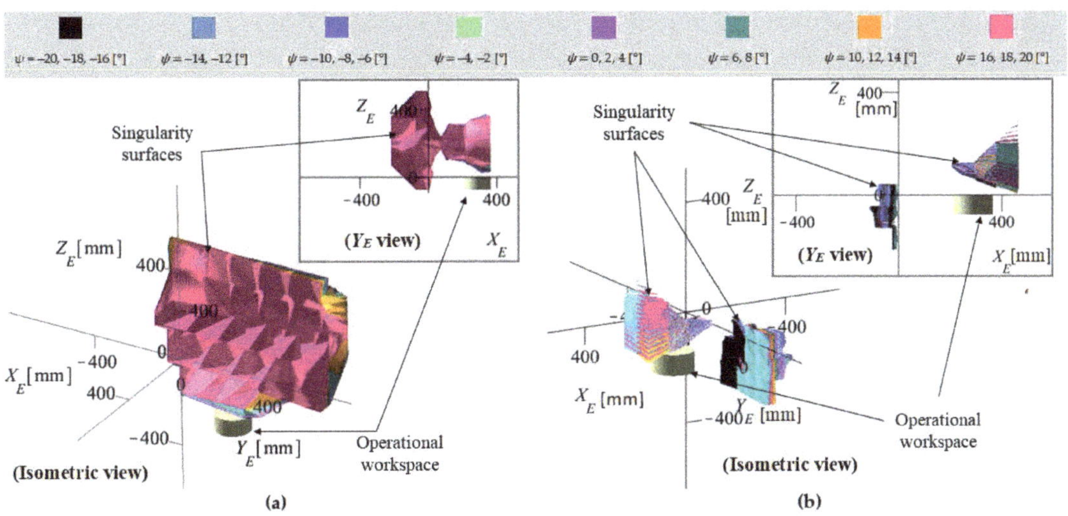

Figure 18. Output singularity surfaces: (**a**) without the inverse kinematic model constraints; (**b**) with the inverse kinematic model constraints.

Some specific singularity surfaces (arbitrarily chosen) are presented next for better detail. Figure 19 shows the surface for the angle values $\{\psi = 0,\ \theta = 0,\ \varphi = 0\}$ [°], illustrating both the output singularity surface (Figure 19a) and the surface which degenerates after applying the inverse kinematic model constraints (Figure 19b). Figure 20 illustrates a possible singular configuration for the surface presented in Figure 19b (which is an output singularity of Case 5). Figures 21 and 22 show the surfaces for the angle values $\{\psi = 20,\ \theta = 0,\ \varphi = 0\}$ [°] and $\{\psi = 20,\ \theta = 20,\ \varphi = 20\}$ [°], respectively.

Following the singularity analysis for the optimized model, a conjecture was proposed: Even though there exist output singularities in the robot workspace, there are none in the operational workspace. Based on this, the factor F (Equation (A1)—Appendix A) must be implemented in the robot control as an avoidance function to (i) avoid losing robot control in the positioning stage (when the robot positions the surgical instruments at the insertion points) and (ii) to avoid damaging the robot (e.g., in homing sequences or laboratory tests). Factor F can be implemented in the robot Programable Logic Controller (PLC) in the motion control functions and tested during the robot motion sequences, when (ideally) F becomes zero or (more often) changes its sign between two consecutive points if a singular pose is reached and the robot will stop. As these cases can appear only outside the operational workspace, when the instruments are not inserted in the body, the user can then move the robot using an alternative trajectory. When a Point to Point (PTP) algorithm is used, the entire trajectory can be checked before the actual robot motion and validated (using the F factor value as discussed before).

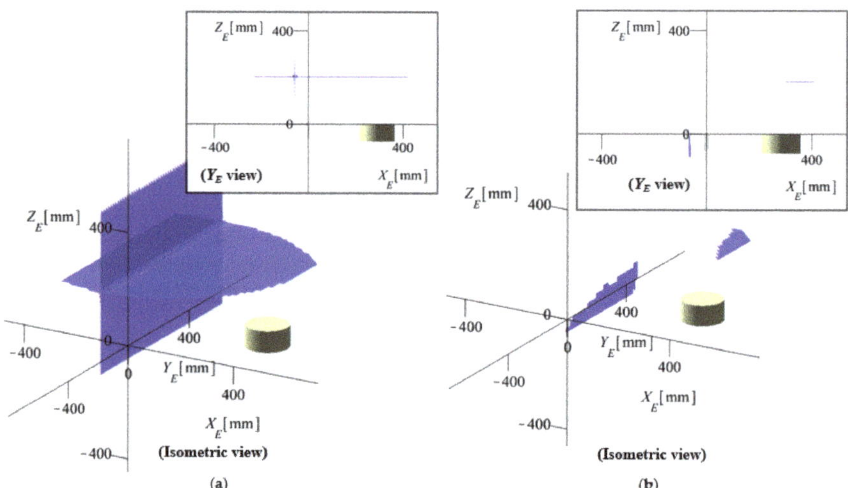

Figure 19. Output singularity surface for $\{\psi = 0,\ \theta = 0,\ \varphi = 0\}$ [°]: (**a**) without the inverse kinematic model constraints; (**b**) with the inverse kinematic model constraints.

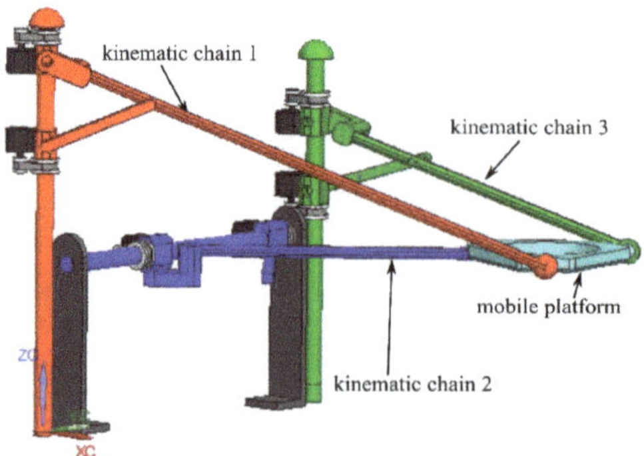

Figure 20. Output singularity configuration for the angle values $\{\psi = 0,\ \theta = 0,\ \varphi = 0\}$.

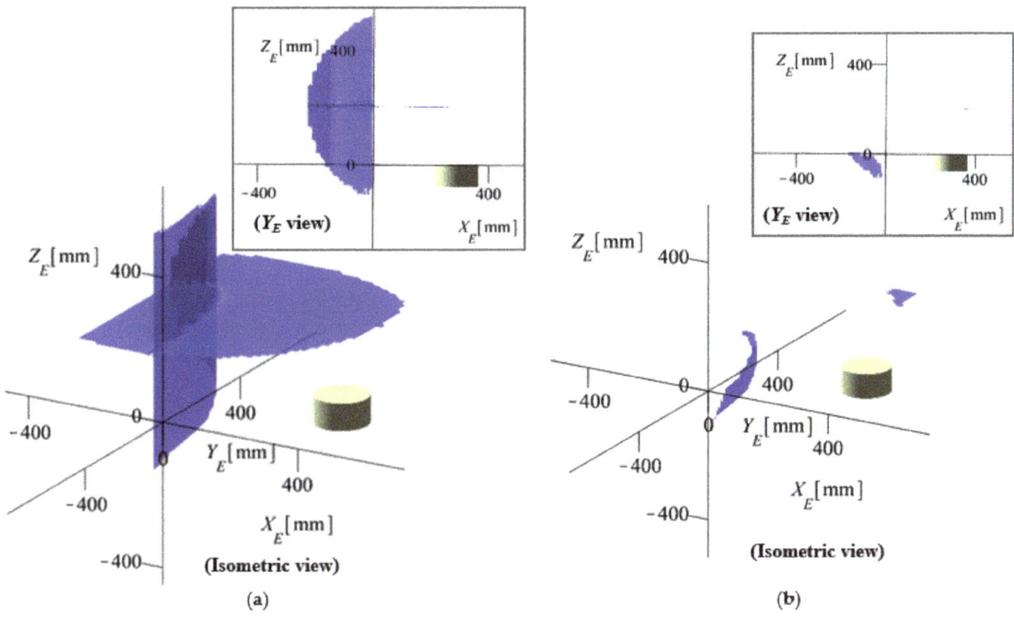

Figure 21. Output singularity surface for $\{\psi = 20, \theta = 0, \varphi = 0\}$ [°]: (**a**) without the inverse kinematic model constraints; (**b**) with the inverse kinematic model constraints.

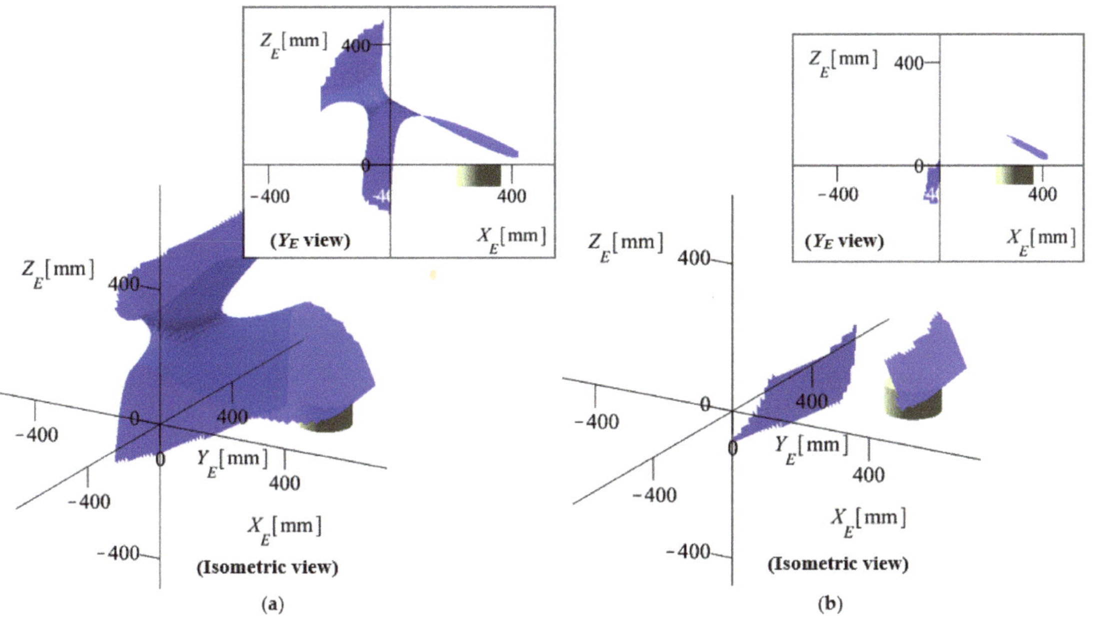

Figure 22. Output singularity surface for $\{\psi = 20, \theta = 20, \varphi = 20\}$ [°]: (**a**) without the inverse kinematic model constraints; (**b**) with the inverse kinematic model constraints.

6.4. Numerical Simulations

The kinematic models of the 6-DOF parallel robot for SILS (the 3-R-$\underline{P}$RR-$\underline{P}$RS parallel robot) were derived in [27]. Considering the inverse kinematic models:

$$\dot{\mathbf{Q}} = -\mathbf{B}^{-1} \cdot \mathbf{A} \cdot \dot{\mathbf{X}}$$
$$\ddot{\mathbf{Q}} = -\mathbf{B}^{-1} \cdot \left(\dot{\mathbf{A}} \cdot \dot{\mathbf{X}} + \mathbf{A} \cdot \ddot{\mathbf{X}} + \dot{\mathbf{B}} \cdot \dot{\mathbf{Q}} \right) \tag{40}$$

With the Jacobian matrices, **A** and **B**, computed from Equation (7) and the inverse geometric model from Equation (5). An input trajectory was provided for the mobile platform coordinates and their velocity $\dot{\mathbf{Q}}$ and acceleration $\ddot{\mathbf{Q}}$ vectors, respectively. The trajectory was defined as a sequence of motions in accordance with the medical protocol:

➤ Stage 1 (align the medical instruments with the insertion points): linear motions from point E_1 [$X_1 = 300$, $Y_1 = 400$, $Z_1 = 20$, $\psi_1 = 0$, $\theta_1 = 0$, $\psi_1 = 0$] [mm, °] to point E_2 [$X_2 = 350$, $Y_2 = 410$, $Z_2 = 0$, $\psi_2 = 0$, $\theta_2 = 0$, $\psi_2 = 0$] [mm, °], with maximum velocity v_max = 10 mm/s and maximum acceleration a_max = 5 mm/s^2;

➤ Stage 2 (insert the instruments—the mobile platform positions the orientation platform RCM's at the insertion points): linear motions from point E_2 [$X_2 = 340$, $Y_2 = 410$, $Z_2 = 0$, $\psi_2 = 0$, $\theta_2 = 0$, $\psi_2 = 0$] [mm, °] to point E_3 [$X_3 = 350$, $Y_3 = 410$, $Z_3 = -50$, $\psi_3 = 0$, $\theta_3 = 0$, $\psi_3 = 0$] [mm, °], with maximum velocity v_max = 10 mm/s and maximum acceleration a_max = 5 mm/s^2;

➤ Stage 3 (reorient the mobile platform): orientation motions from point E_3 [$X_3 = 350$, $Y_3 = 410$, $Z_3 = -50$, $\psi_3 = 0$, $\theta_3 = 0$, $\psi_3 = 0$] [mm, °] to point E_4 [$X_4 = 350$, $Y_4 = 410$, $Z_4 = -50$, $\psi_4 = 0$, $\theta_4 = 10$, $\psi_4 = 20$] [mm, °], with maximum v_max = 4 °/s and maximum acceleration a_max = 2 °/s^2.

Figure 23 shows the time-dependent diagrams for the input trajectories, whereas Figure 24 shows the time-dependent diagrams for the active joints (computed via the inverse kinematic models). The results show no spikes and large (inadequate) values in the velocity and acceleration fields (which represents advantages in the further design stages when actuators must be chosen) and no violation of the active joint boundaries shown in Table 2.

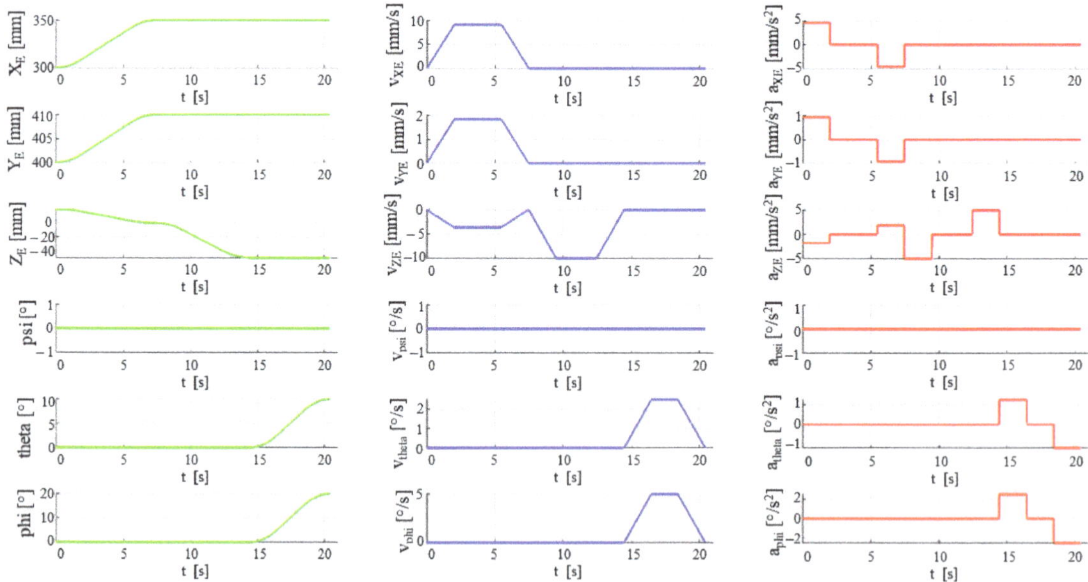

Figure 23. Time history diagrams of the input trajectory (position—green, velocity—blue, accelerations—red).

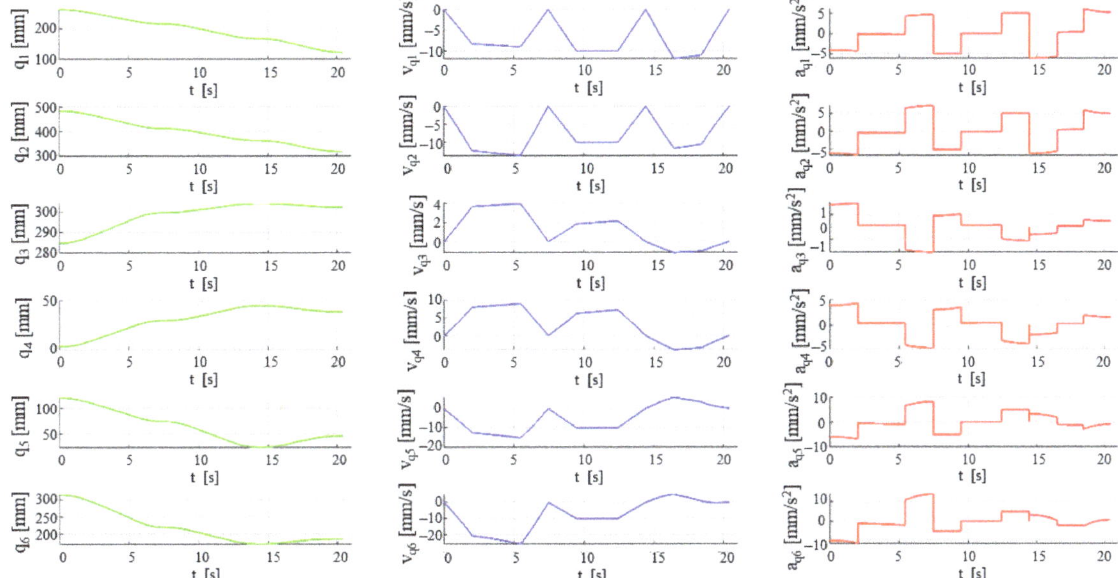

Figure 24. Time history diagrams of the active joints (position—green, velocity—blue, accelerations—red).

7. Conclusions

This paper presented a singularity analysis using the vanishing points of the determinants of the Jacobian matrices and the geometric optimization of a 6-DOF parallel robot for SILS. Numerical analysis for the singularities, where slices of the 6-dimensional singularity hyperspace were studied for imposed orientation angles (correlated with the SILS task), showed that no singularity surface intersects the cylindrical operational workspace. The conjecture is that there are no singularities in the operational workspace (within the boundary of the maximum orientation values for the SILS task). However, the numerical analysis of the singularities also showed that there exist singularity configurations outside the operational workspace. Consequently, the singularity factor (of det(A)) must be implemented in the robotic system control to avoid these configurations. Numerical simulations based on the optimized parallel robot for SILS were performed to validate the proposed solution for the medical task.

Further work is intended for the next development stages of the robotic system, such as designing (prototyping and CAD design), simulating (motion simulations and finite element analysis), and testing the experimental model in laboratory medically-relevant conditions.

Author Contributions: Conceptualization, D.P. and I.B.; data curation, A.P., P.T., N.C., I.A., C.R. and B.G.; formal analysis, D.P., B.G. and C.V.; funding acquisition, D.P.; investigation, I.B., A.P., P.T. and C.V.; methodology, I.B. and C.V.; supervision, D.P.; validation, D.P.; visualization, P.T.; writing—original draft, I.B., A.P. and C.V.; writing—review & editing, D.P., P.T. and B.G. All authors have read and agreed to the published version of the manuscript.

Funding: This work was supported by a grant of the Ministry of Research, Innovation and Digitization, CNCS/CCCDI—UEFISCDI, project number PCE171/2021—Challenge within PNCDI III, and project POCU/380/6/13/123927–ANTREDOC, "Entrepreneurial competencies and excellence research in doctoral and postdoctoral studies programs", a project co-funded by the European Social Fund through the Human Capital Operational.

Institutional Review Board Statement: Not applicable.

Informed Consent Statement: Not applicable.

Data Availability Statement: Not applicable.

Conflicts of Interest: The authors declare no conflict of interest.

Appendix A

$$
\begin{aligned}
F &= 24X_E^3 c_\psi^2 c_\varphi^2 s_\theta + 12X_E Y_E^2 c_\psi^2 s_\theta + 12X_E Y_E^2 c_\varphi^2 s_\theta - 12L_H X_E^2 c_\varphi s_\varphi + 24X_E^2 Y_E c_\varphi s_\varphi - 12X_E^3 c_\psi^2 s_\theta - 12X_E^3 c_\varphi^2 s_\theta + 12X_E^3 s_\theta \\
&+ 24X_E^3 c_\varphi s_\varphi c_\psi s_\psi - 12X_E^2 Y_E c_\theta^2 c_\varphi s_\varphi - 12L_H X_E Y_E c_\psi^2 s_\theta + 12L_H X_E^2 c_\psi s_\psi s_\theta - 12L_H X_E Y_E c_\varphi^2 s_\theta - 24X_E^2 Y_E c_\psi s_\psi s_\theta \\
&+ 2l_p^2 L_V c_\theta^3 c_\varphi^2 c_\psi - 2l_p^2 Z_E c_\theta^3 c_\varphi^2 c_\psi - 24X_E Y_E^2 c_\psi^2 c_\varphi^2 s_\theta + 3l_p L_H L_V c_\varphi^3 c_\theta + 24L_H X_E^2 c_\psi^2 c_\varphi s_\varphi - 3l_p L_H Z_E c_\varphi^3 c_\theta + 6L_H X_E^2 c_\theta^2 c_\varphi s_\varphi \\
&- 12L_V X_E^2 c_\varphi^2 c_\psi c_\theta - 48X_E^2 Y_E c_\psi^2 c_\theta s_\theta + 12X_E^2 Z_E c_\varphi^2 c_\psi c_\theta + 4\sqrt{3}l_p L_H Y_E c_\theta^3 c_\psi^2 s_\psi c_\varphi + 2\sqrt{3}l_p L_H L_V c_\psi^2 c_\theta^2 s_\theta s_\varphi \\
&+ 8\sqrt{3}l_p L_H X_E c_\psi^2 c_\varphi^2 s_\psi c_\theta s_\theta - 2\sqrt{3}l_p L_H Z_E c_\psi^2 c_\theta^2 s_\theta c_\varphi s_\varphi - 4\sqrt{3}l_p L_V Y_E c_\psi^2 c_\theta^2 s_\theta c_\varphi s_\varphi - 4\sqrt{3}l_p X_E^2 c_\psi c_\theta s_\theta - \sqrt{3}l_p^2 L_H c_\varphi^3 c_\theta s_\theta \\
&+ 4\sqrt{3}l_p X_E^2 c_\psi^3 c_\theta s_\theta - 4\sqrt{3}l_p Y_E^2 c_\psi^3 c_\theta s_\theta - 2\sqrt{3}l_p L_V X_E c_\varphi^2 c_\theta^2 + 2\sqrt{3}X_E Z_E c_\varphi^2 c_\theta^2 + 6\sqrt{3}L_H X_E^2 c_\varphi c_\theta s_\theta - \sqrt{3}l_p L_H X_E c_\varphi^2 s_\psi c_\theta s_\theta \\
&+ 8\sqrt{3}l_p X_E Y_E c_\psi^2 s_\psi c_\theta s_\theta + 2\sqrt{3}l_p X_E Y_E c_\varphi^2 s_\psi c_\theta s_\theta + 5\sqrt{3}l_p L_H X_E c_\psi c_\theta c_\varphi s_\varphi + 2\sqrt{3}l_p^2 L_H c_\varphi^2 s_\varphi c_\psi s_\psi c_\theta - 2\sqrt{3}l_p L_H X_E c_\theta^3 c_\psi c_\varphi s_\varphi \\
&+ 4\sqrt{3}l_p L_V Y_E c_\varphi^2 c_\theta^2 c_\psi s_\psi + 16\sqrt{3}l_p X_E Y_E c_\psi^3 c_\theta c_\varphi s_\varphi - 8\sqrt{3}l_p X_E^2 c_\psi^2 c_\theta s_\theta c_\varphi s_\psi + 8\sqrt{3}l_p Y_E^2 c_\psi^2 c_\theta s_\theta c_\varphi s_\psi - 4\sqrt{3}l_p X_E Z_E c_\varphi^2 c_\theta^2 c_\psi s_\psi \\
&+ 4\sqrt{3}l_p X_E Y_E c_\theta^3 c_\psi c_\varphi s_\varphi - 4\sqrt{3}l_p L_H X_E c_\psi^2 s_\psi c_\theta s_\theta - 10\sqrt{3}l_p X_E Y_E c_\psi c_\theta c_\varphi s_\varphi + 4\sqrt{3}l_p L_H X_E c_\psi^3 c_\theta^3 c_\varphi s_\varphi - \sqrt{3}l_p^2 L_H c_\theta^3 c_\varphi^2 s_\varphi c_\psi s_\psi \\
&- 4\sqrt{3}l_p Y_E^2 c_\theta^3 c_\psi^2 s_\psi c_\varphi s_\varphi - 8\sqrt{3}l_p L_H Y_E c_\psi^3 c_\varphi^2 c_\theta s_\theta - 2\sqrt{3}l_p L_H L_V c_\varphi^2 c_\theta^2 c_\psi s_\psi - 8\sqrt{3}l_p L_H X_E c_\psi^3 c_\theta c_\varphi s_\varphi + 2\sqrt{3}l_p L_H Z_E c_\varphi^2 c_\theta^2 c_\psi s_\psi \\
&+ 4\sqrt{3}l_p L_H Y_E c_\varphi^2 c_\psi c_\theta s_\theta + 4\sqrt{3}l_p L_V X_E c_\theta^2 s_\theta c_\varphi s_\varphi c_\psi s_\psi - 4\sqrt{3}l_p X_E Z_E c_\theta^2 s_\theta c_\varphi s_\varphi c_\psi s_\psi + 4\sqrt{3}l_p L_V X_E c_\psi^2 c_\varphi^2 c_\theta^2 \\
&- 4\sqrt{3}l_p X_E Z_E c_\psi^2 c_\varphi^2 c_\theta^2 + 4\sqrt{3}l_p L_H Y_E c_\psi^3 c_\theta s_\theta - \sqrt{3}l_p^2 L_H c_\psi^2 c_\varphi c_\theta s_\theta + 6\sqrt{3}l_p X_E^2 c_\varphi^2 c_\psi c_\theta s_\theta - 4\sqrt{3}l_p Y_E^2 c_\varphi^2 c_\psi c_\theta s_\theta \\
&- 6\sqrt{3}L_H L_V X_E c_\theta^2 c_\psi c_\varphi + 6\sqrt{3}L_H X_E Z_E c_\theta^2 c_\psi c_\varphi + 2\sqrt{3}l_p X_E^2 c_\psi c_\theta c_\varphi s_\varphi + 2\sqrt{3}l_p^2 L_H c_\varphi^3 c_\psi^2 c_\theta s_\theta - 8\sqrt{3}l_p X_E^2 c_\psi^3 c_\varphi^2 c_\theta s_\theta \\
&+ 8\sqrt{3}l_p Y_E^2 c_\psi^3 c_\varphi^2 c_\theta s_\theta - 16\sqrt{3}l_p X_E Y_E c_\psi^2 c_\varphi^2 s_\psi c_\theta s_\theta + 4\sqrt{3}l_p Y_E Z_E c_\psi^2 c_\theta^2 c_\varphi s_\varphi s_\theta - 8\sqrt{3}l_p L_H Y_E c_\psi^2 s_\psi c_\varphi c_\theta s_\theta - 6l_p L_H L_V c_\varphi^2 s_\varphi c_\psi s_\psi c_\theta s_\theta \\
&+ 6l_p L_H Z_E c_\varphi^2 s_\varphi c_\psi s_\psi c_\theta s_\theta + 3l_p L_H L_V c_\varphi^3 c_\theta^3 c_\psi^2 - 3l_p L_H Z_E c_\varphi^3 c_\theta^3 c_\psi^2 - 2l_p^2 X_E c_\varphi^2 c_\theta^2 c_\psi^2 s_\theta - 6l_p L_H L_V c_\varphi^3 c_\psi^2 c_\theta - 6l_p L_H L_V c_\theta^3 c_\psi^2 c_\varphi \\
&+ 6l_p L_H Z_E c_\varphi^3 c_\psi^2 c_\theta + 6l_p L_H Z_E c_\theta^3 c_\psi^2 c_\varphi - 12L_H X_E^2 c_\psi^2 c_\theta^2 c_\varphi s_\varphi - l_p^2 L_H c_\psi^2 c_\theta^2 c_\varphi s_\varphi + 24X_E^2 Y_E c_\psi^2 c_\theta^2 c_\varphi s_\varphi + 2l_p^2 Y_E c_\psi^2 c_\theta^2 c_\varphi s_\varphi \\
&- 12X_E^3 c_\theta^2 c_\psi s_\psi c_\varphi s_\varphi + 24L_H X_E Y_E c_\psi^2 c_\varphi^2 s_\theta - 24L_H X_E^2 c_\varphi^2 s_\theta c_\psi s_\psi + 48X_E^2 Y_E c_\varphi^2 s_\theta c_\psi s_\psi + 3l_p L_H L_V c_\psi^2 c_\varphi c_\theta + 6L_H L_V X_E c_\varphi^2 s_\psi c_\theta \\
&- 3l_p L_H Z_E c_\psi^2 c_\varphi c_\theta - 6L_H X_E Z_E c_\varphi^2 s_\psi c_\theta - 12L_V X_E Y_E c_\varphi^2 s_\psi c_\theta - 24X_E Y_E^2 c_\psi s_\psi c_\theta s_\theta + 12X_E Y_E Z_E c_\varphi^2 s_\psi c_\theta - 12L_V X_E^2 s_\psi c_\varphi s_\varphi c_\theta s_\theta \\
&+ 12X_E^2 Z_E s_\psi c_\varphi s_\varphi c_\theta s_\theta + 24L_H X_E Y_E c_\psi s_\psi c_\varphi s_\varphi - 3l_p L_H X_E c_\varphi^3 c_\theta^2 s_\theta s_\psi + l_p^2 L_H c_\varphi^2 c_\theta^2 s_\theta c_\psi s_\psi - 2l_p^2 Y_E c_\varphi^2 c_\theta^2 s_\theta c_\psi s_\psi \\
&- 3l_p L_H X_E c_\theta^2 c_\varphi^2 s_\varphi s_\psi + 12X_E Y_E^2 c_\theta^2 c_\psi s_\psi c_\varphi s_\varphi - 2l_p^2 X_E c_\theta^2 c_\psi s_\psi c_\varphi s_\varphi + 6l_p L_H X_E c_\varphi^2 s_\theta c_\psi c_\varphi - 12L_H X_E Y_E c_\theta^2 c_\psi s_\psi c_\varphi s_\varphi \\
&- 6L_H L_V X_E c_\theta s_\theta c_\varphi s_\varphi c_\psi + 6L_H X_E Z_E c_\theta s_\theta c_\varphi s_\varphi c_\psi + 12L_V X_E Y_E c_\theta s_\theta c_\varphi s_\varphi c_\psi - 12X_E Y_E Z_E c_\theta s_\theta c_\varphi s_\varphi c_\psi
\end{aligned}
\tag{A1}
$$

References

1. Saidy, M.N.; Tessier, M.; Tessier, D. Single-Incision Laparoscopic Surgery—Hype or Reality: A Historical Control Study. *Perm. J.* **2012**, *16*, 47–50. [CrossRef] [PubMed]
2. Ghezzi, T.L.; Corleta, O.C. 30 Years of Robotic Surgery. *World J. Surg.* **2016**, *40*, 2550–2557. [CrossRef] [PubMed]
3. Pisla, D.; Andras, I.; Vaida, C.; Crisan, N.; Ulinici, I.; Birlescu, I.; Plitea, N. New approach to hybrid robotic system application in Single Incision Laparoscopic Surgery. *Acta Teh. Napoc.* **2021**, *64*, 369–378.
4. Peters, B.S.; Armijo, P.R.; Krause, C.; Choudhury, S.A.; Oleynikov, D. Review of emerging surgical robotic technology. *Surg. Endosc.* **2018**, *32*, 1636–1655. [CrossRef] [PubMed]
5. Kuo, C.-H.; Dai, J.S. Robotics for Minimally Invasive Surgery: A Historical Review from the Perspective of Kinematics. In *International Symposium on History of Machines and Mechanisms*; Yan, H.S., Ceccarelli, M., Eds.; Springer: Dordrecht, The Netherlands, 2009; pp. 337–354.
6. Horgan, S.; Vanuno, D. Robots in laparoscopic surgery. *J. Laparoendosc. Adv. Surg. Tech.* **2001**, *11*, 415–419. [CrossRef]
7. Beatrice, A.; Matthew, J.C.; Danail, S. Gesture Recognition in Robotic Surgery: A Review. *IEEE Trans. Biomed. Eng.* **2021**, *68*, 2021–2035.
8. Saeidi, H.; Opfermann, J.D.; Kam, M.; Raghunathan, S.; Leonard, S.; Krieger, A. A Confidence-Based Shared Control Strategy for the Smart Tissue Autonomous Robot (STAR). In Proceedings of the 2018 IEEE/RSJ International Conference on Intelligent Robots and Systems (IROS), Madrid, Spain, 1–5 October 2018; pp. 1268–1275.
9. Bhandari, M.; Zeffiro, T.; Reddiboina, M. Artificial intelligence and robotic surgery. *Curr. Opin. Urol.* **2020**, *30*, 48–54. [CrossRef]
10. Harky, A.; Chaplin, G.; Chan, J.S.K.; Eriksen, P.; MacCarthy-Ofosu, B.; Theologou, T.; Muir, A.D. The Future of Open Heart Surgery in the Era of Robotic and Minimal Surgical Interventions. *Hear. Lung Circ.* **2019**, *29*, 49–61. [CrossRef]
11. Wang, W.; Sun, X.; Wei, F. Laparoscopic surgery and robotic surgery for single-incision cholecystectomy: An updated systematic review. *Updat. Surg.* **2021**, *73*, 2039–2046. [CrossRef]
12. Kwak, Y.H.; Lee, H.; Seon, K.; Lee, Y.J.; Lee, Y.J.; Kim, S.W. Da Vinci SP Single-Port Robotic Surgery in Gynecologic Tumors: Single Surgeon's Initial Experience with 100 Cases. *Yonsei Med. J.* **2022**, *63*, 179–186. [CrossRef]
13. McCarus, S.D. Senhance Robotic Platform System for Gynecological Surgery. *JSLS J. Soc. Laparoendosc. Surg.* **2021**, *25*, e2020.00075. [CrossRef]
14. Laribi, M.A.; Mlika, A.; Romdhane, L.; Zeghloul, S. Robust Optimization of the RAF Parallel Robot for a Prescribed Workspace. In *Computational Kinematics. Mechanisms and Machine Science*; Zeghloul, S., Romdhane, L., Laribi, M., Eds.; Springer: Cham, Switzerland, 2018; Volume 50, pp. 383–393. [CrossRef]
15. Hay, A.M.; Snyman, J.A. Optimal synthesis for a continuous prescribed dexterity interval of a 3-dof parallel planar manipulator for different prescribed output workspaces. *Int. J. Numer. Methods Eng.* **2006**, *68*, 1–12. [CrossRef]
16. Merlet, J.P. Jacobian, manipulability, condition number, and accuracy of parallel robots. *ASME. J. Mech. Des.* **2006**, *128*, 199–206. [CrossRef]

17. Gosselin, C.; Angeles, J. A global performance index for the kinematic optimization of robotic manipulators. *ASME J. Mech. Des.* **1991**, *113*, 220–226. [CrossRef]
18. Itul, T.P.; Galdau, B.; Pisla, D.L. Static analysis and stiffness of a 2-dof parallel device for orientation. *Acta Tech. Napoc. Ser. Appl. Math. Mech. Eng.* **2012**, *55*, 615–620.
19. Chablat, D.; Wenger, P. Architecture optimization of a 3-DOF translational parallel mechanism for machining applications, the orthoglide. *IEEE Trans. Robot. Autom.* **2003**, *19*, 403–410. [CrossRef]
20. Dastjerdi, A.H.; Sheikhi, M.M.; Masouleh, M.T. A complete analytical solution for the dimensional synthesis of 3-DOF delta parallel robot for a prescribed workspace. *Mech. Mach. Theory* **2020**, *153*, 103991. [CrossRef]
21. Courteille, E.; Deblaise, D.; Maurine, P. Design optimization of a Delta-like parallel robot through global stiffness performance evaluation. In Proceedings of the 2009 IEEE/RSJ International Conference on Intelligent Robots and Systems, St. Louis, MO, USA, 10–15 October 2009; pp. 5159–5166. [CrossRef]
22. Wang, A.U.; Rong, W.B.; Qi, L.M.; Qin, Z.G. Optimal design of a 6-DOF parallel robot for precise manipulation. In Proceedings of the 2009 International Conference on Mechatronics and Automation, Changchun, China, 9–12 August 2009.
23. Qiang, H.; Wang, L.; Ding, J.; Zhang, L. Multiobjective Optimization of 6-DOF parallel manipulator for desired total orientation workspace. *Math. Probl. Eng.* **2019**, *2019*, 5353825. [CrossRef]
24. Aboulissane, B.; Haiek, D.E.; Bakkali, L.E.; Bahaoui, J.E. On the workspace optimization of parallel robots based on CAD approach. *Procedia Manuf.* **2019**, *32*, 1085–1092. [CrossRef]
25. Surbhil, G.; Sankho, T.S.; Amod, K. Design optimization of minimally invasive surgical robot. *Appl. Soft Comput.* **2015**, *32*, 241–249.
26. Lessard, S.; Bigras, P.; Bonev, I.A. A New Medical Parallel Robot and Its Static Balancing Optimization. *J. Med. Devices* **2007**, *1*, 272–278. [CrossRef]
27. Pisla, D.; Pusca, A.; Tucan, P.; Gherman, B.; Vaida, C. Kinematics and workspace analysis of an innovative 6-dof parallel robot for SILS. *Proc. Rom. Acad. Ser. A* 2022, *in press*.
28. Pisla, D.; Carami, D.; Gherman, B.; Soleti, G.; Ulinici, I.; Vaida, C. A novel control architecture for robotic-assisted single incision laparoscopic surgery. *Rom. J. Tech. Sciences. Appl. Mech.* **2021**, *66*, 141–162.
29. Birlescu, I.; Vaida, C.; Pusca, A.; Rus, G.; Pisla, D. A New Approach to Forward Kinematics for a SILS Robotic Orientation Platform Based on Perturbation Theory. In *Advances in Robot Kinematics 2022*; Altuzarra, O., Kecskeméthy, A., Eds.; Springer: Cham, Switzerland, 2022; Volume 24, pp. 171–178. [CrossRef]
30. Ebert-Uphoff, I.; Lee, J.-K.; Lipkin, H. Characteristic tetrahedron of wrench singularities for parallel manipulators with three legs. *Proc. Inst. Mech. Eng. Part C: J. Mech. Eng. Sci.* **2002**, *216*, 81–93. [CrossRef]
31. MathWorks. gamultiobj. Available online: https://ch.mathworks.com/help/gads/gamultiobj.html (accessed on 1 July 2022).
32. Pisla, D.; Birlescu, I.; Vaida, C.; Tucan, P.; Gherman, B.; Plitea, N. Family of Modular Parallel Robots with Active Translational Joints for Single Incision Laparoscopic Surgery. OSIM A00733, 3 December 2021.
33. Gosselin, C.; Angeles, J. Singularity analysis of closed-loop kinematic chains. *IEEE Trans. Robot. Autom.* **1990**, *6*, 281–290. [CrossRef]
34. Zlatanov, D.; Fenton, R.G.; Benhabib, B. A Unifying Framework for Classification and Interpretation of Mechanism Singularities. *J. Mech. Des.* **1995**, *117*, 566–572. [CrossRef]
35. Ma, O.; Angeles, J. Architecture singularities of platform manipulators. In Proceedings of the 1991 IEEE International Conference on Robotics and Automation, Sacramento, CA, USA, 9–11 April 1991; pp. 1542–1547.
36. Spring, I.W. Euler parameters and the use of quaternion algebra in the manipulation of finite rotations. *Mech. Mach. Theory* **1986**, *21*, 365–373. [CrossRef]
37. Deb, K.; Agrawal, S.; Pratap, A.; Meyarivan, T. A Fast Elitist Non-dominated Sorting Genetic Algorithm for Multi-objective Optimization: NSGA-II. In *PPSN 2000: Parallel Problem Solving from Nature PPSN VI.*; Lecture Notes in Computer Science; Springer: Berlin/Heidelberg, Germany, 2000; Volume 1917, pp. 849–858.
38. Wang, W.; Wang, W.; Dong, W.; Yu, H.; Yan, Z.; Du, Z. Dimensional optimization of a minimally invasive surgical robot system based on NSGA-II algorithm. *Adv. Mech. Eng.* **2015**, *7*, 1687814014568541. [CrossRef]

Article

Pick–and–Place Trajectory Planning and Robust Adaptive Fuzzy Tracking Control for Cable–Based Gangue–Sorting Robots with Model Uncertainties and External Disturbances

Peng Liu [1,2,*], Haibo Tian [1], Xiangang Cao [1], Xinzhou Qiao [1], Li Gong [1], Xuechao Duan [2], Yuanying Qiu [2] and Yu Su [3]

[1] School of Mechanical Engineering, Xi'an University of Science and Technology, Xi'an 710054, China
[2] Key Laboratory of Ministry of Education for Electronic Equipment Structure Design, Xidian University, Xi'an 710000, China
[3] School of Mechatronic Engineering, Xi'an Technological University, Xi'an 710021, China
* Correspondence: liupeng@xust.edu.cn

Abstract: A suspended cable–based parallel robot (CBPR) composed of four cables and an end–grab is employed in a pick–and–place operation of moving target gangues (MTGs) with different shapes, sizes, and masses. This paper focuses on two special problems of pick–and–place trajectory planning and trajectory tracking control of the cable–based gangue–sorting robot in the operation space. First, the kinematic and dynamic models for the cable–based gangue–sorting robots are presented in the presence of model uncertainties and unknown external disturbances. Second, to improve the sorting accuracy and efficiency of sorting system with cable–based gangue–sorting robot, a four-phase pick–and–place trajectory planning scheme based on S-shaped acceleration/deceleration algorithm and quintic polynomial trajectory planning method is proposed, and moreover, a robust adaptive fuzzy tracking control strategy is presented against inevitable uncertainties and unknown external disturbances for trajectory tracking control of the cable–based gangue–sorting robot, where the stability of a closed-loop control scheme is proved with Lyapunov stability theory. Finally, the performances of pick–and–place trajectory planning scheme and robust adaptive tracking control strategy are evaluated through different numerical simulations within Matlab software. The simulation results show smoothness and continuity of pick–and–place trajectory for the end–grab as well as the effectiveness and efficiency to guarantee a stable and accurate pick–and–place trajectory tracking process even in the presence of various uncertainties and external disturbances. The pick–and–place trajectory generation scheme and robust adaptive tracking control strategy proposed in this paper lay the foundation for accurate sorting of MTGs with the robot.

Keywords: cable–based parallel robot; gangue sorting robot; pick–and–place operations; trajectory planning; tracking control; robustness

Citation: Liu, P.; Tian, H.; Cao, X.; Qiao, X.; Gong, L.; Duan, X.; Qiu, Y.; Su, Y. Pick–and–Place Trajectory Planning and Robust Adaptive Fuzzy Tracking Control for Cable–Based Gangue–Sorting Robots with Model Uncertainties and External Disturbances. *Machines* **2022**, *10*, 714. https://doi.org/10.3390/machines10080714

Academic Editors: Zhufeng Shao, Dan Zhang and Stéphane Caro

Received: 18 July 2022
Accepted: 19 August 2022
Published: 20 August 2022

1. Introduction

Robotic systems have played an increasingly important role in the intelligent activity of coal mining. One practical and important area of application for robotic systems is in the intelligent identification and roboticized separation of coals and gangues with the machine vision system [1]. The separation of gangues from coals is an extremely critical link for the rational utilization of coal resources. cable–based parallel robots (CBPRs), which have a number of desirable properties, such as simple structure, heavy payload capabilities, large workspace, low energy consumption, and so on [2–4], have been widely used in astronomical observation [5], aerial photography [6], multiple mobile cranes [7], rehabilitation and training [8], and wind tunnel experiments [9]. There has been plenty of prior work in the aspects of workspace generation and analysis [10–12], stability evaluation and stability sensitivity analysis [1,13,14], cable tension optimal distribution [15–17], optimization design [18–20], and so on. The most remarkable characteristic of CBPRs is that it employs flexible cables instead of rigid links, while the main advantage of the robots is their high load-carrying capacity, which makes the robots suitable to be employed in pick–and–place task of the moving target gangues (MTGs). The cable–based parallel robots, according to the number of cables and degrees of freedom of the end-effector, are classified into redundant actuated CBPRs and underactuated CBPRs [21]. It should be noticed that redundant actuated CBPRs are more appropriate than underactuated ones for accurate pick–and–place operations of the heavy loads, where a high payload-to-weight ratio and a high positioning accuracy are required. Therefore, a redundantly cable–based gangue–sorting robot with an end–grab is supposed to perform pick–and–place operation of MTGs with different shapes, sizes and masses. The track, approach, pick, carry, and place operation of MTGs for the end–grab must be investigated firstly in order to accomplish the separation of coals and gangues.

Generally speaking, there inevitably are frictions between the winches and the cables that are generally time-varying and nonlinear, and therefore, the cable–based gangue–sorting robots have a complicated dynamic model, including frictional uncertainties, modeling uncertainties, and external disturbances. Similarly, the total mass of the unloaded and loaded end–grab may also change while the robot performs pick–and–place operations of MTGs. The robot in this application consists of two coupled subsystems, namely the cable–based architecture and the end–grab. Lastly, it should be pointed out that the robot controller must ensure all the cable tensions are always positive because the cables can only pull the end–grab but not push it. Consequently, the pick–and–place trajectory planning and trajectory tracking control of the robot are confronted with additional problems beyond other cable–based parallel robots.

1.1. Pick–and–Place Trajectory Planning

The pick–and–place trajectory planning problem of the cable–based gangue–sorting robot is a fundamental one, and it is finding a smooth and continuous trajectory from a starting position to a desired terminal position within the workspace of the robot. The aim of trajectory planning is to generate the input for the control system of the cable-based gangue–sorting robot to perform pick–and–place operation of MTGs by smooth and continuous motion of the end–grab. Thus far, there are also many publications on the trajectory planning for CBPRs. Qian [22] proposed a new trajectory planning method based on the improved quintic B-splines curves for a 3-DOF CBPR, and furthermore, the effectiveness of this method was verified through the experiments. Zhao et al. [23] presented point-to-point trajectory planning for UCPR with variable angles and height cable mast by using three algebraic methods, while the effectiveness and feasibility of the method were validated with numerical simulation and experiments. The improved rapidly exploring random tree method was proposed to address moving obstacle avoidance for CDPRs, and the suggested method was verified with the experiment [24]. Jiang et al. [25] proposed a point-to-point dynamic trajectory planning technique for reaching a series of poses with a six-DOF cable-suspended parallel robot. Hwang et al. [26] presented a scheme to suppress unwanted oscillatory motions of the payload of a four-cable-driven CDPR based on a zero-vibration input-shaping scheme, and the advantage of the proposed scheme is that it is possible to generate an oscillation-free trajectory based on a ZV input-shaping scheme, and moreover, a series of computer simulations and experiments to verify the effectiveness of the proposed method were conducted for three-dimensional motions of a CBPR with four cables. A smooth trajectory planning algorithm to enhance the smoothness of the trajectory when used in rehabilitation training was proposed for a cable-driven waist rehabilitation robot by employing an improved quintic non-uniform rational B-splines [27]. As demonstrated in [28], a novel methodology for the identification of the inertial parameters of the end-effector, based on the use of internal-dynamics equations and free-motion excitations, was proposed for the underactuated cable-driven parallel robots, where the optimal free-motion trajectories were investigated to obtain optimal identification results. In addition, in ref. [29], a general framework for the planning of point-to-point motions that extend beyond the static workspace was presented for a six-degree-of-freedom cable-suspended parallel robot, and furthermore, the effectiveness of the proposed method was verified through simulations. Furthermore, in [30], the design and experimental validation of a novel three-DOF pendulum-like cable-driven robot capable of executing point-to-point motions by leveraging partial feedback linearization control and on-line trajectory planning based on adaptive frequency oscillators were presented. The research on the trajectory planning for CBPRs, generally speaking, mainly focuses on trajectory planning in either cable space or Cartesian space. The trajectory planning in the Cartesian space can intuitively obtain the end-effector trajectory for CBPRs. Different from other types of CBPRs, the trajectory generation of the end–grab for the pick–and–place operations of MTGs must be determined according to the movement characteristics of MTGs, the location of the gangue recovery bin, and optimal tension condition of the cables in the workspace of the robots. As a result, planning and generating the end–grab trajectory for the cable–based gangue–sorting robot in Cartesian space is the major objective under consideration in this paper, and furthermore, a four-stage trajectory planning scheme of the end–grab is proposed.

1.2. Pick–and–Place Trajectory Tracking Control

As mentioned above, pick–and–place trajectory tracking control of the cable–based gangue–sorting robots is a challenging problem. Moreover, it is well–known that in practice, the control system design for CBPRs with model uncertainties and external disturbances, to the extent of the authors' knowledge, is also a challenging task, and it has attracted more attention recently [31–34]. In order to reduce and eliminate the effect of nonlinear uncertainties and external disturbances on the controller of the robots, a few of approaches for CBPRs are presented, such as nonlinear adaptive control, sliding mode control, robust model predictive control, time-delay control, computed torque control, fuzzy logic control, and so on [35–39]. Izadbakhsh et al. [40] proposed a robust adaptive controller for cable-driven parallel robots subject to dynamic uncertainties, while the stability of the control system was analyzed with a Lyapunov-based method. Wang et al. [41] obtained a model-free robust adaptive control for the cable-driven parallel robots, which is composed of time-delay estimation, a new PID-NFTSM manifold, and a combined adaptive reaching law, using adaptive proportional-integral-derivative nonsingular fast terminal sliding mode. Oh et al. [42] presented an approach to design positive tension controllers for the cable-suspended parallel robots with redundant cables. Shao et al. [43] established the elastic dynamic model for the cable-driven Stewart manipulator, while the rigid-body dynamic model of the A–B rotator and the rigid Stewart manipulator was obtained in detail, and furthermore, the kinematic and dynamic vibration control strategies for the feed support system in FAST were proposed and evaluated with simulations. Duan et al. [44] presented a PID controller with base acceleration feedforward designed in the operational space of the Stewart platform based on the integrated dynamic model of the Stewart platform, and furthermore, vibration isolation and trajectory following control experiments for the cable-suspended Stewart platform was carried out. Schenk et al. [45] developed a super-twisting controller for a redundant cable-driven parallel robot to track a reference trajectory in presence of uncertainties and disturbances. Jafarlou et al. [46] investigated an adaptive fractional-order finite-time sliding mode control for the cable-suspended parallel robots in the presence of model uncertainties. The stability of the closed-loop system was analyzed with developed Lyapunov theory. Hosseini et al. [47] designed a nonlinear PD controller for cable-driven parallel robots in joint space so that the robot can track the reference trajectory quickly and accurately, while the stability of the closed-loop system was examined through Lyapunov direct method. Aghaseyedabdollah et al. [48] discussed the design of supervisory adaptive fuzzy sliding mode control with the fuzzy PID sliding surface for a planar cable-driven parallel robot. Inel et al. [49] addressed a nonlinear continuous-time generalized predictive control for a planar cable-driven parallel robot.

Zi [50] presented a three-DOF cable-driven parallel robot and its adaptive fuzzy control system design and analysis, and the simulation results showed the satisfactory performance of the proposed adaptive fuzzy control system. As a matter of fact, the dynamic model of the cable–based gangue–sorting robot is always contaminated with uncertainties such as nonlinear and time-varying parameters as well as external disturbances, and this makes the dynamic model of the cable–based gangue–sorting robots complicated, and thus, a robust controller and an adaptive fuzzy control system are required to achieve high-performance pick–and–place trajectory tracking control.

Use of the cable–based gangue–sorting robot in the pick–and–place operation of MTGs presents unique challenges: (i) the major challenge for designing a controller of the robot is that the robot may experience abrupt changes in dynamic parameters while the robot captures and places the MTGs with highly variable payload, which can cause traditional control methods to achieve poor results in practical applications; and (ii) it should be emphasized that the most important limitation of the robot is that the cables suffer from unidirectional constraints that can only be pulled and not pushed, and therefore, the cables, to perform pick–and–place operation of MTGs effectively and accurately, must be in tension in the whole workspace. For this reason, the main goal of this paper is to propose a suitable control strategy for the cable–based gangue–sorting robots. To achieve this goal, in view of the nonlinear payload variation and external disturbances of the cable–based gangue–sorting robots, a robust adaptive fuzzy tracking control, which can ensure that the cables are always in positive tension, is investigated in this paper for the high-precision tracking of the robots to efficiently and reliably perform the pick–and–place operations of the MTGs. The advantage of the proposed controller in comparison with ref. [50] is its ability to obtain the positive cable tensions along the pick–and–place trajectory in presence of model uncertainties and external disturbances, providing better tracking performance because a robust term is employed to compensate the estimation errors of the fuzzy control system.

From above, without a smooth and continuous pick–and–place trajectory and appropriate trajectory tracking control for the cable–based gangue–sorting robots, the robots might sustain serious damages, and therefore, the objective of the paper is to generate a smooth and continuous pick–and–place trajectory and to implement a robust control scheme suited for the considered pick–and–place application. As a result, the main contributions of this presented paper are summarized as follows:

i.   Proposing a four-stage trajectory planning scheme for the end–grab of the cable–based gangue–sorting robot while taking account the effect of the synchronous movement of the gangues with the belt conveyor as well as the location of the gangue recovery bin;

ii.   Developing a robust adaptive fuzzy control strategy in the task space to track a given trajectory for the cable–based gangue–sorting robot in the presence of model uncertainties, varying payloads, and external disturbances while guaranteeing closed-loop control system stability;

iii.   Demonstrating the validity of the proposed pick–and–place trajectory planning scheme and the robust adaptive fuzzy tracking control strategy through numerical simulation.

The structure of this paper is as follows. The second section presents the kinematic and dynamic models of a cable–based gangue–sorting robot. The control system is presented in the third section. The effectiveness of the proposed pick–and–place trajectory planning scheme and robust adaptive tracking control strategy are demonstrated by simulation results in the fourth section. Finally, conclusions are drawn, and future work is presented in the fifth section.

2. Description and Modeling of a Cable-Based Gangue-Sorting Robot

2.1. Description of the Robot

In the scope of this research, the investigated gangue–sorting system with a robot, as shown in Figure 1, is composed of a cable–based gangue–sorting robot, a conveyor belt, a machine vision system, gangue recovery bin, as well as coals and gangues. The robot runs across the belt conveyor and employs four cables to drive the end–grab to move to the local neighborhood where the target gangues are located so as to complete the pick–and–place operation of the MTGs. It should be pointed that a collision-free workspace and pick–and–place trajectory can be obtained for the gangues-sorting system. According to the process and characteristics of pick–and–place operation of the target gangues, the pick–and–place trajectory of the end–grab is separated into four phases in this study, namely the starting phase, preparing phase, picking phase, and placing phase. The sorting process of the target gangues can be described in more detail as follows: in the first step, the target gangues, which move synchronously with the conveyor belt at a constant speed, will enter the visual identification area, and meanwhile, the machine vision system identifies and collects the shape and position information of the selected target gangues and transmits it to the main controller of the robot; in the next step, the target gangues move for a while to reach the picking area, where the robot performs the picking operation of the target gangues; in the final step, MTGs are placed into the gangue recovery bin, and at this point, the pick–and–place operation is completed. The robot returns to the zero position and continues to complete the pick–and–place task of other gangues. As shown in Figure 2, the proposed cable–based gangue–sorting robot consists of mechanical module and control module. The mechanical structure is composed of a frame, some pulleys, four cables and motor driven systems, and an end–grab, while the control module consists of an industrial personal computer (IPC), motion controllers, a laser tracker, encoders, and so on. It should be pointed out that the mass of the end–grab and the mass of MTGs could be available with a machine vision system by their shapes and sizes and force sensors to measure cable tensions and estimate the carried masses, while the laser tracker, which can be employed to measure 3D coordinates of the end–grab, are used in combination with the servomotor encoders to obtain the position of the end–grab. In the presence of measuring systems and equipment, the closed–loop control can be employed for the robot, which leads to performance accuracy in the pick–and–place operation of the MTGs.

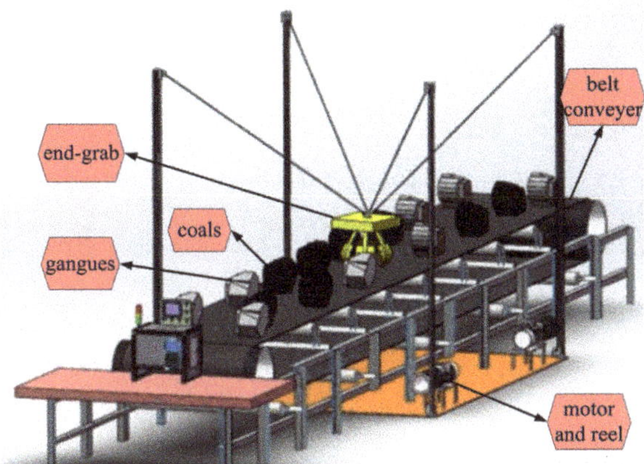

Figure 1. Three-dimensional CAD model of the robot.

The kinematics model of the cable–based gangue–sorting robot is a prerequisite for the dynamics model and fundamental for practical aspects such as motion trajectory planning and control system design. As a result, in this section, a full development of kinematics and dynamic models for a cable–based gangue–sorting robot is established.

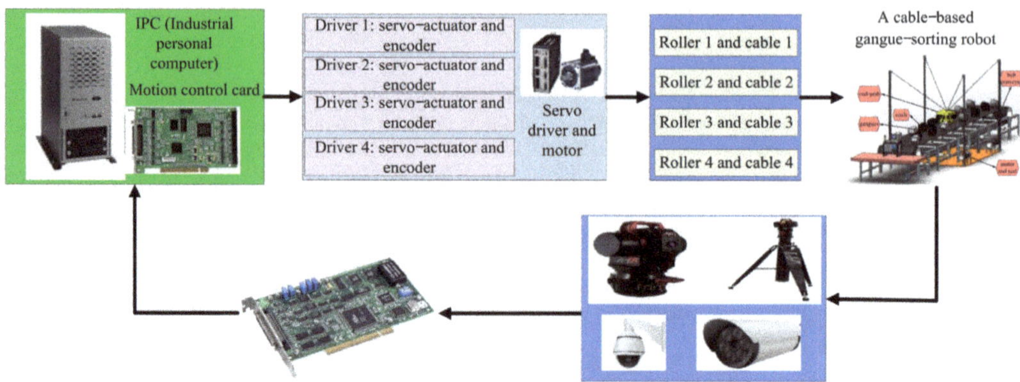

Figure 2. Configuration of the investigated robot.

2.2. Modeling of the Robot

As shown in Figure 3, the fixed coordinate system noted as $OXYZ$ is attached to the fixed base, where O is the origin point. The structure dimensions of the robot are denoted by a, b, and h, respectively. The point A_i (i = 1, 2, 3, 4) represents the position of the fixed pulley in the coordinate system. As in ref. [1], the cable–based gangue–sorting robot can be seen by a CBPR with a point mass, and therefore, the position of the end–grab is denoted by P. With regard to the cable–based gangue–sorting robot, the cables, which are made of lighter materials, can be treated as a kind of massless straight line that can only sustain tension, and therefore, the length of the ith cable can be denoted by L_i. It can be easily obtained as follows:

$$L_i = P - A_i (i = 1, 2, 3, 4) \tag{1}$$

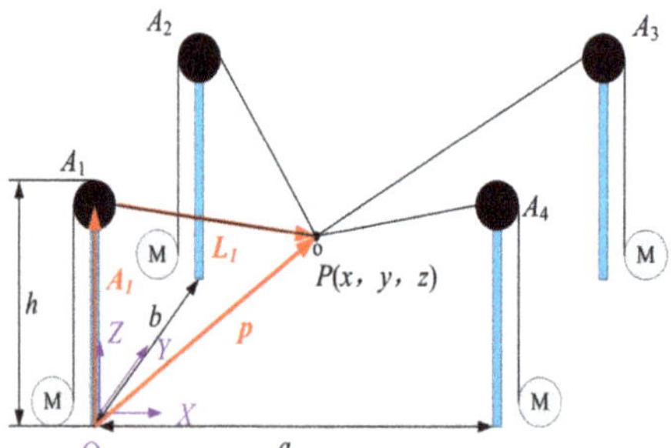

Figure 3. Kinematic model of the robot.

For more detail, the cable length of the four cables can be written as:

$$\begin{cases} L_1 = \sqrt{(x - x_1)^2 + (y - y_1)^2 + (z - z_1)^2} \\ L_2 = \sqrt{(x - x_2)^2 + (y - y_2)^2 + (z - z_2)^2} \\ L_3 = \sqrt{(x - x_3)^2 + (y - y_3)^2 + (z - z_3)^2} \\ L_4 = \sqrt{(x - x_4)^2 + (y - y_4)^2 + (z - z_4)^2} \end{cases} \tag{2}$$

Furthermore, the unit vector along the ith cable is denoted by u_i, and it can be denoted by:

$$u_i = (P - A_i)/L_i \tag{3}$$

A mathematical dynamic model of the cable–based gangue–sorting robot is essential for good control design and analysis to realize high-performance pick–and–place trajectory tracking control. This section presents the dynamic equations for the robot. These equations will be used later to ensure that the cable tensions remain positive during the pick–and–place operation of MTGs. There are different approaches for solving the dynamics of the robots, such as the Newton–Euler equation, Kane equation, and Lagrange equation [51–54]. In this article, the Newton–Euler equation is adopted to solve the dynamics of the robot, and thus, the dynamic equation of the robot can be expressed as:

$$
m\begin{bmatrix} \ddot{x} \\ \ddot{y} \\ \ddot{z} \end{bmatrix} = \begin{bmatrix} \dfrac{x_1-x}{L_1} & \dfrac{x_2-x}{L_2} & \dfrac{x_3-x}{L_3} & \dfrac{x_4-x}{L_4} \\ \dfrac{y_1-y}{L_1} & \dfrac{y_2-y}{L_2} & \dfrac{y_2-y}{L_3} & \dfrac{y_2-y}{L_4} \\ \dfrac{h-z}{L_1} & \dfrac{h-z}{L_2} & \dfrac{h-z}{L_3} & \dfrac{h-z}{L_4} \end{bmatrix}\begin{bmatrix} T_1 \\ T_2 \\ T_3 \\ T_4 \end{bmatrix} + \begin{bmatrix} 0 \\ 0 \\ -mg \end{bmatrix} \tag{4}
$$

where m is the mass of the end–grab; g is gravity acceleration; T is the vector consisting of all cable tensions; $\ddot{x}$, $\ddot{y}$, and $\ddot{z}$ are acceleration of the end–grab, respectively.

For the sake of simplicity, Equation (4) can be rewritten in the following matrix form:

$$
M(X)\ddot{X} + H\left(X, \dot{X}\right) = \tau \tag{5}
$$

where $M(X) = \begin{bmatrix} m & & \\ & m & \\ & & m \end{bmatrix}$ is the inertia matrix, which is defined as symmetric and pos-

itive; $H\left(X, \dot{X}\right) = C(X, \dot{X})\dot{X} + G(X)$ is the vector of Coriolis, centripetal, and gravity terms;

$C(X, \dot{X}) = \begin{bmatrix} 0 & & \\ & 0 & \\ & & 0 \end{bmatrix}$ is a nonlinear Coriolis and centripetal vector terms; $G(X) = \begin{bmatrix} 0 \\ 0 \\ mg \end{bmatrix}$

is gravity vector; τ is the input torque vector; and $\tau = J^T T$ and $T = \begin{bmatrix} T_1 \\ T_2 \\ T_3 \\ T_4 \end{bmatrix}$ are the cable

tension vectors; $J^T = \begin{bmatrix} \dfrac{x_1-x}{L_1} & \dfrac{x_2-x}{L_2} & \dfrac{x_3-x}{L_3} & \dfrac{x_4-x}{L_4} \\ \dfrac{y_1-y}{L_1} & \dfrac{y_2-y}{L_2} & \dfrac{y_2-y}{L_3} & \dfrac{y_2-y}{L_4} \\ \dfrac{h-z}{L_1} & \dfrac{h-z}{L_2} & \dfrac{h-z}{L_3} & \dfrac{h-z}{L_4} \end{bmatrix}$ is the wrench matrix of the

robot; $\ddot{X} = \begin{bmatrix} \ddot{x} \\ \ddot{y} \\ \ddot{z} \end{bmatrix}$ is the acceleration vector of the end–grab.

As mentioned above, when the cable–based gangue–sorting robot performs the pick–and–place operations of MTGs, its total mass of the unloaded and loaded end–grab inevitably changes. These parameter uncertainties and load variations of the end–grab will introduce disturbances to the closed-loop system for the robot and greatly affect the control performance. In practical engineering application, it is difficult to acquire complete information of the cable–based gangue–sorting robot because of parameter uncertainties and external disturbances. As a result, in the presence of the inertial parameter uncertainties and external disturbances, we can express the actual dynamic model of the robot by the following equation:

$$M(X)\ddot{X} + H(X, \dot{X}) + f + \tau_d = \tau. \tag{6}$$

where $M(X) = M_0 + \Delta M$ and $H(X) = H_0 + \Delta H$ are the actual dynamic parameters of the robot; M_0 is the estimated inertia matrix and H_0 is the estimated Coriolis, centrifugal force, and gravity matrix, while ΔM and ΔH are dynamic modeling errors, respectively; τ_d is the vector containing the inertial parameter uncertainties and external disturbances effects; f is the viscous and Coulomb friction matrix.

Finally, we obtain the actual dynamic equation of the cable–based gangue picking robot as follows:

$$M_0(X)\ddot{X} + H_0(X, \dot{X}) + D = \tau \tag{7}$$

where $D = \Delta M\ddot{X} + \Delta H + f + \tau_d$ is the lumped composite disturbance including modeling errors, friction forces, and external disturbances.

It should be pointed out that the Equation (7) is a non-homogeneous linear equation, and therefore, the cable tensions may exist in infinite solutions. It must be emphasized that the equations of motion are valid only if the cables are all in tension. The suitable solution can be obtained by the cable tension optimization model with the pseudo-inverse method [16,17].

3. Control System

The control system of the cable–based gangue–sorting robot is responsible for: (i) planning the pick–and–place trajectory of the end–grab to track, approach, capture, carry, and place operation of MTGs and (ii) assuring realization of the designed and selected pick–and–place trajectory of the end–grab despite the parameters' uncertainty and disturbances. Pick–and–place trajectory planning can be performed while the position and dimension information of the MTG is acquired, while any controller, which is responsible for realization of the designed and selected pick–and–place trajectory of the end–grab, must work in real time. As a result, this paper proposes a new control system for the cable–based gangue–sorting robot that consists of two modules: a pick–and–place trajectory planning module and a robust adaptive fuzzy tracking controller. As mentioned above, the pick–and–place trajectory planning and control of the robot can be performed in the cable space or in the Cartesian space of the end–grab. The pick–and–place trajectory planning and trajectory track control in the Cartesian space is carried out for performing the pick–and–place task of the MTGs in the paper. The position and dimension information of the MTG is obtained by using the machine vision system, and furthermore, the pick–and–place operation of the MTGs is performed autonomously after the pick–and–place trajectory is planned and generated. During execution of the pick–and–place trajectory of the end–grab, the position error of the end–grab is measured and used as a feedback for the control system.

3.1. Proposed Trajectory Planner

In the process of gangue sorting, the target gangue moves synchronously with the belt conveyor, and the flexible cable drives the gangue–sorting robot to complete the task of picking the gangue. The position, speed, and acceleration of the end–grab during the working process need to be set manually according to the specific requirements of the task of picking the gangue so as to achieve accurate grasping of the gangue. It is desirable to design a continuous and smooth tracking and approach trajectory for the end–grab of the cable–based gangue–sorting robot to perform the pick–and–place task of MTGs. Therefore, generation of smooth and continuous trajectory for performing the pick–and–place operation of MTGs is the major objective under consideration in this section.

3.1.1. Requirements for Trajectory Planning

The location and distribution of the gangues on the belt conveyor are shown in Figure 4. The geometric center of the cable–based gangue–sorting robot is located at the center line of the belt conveyor. Gangue recovery bins are arranged on both sides of the belt conveyor.

Therefore, the gangues to be sorted on the belt conveyor can be considered to be distributed in zone *A* and zone *B*, which are symmetric about the center line denoted by *b*. As a result, it is reasonable that we can take the sorting of the gangues within either zone *A* or zone *B* as an example to plan the pick–and–place trajectory of the end–grab for the robot, while the other side can be solved by using the symmetry relationship. We take zone *B* as an example to illustrate the proposed pick–and–place trajectory planning scheme for the robot in this section.

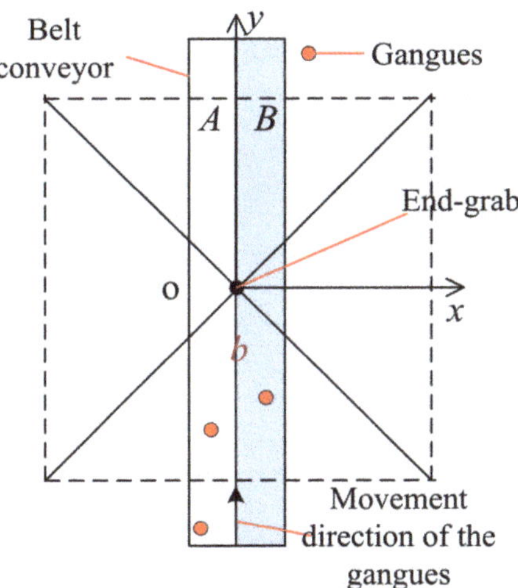

Figure 4. The location and distribution of the gangues.

According to the synchronous movement characteristics of MTGs with the belt conveyor, the location of the gangue recovery bin, and optimal stress condition of the cables in the workspace of the cable–based gangue–sorting robot, we proposes a four-stage pick–and–place trajectory planning scheme for the end–grab of robot, which comprises the following four periods: the starting period, preparing period, picking period, and placing period. As a result, the pick–and–place trajectory of the end–grab must meet the following requirements:

(1) Starting period: the movement velocity of the end–grab increases from 0 to a predetermined constant speed while the robot starts for the first time. There inevitably is an acceleration stage for the end–grab, and therefore, the motion state of the end–grab should be continuous and smooth.

(2) Preparation period: in order to avoid impact during the process of picking the target gangue, the end–grab and the target gangue to be grabbed should be in a relatively static state. Therefore, there needs to be a constant linear motion along the movement direction of the target gangue.

(3) Picking period: no impact can occur during the operation of carrying the picked gangue after the target gangue is captured by the end–grab. In addition, in order to ensure that the captured gangue can fall into the gangue bin at a fixed direction and speed, the end–grab should also have a uniform linear motion at this stage.

(4) Placing period: in order to avoid repeated acceleration leading to a discontinuous trajectory for the end–grab, the end–grab directly enters stage (2) to perform the pick–and–place operation of the next target gangue after the current picked gangue is placed in the gangue recovery bin.

In addition, each trajectory for the four periods mentioned above should be smoothly connected to the next one to avoid the motion state discontinuity in the neighborhood of the trajectory transition point, which can lead to dynamic impact.

3.1.2. Trajectory Planning Scheme

(1)   Determination of the end-grab position and zero position

As shown in Figure 5, considering the unidirectional characteristics of the cables, optimal tension condition of the cables, and the workspace where the end–grab is located, the tension of each cable is equal to each other while the end–grab is located at the geometric center of the workspace, so the ideal position of the end–grab for grabbing the target gangue should be located on the vertical axis of the workspace. In addition, the target gangue moves along the positive direction of y–axis. Considering the tension condition of the cable, a certain position along the x–axis can be selected as the most appropriate grabbing point, and thus, point E is selected as the grabbing point of the target gangue. Then, the straight line is perpendicular to the movement direction of the target gangue and the belt through E point, which can be obtained, and the straight line crosses the belt at points Q and R on both sides. As a result, the grabbing point of the target gangue must be on the QR while the position coordinate of x–direction deviates from the center line b. As requirement (2) states, the end–grab and the target gangue to be grabbed should be in a relatively static state, and therefore, the zero point the end–grab should be ahead of the grab point to achieve the synchronous movement of the end–grab and the target gangue. As a result, point C is chosen as the zero point of the end–grab for the cable–based gangue–sorting robot.

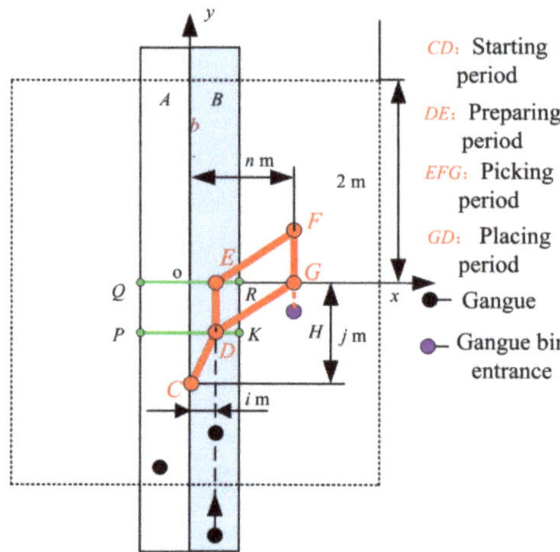

Figure 5. Pick–and–place trajectory planning scheme.

(2)   Determination of the starting period and preparation period

According to the requirement (2) of trajectory planning, point E is the terminal point of the uniform motion in a straight line of the end–grab, while the starting point of the uniform linear motion of the end–grab should be reasonably selected. Point D is selected as the starting point of the uniform linear motion. Then the straight line is perpendicular to the movement direction of the target gangue and the belt through D point, which can be obtained, and the straight line crosses the belt at points P and K on both sides. As a result, the starting point of the uniform linear motion is on PK no matter where the target gangues are located. Therefore, the CD segment is the starting period while the DE segment is the preparation period. It should be noted that the starting section CD and preparing section DE coincide on the same straight line when the target gangue is located at the center line b of the conveyor belt, and thus, the end–grab does not move in the x–direction.

(3)   Determination of the picking period

According to the requirement (3) of the trajectory planning, the end–grab should also move in a straight line at a constant speed at this stage. The cable tensions will change while the target gangue is placed, so the reasonable point should be selected as the placing point of the target gangues. Considering the fact that the cable tensions are relatively reasonable when the end–grab is on the x–axis, point G is selected as the terminal point of the straight line. Connecting the line segment HG and the starting point F of the uniform linear motion must happen on the extension line of HG along the positive direction of the y–axis, and the position of point F can be determined while the operation time of the end–grab on FG is consistent with DE. As a result, EFG section is the picking period when the target gangue is carried by the end–grab from point E to point G.

(4)   Determination of the placing period

According to the requirement (4) of trajectory planning, the end–grab will enter the placing period after the target gangue is placed by the end–grab at G point to avoid repeatedly returning to zero to accelerate. The end–grab can directly transition to the preparation section after the current picked gangue is placed in order to avoid repeated acceleration, leading to a discontinuous trajectory for the end–grab. Thus, GD is the placing period of the pick–and–place trajectory for the end–grab.

To sum up, as shown in Figure 5, point C is selected as the zero position of the end–grab; point D and point F are the starting points of the uniform linear motion of the end grab, respectively, while point E is the grabbing point of the target gangues, and point G is the placing point of the target gangues. The CD segment, the DE segment, the EFD segment, and the GD segment are the starting period, the preparation period, the picking period, and the gangue-placing period, respectively. The four periods above shall be smoothly connected to ensure that the end–grab will not be impacted during the process of the pick–and–place operation of target gangues.

3.1.3. Implementation Methods of Each Trajectory for the Four Periods

(1)   S-shaped acceleration/deceleration algorithm

The S-shaped acceleration/deceleration algorithm is optimized on the basis of T-shaped velocity programming. The planned trajectory, velocity, and acceleration are continuous, which can ensure smooth acceleration and deceleration of the end-effector without impact [55,56]. There are four types of S-shaped velocity curve planning: seven-stage, six-stage, five–stage, and four-stage, respectively. Considering the fact that the motion planning of the end–grab meets the conditions of five–stage planning, the five–stage S-shaped velocity curve planning method is introduced here, whose velocity and acceleration curves are shown in Figure 6. As shown in Figure 6, $v_{\max}$ represents the maximum planned speed; t_i represents the time; $T_i = T_{i+1} - T_i$ is the time of each segment; $a_{\max}$ represents the maximum acceleration; and J represents the jerk. As a result, the relationship between the total displacement of the end–grab s, displacement of the acceleration stage s_1, the jerk J, and acceleration of the end–grab $a_{\max}$ is as follows:

$$s_1 = \frac{J v_{\max}^2 + v_{\max} a_{\max}^2}{2 J a_{\max}} \tag{8}$$

If the above parameters meet $v_{\max} \leq \dfrac{a_{\max}^2}{J}$ and $s > 2v_{\max}\sqrt{\dfrac{v_{\max}}{J}}$ in the given range of the total displacement of the end–grab s, the acceleration cannot reach the maximum value, but the speed can reach the maximum value, and in this case, the maximum acceleration needs to be adjusted as follows:

$$a_{new} = \sqrt{v_{\max} J} \tag{9}$$

Further, the time of each segment can be obtained as follows:

$$T_1 = T_2 = T_4 = T_5 = \frac{a_{new}}{J} \tag{10}$$

$$T_3 = \frac{s - 2v_{max}\sqrt{\frac{v_{max}}{J}}}{v_{max}} = \frac{s}{v_{max}} - 2\sqrt{\frac{v_{max}}{J}} \tag{11}$$

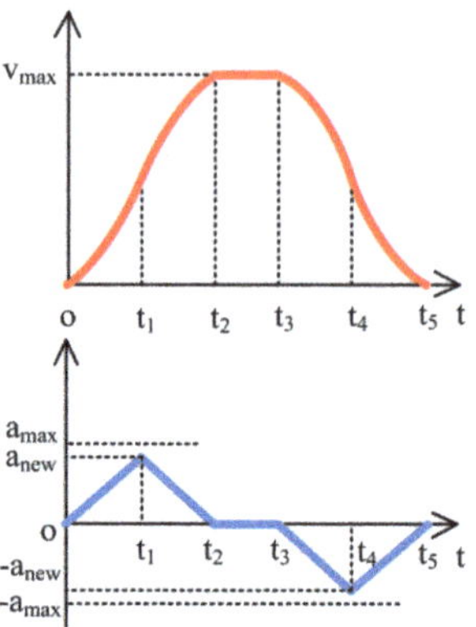

Figure 6. Velocity and acceleration curves of the five−stage.

In this paper, this method is only applied to the acceleration section of *y*-direction, and the expressions of its speed and displacement acceleration section are as follows:

$$v(t) = \begin{cases} \dfrac{1}{2}Jt^2 & 0 < t < T_1 \\ \dfrac{1}{2}JT_1{}^2 + JT_1t - \dfrac{1}{2}Jt^2 & T_1 < t < T_2 \end{cases} \tag{12}$$

$$s(t) = \begin{cases} \dfrac{1}{6}Jt^3 & 0 < t < T_1 \\ \dfrac{1}{6}JT_1{}^3 + \dfrac{1}{2}JT^2{}_1t + \dfrac{1}{2}JT_1t^2 - \dfrac{1}{6}Jt^3 & T_1 < t < T_2 \end{cases} \tag{13}$$

(2) Quintic polynomials

The four trajectories of the four periods should be smoothly connected to each other. Therefore, the sharp points, which will cause local discontinuity of the speed curve and impact of the end–grab, are avoided in the trajectory. The quintic polynomial trajectory planning, which can ensure the continuity up to the acceleration level, is employed to plan the pick–and–place trajectory of the end–grab [57]. The position, velocity, and acceleration equations of quintic polynomial trajectory planning can be expressed as:

$$\begin{bmatrix} q(0) \\ \dot{q}(0) \\ \ddot{q}(0) \\ q(t) \\ \dot{q}(t) \\ \ddot{q}(t) \end{bmatrix} = \begin{bmatrix} 1 & 0 & 0 & 0 & 0 & 0 \\ 0 & 1 & 0 & 0 & 0 & 0 \\ 0 & 0 & 2 & 0 & 0 & 0 \\ 1 & t & t^2 & t^3 & t^4 & t^5 \\ 0 & 1 & t & t^2 & t^3 & t^4 \\ 0 & 0 & 1 & t & t^2 & t^3 \end{bmatrix} \begin{bmatrix} a_0 \\ a_1 \\ a_2 \\ a_3 \\ a_4 \\ a_5 \end{bmatrix} \tag{14}$$

where $q = [q(0), \dot{q}(0), \ddot{q}(0), q(t), \dot{q}(t), \ddot{q}(t)]^{\mathrm{T}}$ is a vector composed of generalized coordinates, generalized velocity, and generalized acceleration from the starting point to the terminal point. a_i represents the coefficients of the quintic polynomial.

From above, the trajectory planning methods adopted by each period are shown in Table 1.

Table 1. Trajectory planning method of each section.

Periods	Direction	Used Trajectory Planning Method
	x	Quintic polynomials
Starting period CD	y	S-shaped acceleration/deceleration algorithm
	z	No displacement
	x	No displacement
Preparation period DE	y	S-shaped acceleration/deceleration
	z	No displacement
	x	Quintic polynomials
Picking period EF	y	Quintic polynomials
	z	Quintic polynomials
	x	No displacement
Picking period FG	y	S-shaped acceleration/deceleration
	z	Quintic polynomials
	x	Quintic polynomials
Placing period GD	y	Quintic polynomials
	z	No displacement

3.2. Proposed Controller

For a cable–based gangue–sorting robot, the external disturbances, frictional terms, internal uncertainties, and payload variation are considered as a composite disturbance, which mostly affects the control performance and causes inaccuracy in the control process. The main objective of control for the cable–based gangue–sorting robot is to achieve high-performance pick–and–place trajectory tracking accuracy in the pick–and–place operation of MTGs. As a result, considering the structural characteristics and control difficulties of the cable–based gangue–sorting robot, in the presented control system, a robust adaptive fuzzy tracking control is used for assuring realization of the selected trajectory of the end–grab to perform the pick–and–place operation of MTGs for the cable–based gangue–sorting robot. The uncertainties of the control system are adaptively compensated by fuzzy control system, while a robust term is employed to compensate the estimation errors of the fuzzy control system. Meanwhile, the stability of the whole closed-loop system is guaranteed.

As an intelligent control method, fuzzy control is based on fuzzy logic inference, and the microcomputer control method has also been widely applied to generate auxiliary joint torques to compensate these uncertainties [58,59]. Fuzzy logic system can be employed to approximate the unknown nonlinear functions as well as external disturbances. It is especially suitable for the control of nonlinear, time-varying systems, such as robotics. The approximation characteristics of the fuzzy logic system are used to compensate for the composite disturbance for the cable–based gangue–sorting robot. A robust term is designed to eliminate the estimation errors and external disturbance of fuzzy logic system.

A multi-input and multi-output fuzzy logic system performs mapping from fuzzy sets in $U \in R^n$ to fuzzy sets in $V \in R^m$, based on the fuzzy IF–THEN rules. The output of a multi-input and multi-output fuzzy logic system with center-average defuzzifier, product inference, and singleton fuzzifier takes the following form:

$$y_j = \frac{\sum\limits_{l=1}^{M} \bar{y}_j^l \left(\prod\limits_{i=1}^{n} \mu_{A_i^l}(x_i) \right)}{\sum\limits_{l}^{M} \left(\prod\limits_{i=1}^{n} \mu_{A_i^l}(x_i) \right)}, (j = 1, 2, \ldots, m) \tag{15}$$

where $\bar{y}_j^l$ is a value at which the membership function for output fuzzy set reaches its maximum; $\mu_{A_i^l}(x_i)$ is the membership function of the linguistic variable x_i and can be defined as follows:

$$\mu_{A_i^l}(x_i) = \exp\left(\frac{-(x_i - \bar{x}_i^l)}{\sigma^2} \right)^2 \tag{16}$$

where $\bar{x}_i^l$ and σ are the mean and the deviation of the Gaussian membership function, respectively.

The fuzzy basis function can be defined as:

$$\varepsilon_l(x) = \frac{\prod\limits_{i=1}^{n} \mu_{A_i^l}(x_i)}{\sum\limits_{l}^{M}\left(\prod\limits_{i=1}^{n} \mu_{A_i^l}(x_i)\right)} \tag{17}$$

As a result, Equation (15) can be rewritten as follows:

$$y_j = \Theta_j^{\mathrm{T}}\varepsilon(x) \tag{18}$$

where $\varepsilon\left(x\right) = [e_1\left(x\right), e_2\left(x\right), \ldots, e_M\left(x\right)]^{\mathrm{T}}$ is the fuzzy basis function vector, while $\Theta_j = [\bar{y}_j^1, \bar{y}_j^2, \ldots, \bar{y}_j^M]^{\mathrm{T}}$ is the center of the fuzzy subset.

Furthermore, the overall output of a MIMO fuzzy logic system can be represented as:

$$y = \Theta^{\mathrm{T}}\varepsilon(x) \tag{19}$$

The control problem for the cable–based gangue–sorting robot is to design a controller to compute the cable tensions T applied to the end–grab such that the end–grab positions X tend asymptotically toward the constant desired end–grab positions X_d. Therefore, the tracking error is defined as:

$$e = X(t) - X_{\mathrm{d}}(t) \tag{20}$$

where $X_{\mathrm{d}}(t)$ is the desired trajectory of the end–grab, while $X(t)$ is the actual trajectory of the end–grab.

Moreover, the sliding surface s is defined as follows:

$$s = \dot{e} + \Lambda e \tag{21}$$

where Λ is a positive definite parameter matrix.

The reference tracking velocity can be defined as follows:

$$\dot{X}_r(t) = \dot{X}_d(t) - \Lambda X(t) \tag{22}$$

The proposed controller, which can counteract the external disturbances, frictional terms, and internal uncertainties, can be expressed by the following equation:

$$\begin{aligned}
T &= T_n + T_h \\
T_n &= (J^{\mathrm{T}})^{+}\left(M(X)\ddot{X}_r + H\left(X, \dot{X}\right) + \hat{F}\left(X, \dot{X}, \ddot{X}|\tilde{\Theta}\right) - K_D s - W\mathrm{sgn}(s)\right) \\
T_h &= \left(I - (J^{\mathrm{T}})^{+}J^{\mathrm{T}}\right)\lambda \\
\hat{F}\left(X, \dot{X}, \ddot{X} \mid \tilde{\Theta}\right) &= \Theta_i^{\mathrm{T}}\varepsilon\left(X, \dot{X}, \ddot{X}\right)
\end{aligned} \tag{23}$$

in which T_n is the special solution to the vector T, while T_h is the homogeneous solution to the vector T and $J^{\mathrm{T}}T_h = 0$; $(\bullet)^{+}$ denotes the pseudo inverse; $K_D = \mathrm{diag}(K_i), K_i > 0$; λ is an arbitrary scalar; $W = \mathrm{diag}\left[\omega_{M_1}, \omega_{M_2}, \ldots, \omega_{M_n}\right], \omega_{M_i} \geq |\omega_i|, i = 1, 2, \ldots, n$; $\hat{F}\left(X, \dot{X}, \ddot{X}|\tilde{\Theta}\right)$

is the fuzzy logic compensation control for the lumped composite disturbance, and it is

represented as
$$\hat{F}(X,\dot{X},\ddot{X}\mid\tilde{\Theta}) = \begin{bmatrix} \hat{F}\left(X,\dot{X},\ddot{X}\mid\tilde{\Theta}_1\right) \\ \hat{F}\left(X,\dot{X},\ddot{X}\mid\tilde{\Theta}_2\right) \\ \hat{F}\left(X,\dot{X},\ddot{X}\mid\tilde{\Theta}_3\right) \end{bmatrix} = \begin{bmatrix} \Theta_1^{\mathrm{T}}\varepsilon\left(X,\dot{X},\ddot{X}\right) \\ \Theta_2^{\mathrm{T}}\varepsilon\left(X,\dot{X},\ddot{X}\right) \\ \Theta_3^{\mathrm{T}}\varepsilon\left(X,\dot{X},\ddot{X}\right) \end{bmatrix}.$$

As stated in the introduction, the proposed robust adaptive fuzzy tracking controller has an advantage over the one in ref. [50]. It is assumed that the cables, however, are ideally massless and nonelastic elements in this paper. It should be pointed out that the cables may behave as elastic elements in practice. This elasticity of the cables inevitably causes unwanted vibrations, leading to degradation of the positioning accuracy of the end–grab for the cable–based gangue–sorting robot. For this reason, the proposed controller for the cable–based gangue–sorting robot should efficiently dampen the vibrations of the cables, leading to enhancement of the motion accuracy of the end–grab [60]. Future research will be dedicated to investigate the effect of the cable vibration on the controller for the robots using the singular perturbation theory.

By using Lyapunov theory, the stability of the cable–based gangue–sorting robot according the dynamics in the presence of the disturbances expressed in Equation (7) with the robust adaptive fuzzy tracking control in Equation (23) is proven. As a result, the Lyapunov function candidate is defined as:

$$V(t) = \frac{1}{2}\left(s^T M s + \sum_{i=1}^{n} \tilde{\Theta}_i^{\mathrm{T}}\Gamma_i\tilde{\Theta}_i\right) \tag{24}$$

$$\begin{aligned}
\dot{V}(t) &= -s^{\mathrm{T}}\left(M\ddot{X}_r + H\dot{X}_r + F(X,\dot{X},\ddot{X}\mid\tilde{\Theta}) - J^{\mathrm{T}}T\right) + \sum_{i=1}^{n}\tilde{\Theta}_i^{\mathrm{T}}\Gamma_i\dot{\tilde{\Theta}}_i \\
&= -s^{\mathrm{T}}\left(F(X,\dot{X},\ddot{X}) - \hat{F}(X,\dot{X},\ddot{X}\mid\tilde{\Theta}^{*}) + K_D s + W\mathrm{sgn}(s)\right) + \sum_{i=1}^{n}\tilde{\Theta}_i^{\mathrm{T}}\Gamma_i\dot{\tilde{\Theta}}_i \\
&= -s^{\mathrm{T}}\left(F(X,\dot{X},\ddot{X}) - \hat{F}(X,\dot{X},\ddot{X}\mid\Theta) + \hat{F}\left(X,\dot{X},\ddot{X}\mid\tilde{\Theta}^{*}\right) - \hat{F}\left(X,\dot{X},\ddot{X}\mid\tilde{\Theta}^{*}\right) + K_D s + W\mathrm{sgn}(s)\right) + \sum_{i=1}^{n}\tilde{\Theta}_i^{\mathrm{T}}\Gamma_i\dot{\tilde{\Theta}}_i \\
&= -s^{\mathrm{T}}\left(\tilde{\Theta}^{\mathrm{T}}\varepsilon(X,\dot{X},\ddot{X}) + \omega + K_D s + W\mathrm{Sgn}(s)\right) + \sum_{i=1}^{n}\tilde{\Theta}_i^{\mathrm{T}}\Gamma_i\dot{\tilde{\Theta}}_i \\
&= -s^{\mathrm{T}}K_D s - s^{\mathrm{T}}\omega - s^{\mathrm{T}}W\mathrm{sgn}(s) + \sum_{i=1}^{n}\left(\tilde{\Theta}_i^{\mathrm{T}}\Gamma_i\dot{\tilde{\Theta}}_i - s_i\tilde{\Theta}\varepsilon(X,\dot{X},\ddot{X})\right)
\end{aligned} \tag{25}$$

where ω is the fuzzy approximation error; $\tilde{\Theta}_i = \Theta_i^{*} - \Theta_i$; $\varepsilon(X,\dot{X},\ddot{X})$ is fuzzy basis function. Moreover, the adaptive law based on Equation (25) is defined as:

$$\dot{\tilde{\Theta}}_i = -\Gamma_i^{-1}s_i\varepsilon\left(X,\dot{X},\ddot{X}\right) \tag{26}$$

Consequently, the following equation can be obtained:

$$\dot{V}(t) \leq -s^{\mathrm{T}}K_D s \leq 0 \tag{27}$$

It can be seen from Equations (24) and (27) that, since function V is a positive definite function, and $\dot{V}$ is a negative definite function, the cable–based gangue–sorting robot in Equation (7) controlled by the proposed robust adaptive fuzzy tracking controller in Equation (23) is globally asymptotically stable with respect to s and Θ based on the Lyapunov method. It means that $\lim_{t\to\infty} s = 0$, and thus, this is equivalent to $\lim_{t\to\infty} e = 0 \Rightarrow \lim_{t\to\infty} X = \lim_{t\to\infty} X_d$. As a result, the control object for the end–grab to track the scheduled trajectory to perform the pick–and–place operation of MTGs can be realized.

4. Simulation Study

4.1. Generation of the Proposed Pick–and–Place Trajectory

The end–grab of the robot is requested to start from the state of rest, merge into four consecutive trajectories in sequence to perform the pick–and–place operation of the current target gangue, and then go back to the point D to begin the pick–and–place operation of the next gangue. Assuming that the distance the target gangue is from the centerline of the belt conveyor is 0.25 m, the pick–and–place trajectory of the end–grab is achieved in this example. As is shown in Figure 3, some parameters of the cable–based gangue–sorting robot are given as follows: a = 4 m, b = 4 m, and h = 3 m. In order to ensure the robustness and the stability of the cable–based gangue–sorting robot, in this paper, all positions of the end–grab from a planned trajectory are required to be completely within the stability workspace, which refers to the set of points meeting certain stability requirements of the end–grab [61]. The stability workspace is composed of the positions of the end–grab with specified stability performance index, which can be calculated using the position and cable tension influencing factors, and furthermore, it can be obtained by the stability workspace generation algorithm as described in the previously published paper, ref. [61], by the author.

Figure 7 shows the pick–and–place trajectory of the end–grab and the positional relationship between the trajectory and the stability workspace. It should be noted that the pick–and–place trajectory is obtained while the distance from the position of the MTG to be grabbed to the center line b, denoted by j, is 0.25 m. It is observed from Figure 7a that the pick–and–place trajectory consists of four periods, and furthermore, each period's connection with each other is smooth, and thus, this leads to no impact for the movement of the end–grab. In addition, it can be seen from the Figure 7b that the pick–and–place trajectory of the end–grab is located completely within the stability workspace, and this can ensure the stability of the end–grab. Indeed, it should be pointed out that all of the positions when the end–grab is located within the surface meet the specified stability requirement. Moreover, its colors are worthy of note, as they represent the elevation of the positions for the end–grab along z–direction. As can be seen from Figure 7c,d, the velocities and accelerations of the end–grab along the proposed pick–and–place trajectory are continuous.

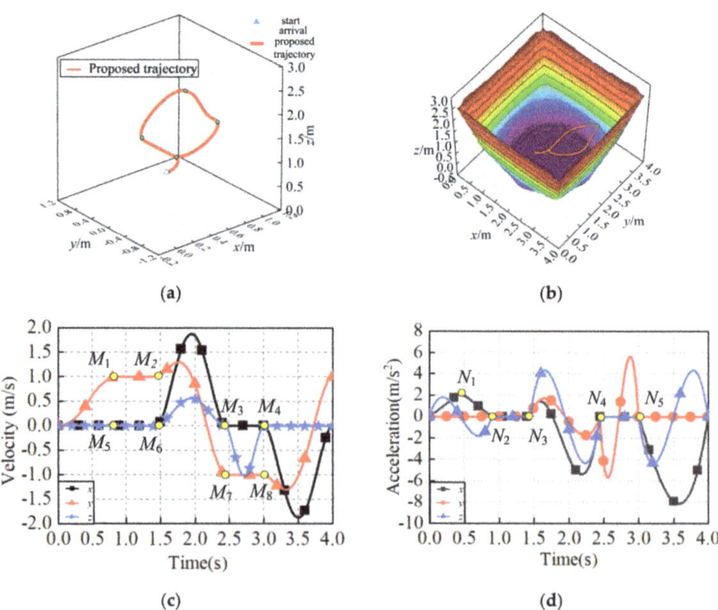

Figure 7. Pick–and–place trajectory of the end–grab. (**a**) pick–and–place trajectory; (**b**) trajectory within the stability workspace; (**c**) velocities of the end–grab; (**d**) accelerations of end–grab.

In addition, the proposed strategy can generate a smooth and continuous pick–and–place trajectory while the MTG is located at an arbitrary position of the belt, such as the center line b, namely $j = 0$, shown in Figure 8. Comparing Figures 7a and 8, it is clear that the trajectories of the starting period CD are slightly different from each other, and in more detail, there is no displacement of the end–grab along the x–direction for the trajectory of the starting period CD in Figure 8. This is because the pick–and–place position of the MTG is on the center line b.

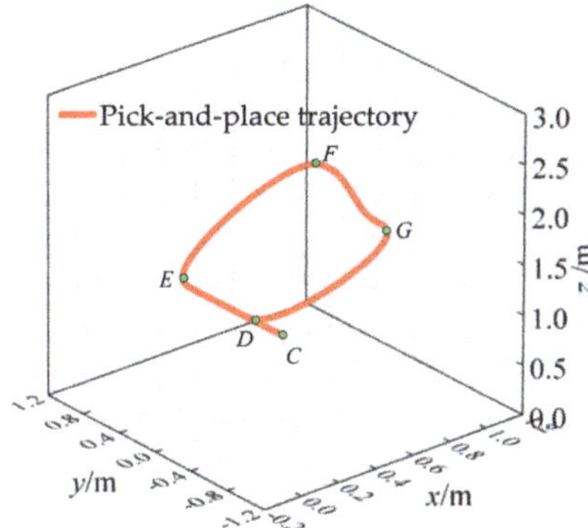

Figure 8. Pick–and–place trajectory of the end–grab while $j = 0$.

Figure 9 displays the length of the four cables. From the simulation results, it may be concluded that cable 1 and cable 3, cable 2, and cable 4 show opposite trends in the whole trajectory curve because of the symmetrical geometric relationship between the cables, and moreover, the length of the four cables change smoothly and continuously.

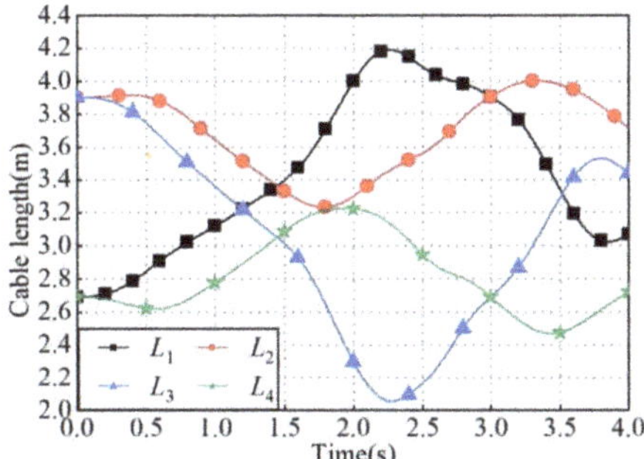

Figure 9. Length of the four cables.

4.2. Control System Validation

In this section, the proposed pick–and–place trajectory control strategies are evaluated and compared with MATLAB software. Two motion trajectories for the end–grab, a spatial circle, and the proposed pick–and–place trajectory generated in this paper are considered in this section to illustrate the efficiency of the designed controller.

To illustrate the effectiveness of the proposed robust adaptive fuzzy tracking controller, we compared it with the fuzzy controller presented in ref. [50]. The controller mentioned above was modified to be implementable for the cable–based gangue–sorting robot considered in this paper. Thus, it can be expressed as:

$$\tau = M(X)\ddot{X}_r + H\left(X, \dot{X}\right) + \hat{f}\left(X, \dot{X} \middle| \tilde{\Theta}\right) - K_D s \tag{28}$$

in which

$$\hat{f}(X,\dot{X}\mid\tilde{\Theta}) = \begin{bmatrix} \hat{f}\left(X,\dot{X}\mid\tilde{\Theta}_1\right) \\ \hat{f}\left(X,\dot{X}\mid\tilde{\Theta}_2\right) \\ \hat{f}\left(X,\dot{X}\mid\tilde{\Theta}_3\right) \end{bmatrix} = \begin{bmatrix} \Theta_1^{\mathrm{T}}\varepsilon\left(X,\dot{X}\right) \\ \Theta_2^{\mathrm{T}}\varepsilon\left(X,\dot{X}\right) \\ \Theta_3^{\mathrm{T}}\varepsilon\left(X,\dot{X}\right) \end{bmatrix} \tag{29}$$

Comparing Eqations (23) and (28), it can be observed that the fuzzy controller is a specific case of the proposed controller in this paper, in which the robust term $W = 0$. Moreover, the proposed control algorithm benefiting from internal force concept can ensure that all cables remain in tension while the end–grab moves along the designed pick–and–place trajectory. As a result, the proposed control scheme in this paper can bring a better comprehensive control performance.

Moreover, in order to have a quantitative evaluation and comparison between the performance of the two controllers, the root mean square error (RMSE) and maximum absolute of tracking errors (MAE) of the end–grab in all directions of the task space are presented, and they can be expressed as follows:

$$\mathrm{RMSE} = \sqrt{\frac{1}{N}\sum_{k=1}^{N}|e(k)|^2} \tag{30}$$

$$\mathrm{MAE} = \max(|e(k)|_{k=1\sim N}) \tag{31}$$

where $e(k)$ denotes the position tracking error of the end–grab; N is the number of samples, and k the current sample.

All dynamic and kinematic parameters of the cable–based gangue–sorting robot and the proposed controller parameters are given in Table 2. Without loss of generality, the lumped composite disturbance vector is set as $D= [4\sin(10t) \quad 2\sin(10t) \quad 4\sin(10t)]^{T}$; the initial motion state of the end–grab is $X_0= [X_1,\dot{X}_1,X_2,\dot{X}_2,X_3,\dot{X}_3] = [1.3, \quad 0, \quad 1.2, \quad 0.03\pi, \quad 1.5, \quad 0]^{T}$; the simulation time is set to 40 s; the membership function of the fuzzy control system is selected as $\mu_{A_i^l}(x_i) = \exp\left(\dfrac{-(x_i - \overline{x}_i^l)}{0.2}\right)^2$, in which $\overline{x}_i^l$ is 1, 2, 3, 4, and A_i is NB, NS, ZO, PS, and PB, respectively.

Table 2. Dynamic and kinematic parameters of the robot and the proposed controller parameters.

Parameter	Symbol	Value
Height of the pillar, Figure 3	h	3 m
Length of the rectangle formed by A_i	b	4 m
Width of the rectangle	a	4 m
Mass of the end–grab	m	5 kg
Acceleration of gravity	g	9.8 m/s^2
Gain matrix	K_D	250 $I_{4\times4}$
Matrix of the sliding surface	Λ	10 $I_{3\times3}$
Adaptation law matrix	Γ	Diag(10,10,10) $\times\ 10^{-4}$
Gain matrix of the robust term	W	Diag(0.2,0.2,0.2)

The spatial circle trajectory for the end–grab can be expressed as:

$$\begin{cases} x = 0.8\cos(0.1\pi t) + 1.8 \\ y = 0.8\sin(0.1\pi t) + 2 \\ z = 1.5 \end{cases} \tag{32}$$

The results of the position trajectory tracking of the end–grab are shown in Figure 10. It is observed that the end–grab tracks the planned spatial circle trajectory relatively well with the proposed robust adaptive fuzzy tracking controller. Figure 11 illustrates the position trajectory and the position errors of the end–grab in x–, y–, and z–directions, respectively. It can be seen that, compared with x–direction and y–directions, z–direction has the better tracking effect, and its absolute error is less than 6‰. Furthermore, the error in z–direction obviously changes in a fixed period, and the errors are always stable in the above range. Additionally, it is concluded from Figure 11d that the absolute value of the error in the x–direction and y–direction are all within 1%, fluctuating in the range of -8–(8‰). Moreover, the absolute value of the fluctuation is less than 3‰.

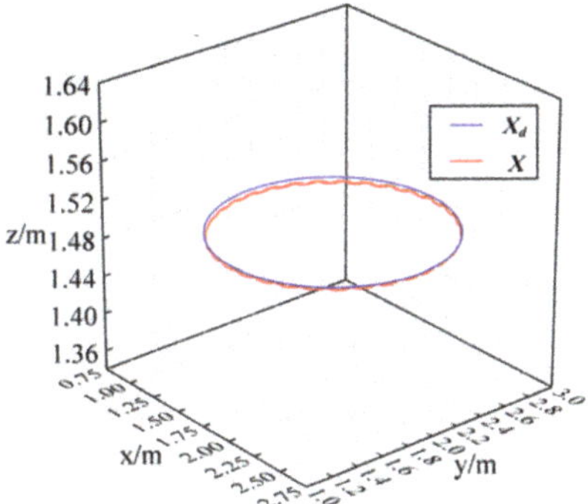

Figure 10. The spatial circle and trajectory tracking.

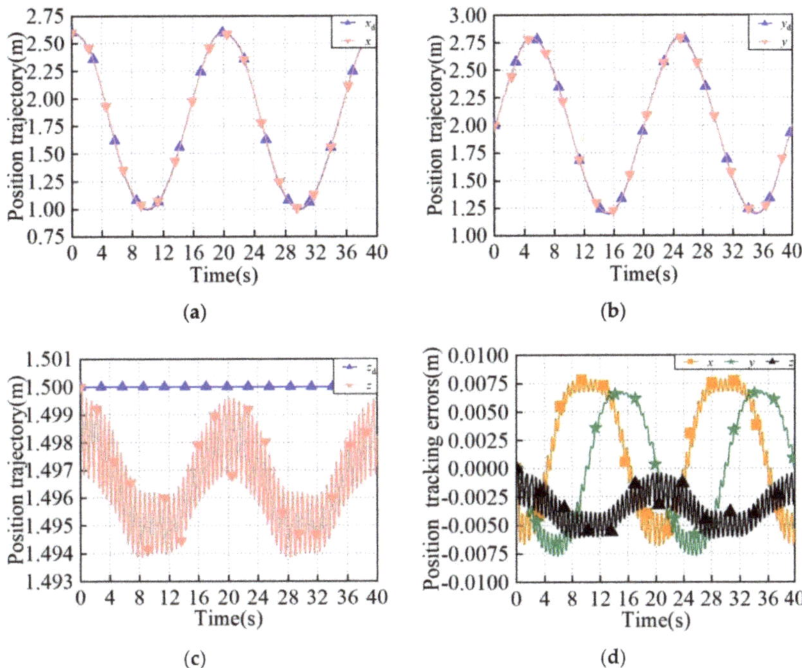

Figure 11. Position trajectory tracking of the end–grab for the spatial circle. (**a**) Position trajectory in x–direction; (**b**) position trajectory in y–direction; (**c**) position trajectory in z–direction; (**d**) trajectory tracking errors.

Figure 12 illustrates the velocity tracking and the velocity tracking errors of the end–grab in x–, y–, and z–directions, respectively. It can be observed that the absolute value of velocity error in x-direction is within 1.6%, and that in y-direction is within 1.5%, while that in z-direction is within 1.6%. Additionally, it can be seen that the velocities in the three directions fluctuate greatly within 0.5 s; however, the velocities in the three directions fluctuate steadily in an acceptable range.

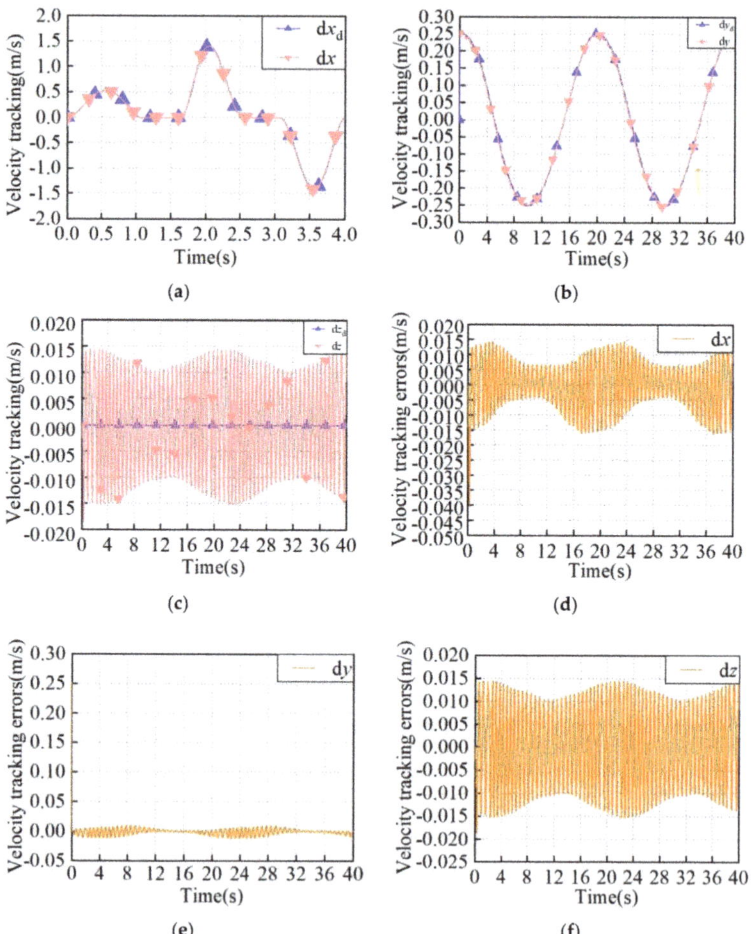

Figure 12. Velocity tracking of the end–grab for the spatial circle. (**a**) Velocity tracking in x–direction; (**b**) velocity tracking in y–direction; (**c**) velocity tracking in z–direction; (**d**) velocity tracking error in x–direction; (**e**) velocity tracking error in y–direction; (**f**) velocity tracking error in z–direction.

The four cable tensions while the end–grab follows the spatial circle defined by Equation (32) are shown in Figure 13. It is obvious that the tension of each cable is greater than 5 N, which satisfies the unidirectional characteristics of the cables. In addition, the phase, period, and fluctuation range of the cable tensions change steadily, which can generate smooth and continuous movement for the end–grab.

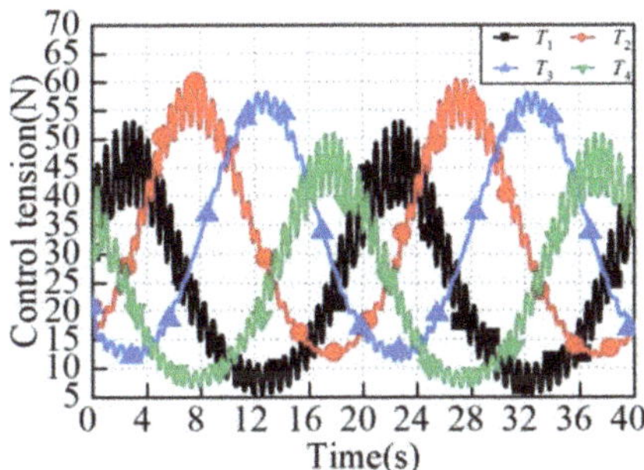

Figure 13. The control input cable tensions for the spatial circle.

While the initial motion state of the end–grab is set as $X_0 = [X_1, \dot{X}_1, X_2, \dot{X}_2, X_3, \dot{X}_3] = [2.25,\ 0,\ 1,\ 0,\ 1.5,\ 0]^{\mathrm{T}}$, the simulation time is set to 4 s. The position tracking of the end–grab for the proposed pick–and–place trajectory is displayed in Figure 14. It is observed that the end–grab, in common with the spatial circle, tracks the proposed pick–and–place trajectory relatively well with the proposed robust adaptive fuzzy tracking controller.

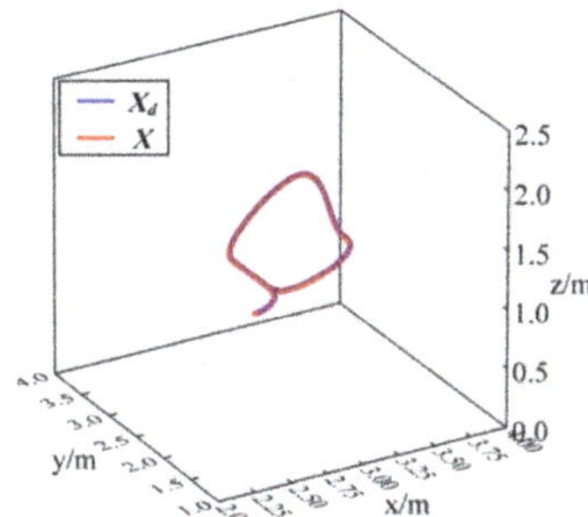

Figure 14. The proposed pick–and–place trajectory and trajectory tracking.

The position trajectory tracking errors of the end-effector in the x–direction, y–direction, and z–direction are shown in Figure 15. As can be seen from the figure, the absolute values of the errors in x–, y–, and z–directions are all within 2‰, and the error in y-direction is the largest. The maximum absolute value of the error in y–direction reaches 1.75‰ around 3.8 s. However, from the pick–and–place trajectory planning in Section 4.1, it can be seen that the end–grab has unloaded the gangues, which will not affect the smooth operation of the end–grab. Thus, the above error is acceptable.

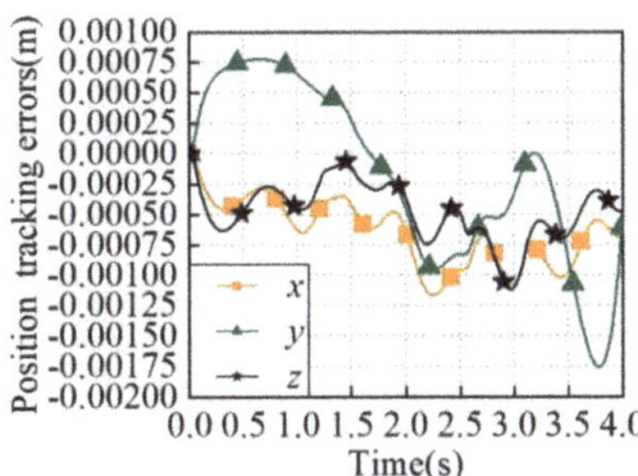

Figure 15. Position trajectory tracking error for the proposed pick–and–place trajectory.

RMSE and MAE for the both controllers in the task space while the end–grab moves along the proposed pick–and–place trajectory are listed in Table 3, respectively. The results obtained for the performance indices of the both controllers indicate the appropriate performance in practice. However, it can be seen from the qualitative analysis that the proposed controller has outperformed the fuzzy controller. As an example, the RMSE and MAE values of the proposed controller are 8.9867×10^{-4} and 2×10^{-2} m, respectively, which are better than that of the fuzzy controller, 9.1872×10^{-4} and 2×10^{-2}. It can be seen from the MAE values of the two controllers that they are equal to each other, and this is because the maximum absolute of tracking errors occurs at the initial position of the pick–and–place trajectory for the end–grab. The initial motion state of the end–grab is set as equal to each other for the two controllers. It should be pointed out that the initial motion state of the end–grab has an effect on the RMSE and MAE.

Table 3. RMSE and MAE for the both controllers.

Controller	RMSE	MAE
Proposed controller	8.9867×10^{-4} m	2×10^{-2} m
Fuzzy controller	9.1872×10^{-4} m	2×10^{-2} m

Velocity tracking of the end–grab for the proposed pick–and–place trajectory is depicted in Figure 16. From Figure 16, we can see that the absolute values of velocity errors in x–, y–, and z–directions are within 1.8%, and thus, the velocity tracking effect is good. Furthermore, the velocity fluctuation is the most complicated from 1.8 s to 3.2 s. However, it can be seen from the figure that the absolute value of the maximum velocity fluctuation error is about 0.75%.

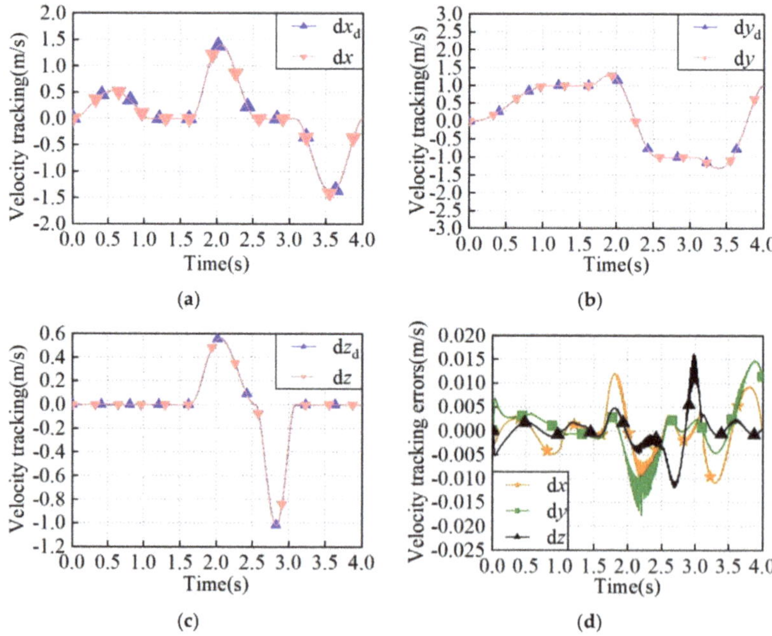

Figure 16. Velocity tracking of the end–grab for the proposed pick–and–place trajectory. (**a**) Velocity tracking in x–direction; (**b**) velocity tracking in y–direction; (**c**) velocity tracking in z–direction; (**d**) velocity tracking errors.

The four cable tensions while the end–grab follows the proposed pick–and–place trajectory are shown in Figure 17. It is obvious that the tension of each cable is also greater than 5 N, which satisfies the unidirectional characteristics of the cables. In addition, the cable tensions are smooth and continuous.

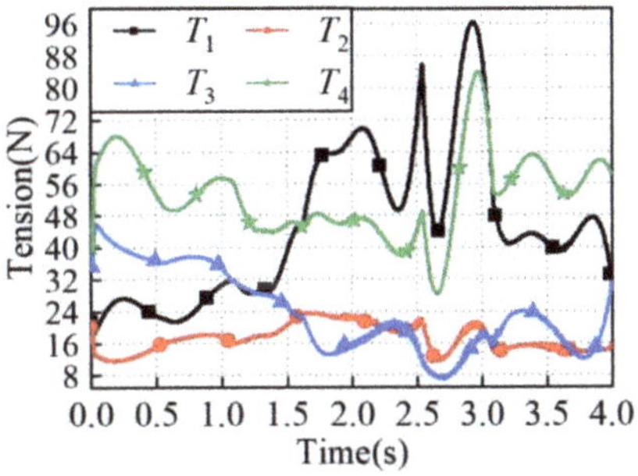

Figure 17. The control input cable tensions for the proposed pick–and–place trajectory.

The performed study clearly shows certain advantages of the proposed control system, which include: first, the presented four−phase pick–and–place trajectory planning scheme can generate a smooth and continuous trajectory for the end–grab to perform the pick–and–place operation of the MTGs; second, the proposed robust adaptive fuzzy tracking control strategy is able to track a trajectory as close as possible to the desired one without violating the cable tension limits.

5. Conclusions and Future Works

In this paper, a control system for the cable–based gangue–sorting robot performing the pick–and–place operation of MTGs with different shapes, sizes, and masses was presented, which consists of two modules: (i) the pick–and–place trajectory planning module and (ii) trajectory tracking control module.

In more detail, a four-stage trajectory planning scheme for the end–grab of the cable–based gangue–sorting robot is proposed in this paper. The pick–and–place trajectory of the end–grab is planned into four periods: the starting period, preparing period, picking period, and placing period. In addition, according to the established dynamic equation of the cable–based gangue–sorting robot, a robust adaptive fuzzy control strategy was designed to track a given trajectory in the presence of model uncertainties, varying payloads, and external disturbances. Based on Lyapunov stability analysis, the stability of the closed-loop control system was theoretically proved. To evaluate the proposed control system, numerical simulations were performed for the cable–based gangue–sorting robot. The simulation analysis of the pick–and–place trajectory planning scheme of the robot shows that the motion trajectory, velocity, and acceleration of the end–grab are smooth and continuous, and moreover, the cable length changes smoothly and continuously. Meanwhile, simulation analysis of the robust adaptive fuzzy control strategy of the robot shows that a high-precision position and velocity tracking performance for the end–grab was obtained with the proposed control strategy, which is robust against the uncertainties.

As a matter of fact, tracking, approaching, picking, and placing the MTGs have become an important mission of the cable–based gangue–sorting robots. In this paper, the authors present the proposed pick–and–place trajectory planning and trajectory tracking control for the robots. In the next phase, we will focus on the following parts as our future work: (i) contact and impact analysis during the picking and placing process of the MTGs; (ii) motion stabilization of the cable–based gangue–sorting robots after picking and placing the MTGs; and (iii) the experimental validation of the four-stage trajectory planning scheme and the proposed pick–and–place trajectory tracking control scheme for the cable–based gangue–sorting robots.

Author Contributions: Conceptualization, P.L. and L.G.; methodology, P.L.; software, Y.S.; validation, P.L. and X.Q.; formal analysis, Y.Q.; investigation, P.L.; resources, X.D.; data curation, X.C.; writing—original draft preparation, P.L.; writing—review and editing, P.L., Y.Q., and X.Q.; supervision, Y.Q.; project administration, X.C.; funding acquisition, P.L. and H.T. All authors have read and agreed to the published version of the manuscript.

Funding: This research was funded by the financial support of National Natural Science Foundation of China (NSFC) under Grant NO. 52174149, Shaanxi Province Natural Science Basic Research Program Project under Grant No. 2019JQ-796, and Key Research and Development Program of Shaanxi Province under Grant NO. 2022GY-241.

Institutional Review Board Statement: Not applicable.

Informed Consent Statement: Not applicable.

Data Availability Statement: The data used to support the findings of this study are available from the corresponding author upon request.

Acknowledgments: The research is supported by Open Fund of Key Laboratory of Electronic Equipment Structure Design (Ministry of Education) in Xidian University. And moreover, the authors are grateful to the guest editor, Zhufeng Shao, Dan Zhang, and Stéphane Caro as well as the anonymous reviewers for their constructive comments and helpful suggestions that greatly improved the quality of this article.

Conflicts of Interest: The authors declared no potential conflict of interest with respect to the research, authorship, and/or publication of this article.

References

1. Liu, P.; Qiao, X.Z.; Zhang, X.H. Stability sensitivity for a cable–based coal–gangue picking robot based on grey relational analysis. *Int. J. Adv. Robot. Syst.* **2021**, *18*, 17298814211059729. [CrossRef]
2. Wang, R.; Xie, Y.; Chen, X.; Li, Y. Kinematic and dynamic modeling and workspace analysis of a suspended cable-driven parallel robot for Schönflies motions. *Machines* **2022**, *10*, 451. [CrossRef]
3. Tang, X.Q. An overview of the development for cable-driven parallel manipulator. *Adv. Mech. Eng.* **2014**, *6*, 823028. [CrossRef]
4. Zarebidoki, M.; Dhupia, J.S.; Xu, W. A review of cable-driven parallel robots: Typical configurations, analysis techniques, and control methods. *IEEE Robot. Autom. Mag.* **2022**, 2–19. [CrossRef]
5. Liu, Z.H.; Tang, X.Q.; Shao, Z. Research on longitudinal vibration characteristic of the six-cable-driven parallel manipulator in FAST. *Adv. Mech. Eng.* **2013**, *2013*, 547416. [CrossRef]
6. Su, Y.; Qiu, Y.; Liu, P.; Tian, J.W.; Wang, Q.; Wang, X.G. Dynamic modeling, workspace analysis and multi-objective structural optimization of the large-span high-speed cable-driven parallel camera robot. *Machines* **2022**, *10*, 565. [CrossRef]
7. Zi, B.; Zhang, L.; Zhang, D.; Qian, S. Modeling, analysis, and co-simulation of cable parallel manipulators for multiple cranes. *Proc. Inst. Mech. Eng. Part C J. Mech. Eng. Sci.* **2015**, *229*, 1693–1707. [CrossRef]
8. Russo, M.; Ceccarelli, M. Analysis of a wearable robotic system for ankle rehabilitation. *Machines* **2020**, *8*, 48. [CrossRef]
9. Tourajizadeh, H.; Yousefzadeh, M.; Khalaji, A.K.; Bamdad, M. Design, modeling and control of a simulator of an aircraft maneuver in the wind tunnel using cable robot. *Int. J. Control Autom. Syst.* **2022**, *20*, 1671–1681. [CrossRef]
10. Bosscher, P.; Riechel, A.T.; Ebert-Uphoff, I. Wrench-feasible workspace generation for cable-driven robots. *IEEE Trans. Robot.* **2006**, *22*, 890–902. [CrossRef]
11. Pham, C.B.; Yeo, S.H.; Yang, G. Workspace analysis of fully restrained cable-driven manipulators. *Robot. Auton. Syst.* **2009**, *57*, 901–912. [CrossRef]
12. Kieu, V.N.D.; Huang, S.C. Dynamic and wrench-feasible workspace analysis of a cable-driven parallel robot considering a nonlinear cable tension model. *Appl. Sci.* **2022**, *12*, 244. [CrossRef]
13. Liu, P.; Qiu, Y.Y.; Su, Y. A new hybrid force-position measure approach on the stability for a camera robot. *Proc. Inst. Mech. Eng. Part C J. Mech. Eng. Sci.* **2016**, *230*, 2508–2516. [CrossRef]
14. Wei, H.L.; Qiu, Y.Y.; Sheng, Y. An approach to evaluate stability for cable-driven parallel camera robots with hybrid tension-stiffness properties. *Int. J. Adv. Robot. Syst.* **2015**, *12*, 185. [CrossRef]
15. Su, Y.; Qiu, Y.Y.; Wei, H.L. Dynamic modeling and tension optimal distribution of cable-driven parallel robots considering cable mass and inertia force effects. *Eng. Mech.* **2016**, *33*, 231–239. [CrossRef]
16. Su, Y.; Qiu, Y.Y.; Sheng, Y. Optimal cable tension distribution of the high-speed redundant driven camera robots considering cable sag and inertia effects. *Adv. Mech. Eng.* **2014**, *2014*, 729020. [CrossRef]
17. Liu, P.; Qiu, Y.Y.; Su, Y.; Chang, J.T. On the minimum cable tensions for the cable–based parallel robots. *J. Appl. Math.* **2014**, *2014*, 350492. [CrossRef]
18. Amine, L.M.; Giuseppe, C.; Said, Z. On the optimal design of cable driven parallel robot with a prescribed workspace for upper limb rehabilitation tasks. *J. Bionic Eng.* **2019**, *16*, 503–513. [CrossRef]

19. Yao, R.; Tang, X.; Wang, J.; Huang, P. Dimensional optimization design of the four-cable-driven parallel manipulator in FAST. *IEEE/ASME Trans. Mechatron.* **2010**, *15*, 932–941. [CrossRef]

20. Zhang, F.; Shang, W.; Zhang, B.; Cong, S. Design optimization of redundantly actuated cable-driven parallel robots for automated warehouse system. *IEEE Access* **2020**, *8*, 56867–56879. [CrossRef]

21. Kawamura, S.; Kino, H.; Won, C. High-speed manipulation by using parallel wire-driven robots. *Robotica* **2000**, *18*, 13–21. [CrossRef]

22. Qian, S.; Bao, K.L.; Zi, B.; Zhu, W.D. Dynamic trajectory planning for a three degrees-of-freedom cable-driven parallel robot using quintic B-splines. *J. Mech. Design* **2020**, *142*, 073301. [CrossRef]

23. Zhao, T.; Zi, B.; Qian, S.; Zhao, J.H. Algebraic method-based point-to-point trajectory planning of an under-constrained cable-suspended parallel robot with variable angle and height cable mast. *Chin. J. Mech. Eng.* **2020**, *33*, 073301. [CrossRef]

24. Xu, J.J.; Park, K.S. Moving obstacle avoidance for cable-driven parallel robots using improved RRT. *Microsyst. Technol.* **2020**, *27*, 2281–2292. [CrossRef]

25. Jiang, X.; Barnett, E.; Gosselin, C. Dynamic point-to-point trajectory planning beyond the static workspace for six-DOF cable-suspended parallel robots. *IEEE Trans. Robot.* **2018**, *34*, 781–793. [CrossRef]

26. Hwang, S.W.; Bak, J.H.; Yoon, J.; Park, J.H.; Park, J.O. Trajectory generation to suppress oscillations in under-constrained cable-driven parallel robots. *J. Mech. Sci. Technol.* **2016**, *30*, 5689–5697. [CrossRef]

27. Jiang, M.; Yang, Z.M.; Li, Y.; Sun, Z.; Zi, B. Smooth Trajectory Planning for a Cable-Driven Waist Rehabilitation Robot Using Quintic NURBS. In *Intelligent Robotics and Applications. ICIRA 2021. Lecture Notes in Computer Science*; Liu, X.J., Nie, Z., Yu, J., Xie, F., Song, R., Eds.; Springer: Cham, Switzerland, 2021; p. 13013. [CrossRef]

28. Idá, E.; Briot, S.; Carricato, M. Identification of the inertial parameters of underactuated cable-driven parallel robots. *Mech. Mach. Theory* **2022**, *167*, 104504. [CrossRef]

29. Dion-Gauvin, P.; Gosselin, C. Beyond-the-static-workspace point-to-point trajectory planning of a 6-DoF cable-suspended mechanism using oscillating SLERP. *Mech. Mach. Theory* **2022**, *174*, 104894. [CrossRef]

30. Scalera, L.; Gasparetto, A.; Zanotto, D. Design and experimental validation of a 3-DOF underactuated pendulum-like robot. *IEEE/ASME Trans. Mechatron.* **2019**, *25*, 217–228. [CrossRef]

31. Zi, B.; Sun, H.H.; Zhang, D. Design, analysis and control of a winding hybrid-driven cable parallel manipulator. *Robot. Comput. Integr. Manuf.* **2017**, *48*, 196–208. [CrossRef]

32. Fang, S.Q.; Franitza, D.; Torlo, M.; Bekes, F.; Hiller, M. Motion control of a tendon-based parallel manipulator using optimal tension distribution. *IEEE/ASME Trans. Mechatron.* **2004**, *9*, 561–568. [CrossRef]

33. Ji, H.; Shang, W.W.; Cong, S. Adaptive synchronization control of cable-driven parallel robots with uncertain kinematics and dynamics. *IEEE T Ind. Electron.* **2021**, *68*, 8444–8454. [CrossRef]

34. Khalilpour, S.A.; Khorrambakht, R.; Taghirad, H.D.; Cardou, P. Robust cascade control of a deployable cable-driven robot. *Mech. Syst. Signal Processing* **2019**, *127*, 513–530. [CrossRef]

35. Jia, H.Y.; Shang, W.W.; Xie, F.; Zhang, B.; Cong, S. Second-order sliding-mode-based synchronization control of cable-driven parallel robots. *IEEE/ASME Trans. Mechatron.* **2020**, *25*, 383–394. [CrossRef]

36. Santos, J.C.; Gouttefarde, M.; Chemori, A. A nonlinear model predictive control for the position tracking of cable-driven parallel robots. *IEEE Trans. Robot.* **2022**, *38*, 2597–2616. [CrossRef]

37. Wang, Y.Y.; Liu, L.F.; Yuan, M.X.; Di, Q.X.; Chen, B. Time-delay control using a novel nonlinear adaptive law for accurate trajectory tracking of cable-driven robots. *IEEE Trans. Ind. Inform.* **2020**, *16*, 5234–5243. [CrossRef]

38. Aflakiyan, A.; Bayani, H.; Masouleh, M.T. Computed torque control of a cable suspended parallel robot. In Proceedings of the 2015 3rd RSI International Conference on Robotics and Mechatronics (ICROM), Tehran, Iran, 7–9 October 2015; pp. 749–754. [CrossRef]

39. Wang, T.C.; Tong, S.C.; Yi, J.Q.; Li, H.Y. Adaptive inverse control of cable-driven parallel system based on type-2 fuzzy logic systems. *IEEE Trans. Fuzzy Syst.* **2015**, *23*, 1803–1816. [CrossRef]

40. Izadbakhsh, A.; Asl, H.J.; Narikiyo, T. Robust adaptive control of over-constrained actuated cable-driven parallel robots. In *Cable-Driven Parallel Robots*; Pott, A., Bruckmann, T., Eds.; Springer: Cham, Switzerland, 2019; Volume 74, pp. 209–220. [CrossRef]

41. Wang, Y.Y.; Liu, L.F.; Yuan, M.X.; Di, Q.X.; Chen, B.; Wu, H.T. A new model-free robust adaptive control of cable-driven robots. *Int. J. Control Autom. Syst.* **2021**, *19*, 3209–3222. [CrossRef]

42. Oh, S.-R.; Agrawal, S.K. Cable suspended planar robots with redundant cables: Controllers with positive tensions. *IEEE Trans. Robot.* **2005**, *21*, 457–465. [CrossRef]

43. Shao, Z.F.; Tang, X.Q.; Wang, L.P.; Chen, X. Dynamic modeling and wind vibration control of the feed support system in FAST. *Nonlinear Dyn.* **2012**, *67*, 965–985. [CrossRef]

44. Duan, X.C.; Mi, J.W.; Zhao, Z. Vibration isolation and trajectory following control of a cable suspended stewart platform. *Machines* **2016**, *4*, 20. [CrossRef]

45. Schenk, C.; Bülthoff, H.H.; Masone, C. Robust adaptive sliding mode control of a redundant cable driven parallel robot. In Proceedings of the 2015 19th International Conference on System Theory, Control and Computing (ICSTCC), Cheile Gradistei, Romania, 14–16 October 2015; pp. 427–434. [CrossRef]

46. Jafarlou, F.; Peimani, M.; Lotfivand, N. Fractional order adaptive sliding-mode finite time control for cable-suspended parallel robots with unknown dynamics. *Int. J. Dynam. Control.* **2022**, 1–11. [CrossRef]

47. Hosseini, M.I.; Khalilpour, S.A.; Taghirad, H.D. Practical robust nonlinear PD controller for cable-driven parallel manipulators. *Nonlinear Dyn.* **2021**, *106*, 405–424. [CrossRef]

48. Aghaseyedabdollah, M.; Abedi, M.; Pourgholi, M. Supervisory adaptive fuzzy sliding mode control with optimal Jaya based fuzzy PID sliding surface for a planer cable robot. *Soft Comput.* **2022**, *26*, 8441–8458. [CrossRef]

49. Inel, F.; Medjbouri, A.; Carbone, G. A non-linear continuous-time generalized predictive control for a planar cable-driven parallel robot. *Actuators* **2021**, *10*, 97. [CrossRef]

50. Zi, B. Fuzzy Control System Design and Analysis for Completely Restrained Cable-Driven Manipulators. In *Fuzzy Logic: Controls, Concepts, Theories and Applications*; Dadios, E., Ed.; IntechOpen: London, UK, 2012. [CrossRef]

51. Miermeister, P.; Pott, A. Modelling and real-time dynamic simulation of the cable-driven parallel robot IPAnema. In *New Trends in Mechanism Science. Mechanisms and Machine Science*; Pisla, D., Ceccarelli, M., Husty, M., Corves, B., Eds.; Springer: Berlin/Heidelberg, Germany, 2010; Volume 5. [CrossRef]

52. Korayem, M.H.; Yousefzadeh, M.; Beyranvand, B. Dynamics and control of a 6-dof cable-driven parallel robot with visco-elastic cables in presence of measurement noise. *J. Intell. Robot. Syst* **2017**, *88*, 73–95. [CrossRef]

53. Korayem, M.H.; Yousefzadeh, M.; Manteghi, S. Dynamics and input–output feedback linearization control of a wheeled mobile cable-driven parallel robot. *Multibody Syst. Dyn.* **2017**, *40*, 55–73. [CrossRef]

54. Caverly, R.J.; Forbes, J.R. Dynamic modeling and noncollocated control of a flexible planar cable-driven manipulator. *IEEE Trans. Robot.* **2014**, *30*, 1386–1397. [CrossRef]

55. Fang, S.X.; Cao, J.G.; Cheng, W.D. Study on high-speed and smooth transfer of robot motion trajectory based on modified S-shaped acceleration/deceleration algorithm. *IEEE Access* **2020**, *8*, 199747–199758. [CrossRef]

56. Ni, H.P.; Ji, S.; Liu, Y.N.; Ye, Y.X.; Zhang, C.R.; Chen, J.W. Velocity planning method for position–velocity–time control based on a modified S-shaped acceleration/deceleration algorithm. *Int. J. Adv. Robot. Syst.* **2022**, *19*, 17298814211072418. [CrossRef]

57. Lu, S.; Ding, B.X.; Li, Y.M. Minimum-jerk trajectory planning pertaining to a translational 3-degree-of-freedom parallel manipulator through piecewise quintic polynomials interpolation. *Adv. Mech. Eng.* **2020**, *12*, 1687814020913667. [CrossRef]

58. Chen, Y.; Ma, G.Y.; Lin, S.X.; Gao, J. Adaptive fuzzy computed-torque control for robot manipulator with uncertain dynamics. *Int. J. Adv. Robot. Syst.* **2012**, *9*, 237. [CrossRef]

59. Zhou, Q.; Wang, L.J.; Wu, C.W.; Li, H.Y.; Du, H.P. Adaptive fuzzy control for nonstrict-feedback systems with input saturation and output constraint. *IEEE Trans. Syst. Man Cybern. Syst.* **2017**, *47*, 1–12. [CrossRef]

60. Babaghasabha, R.; Khosravi, M.A.; Taghirad, H.D. Adaptive robust control of fully constrained cable robots: Singular perturbation approach. *Nonlinear Dyn.* **2016**, *85*, 607–620. [CrossRef]

61. Liu, P.; Qiu, Y.Y. Approach with a hybrid force-position property to assessing the stability for camera robots. *J. Xidian Univ.* **2016**, *43*, 87–93. [CrossRef]

Article

Fractional Order KDHD Impedance Control of the Stewart Platform

Luca Bruzzone * and Alessio Polloni

DIME Department, University of Genoa, 16145 Genoa, Italy; alessio.polloni@elsel.it
* Correspondence: luca.bruzzone@unige.it; Tel.: +39-010-3352967

Abstract: In classical impedance control, KD, the steady-state end-effector forces are imposed to be proportional to the end-effector position errors through the stiffness matrix, K, and a proper damping term is added, proportional to the first-order derivatives of the end-effector position errors according to the damping matrix, D. This paper presents a fractional-order impedance control scheme, named KDHD, in which an additional damping is added, proportional to the half-order derivatives of the end-effector position errors according to the half-derivative damping matrix, HD. Since the finite-order digital filters which implement in real-time the half-order derivatives modify the steady-state stiffness of the end-effector—which should be defined exclusively by the stiffness matrix—a compensation method is proposed (KDHDc). The effectiveness of this approach is validated by multibody simulation on a Stewart platform. The proposed impedance controller represents the extension to multi-input multi-output robotic systems of the $PDD^{1/2}$ controller for single-input single-output systems, which overperforms the PD scheme in the transient behavior.

Keywords: impedance control; fractional calculus; half-order derivative; parallel kinematics machine; KDHD; Stewart platform

Citation: Bruzzone, L.; Polloni, A. Fractional Order KDHD Impedance Control of the Stewart Platform. *Machines* **2022**, *10*, 604. https://doi.org/10.3390/machines10080604

Academic Editors: Zhufeng Shao, Dan Zhang and Stéphane Caro

Received: 30 June 2022
Accepted: 21 July 2022
Published: 24 July 2022

Publisher's Note: MDPI stays neutral with regard to jurisdictional claims in published maps and institutional affiliations.

1. Introduction

Industrial robots are in most applications controlled in their position, independently of their serial or parallel kinematic architecture [1]. Position control is appropriate whenever the task can be correctly performed by regulating only the end-effector trajectory and when the interaction forces with the environment are not crucial. On the contrary, when, for proper task execution, it is fundamental to regulate the force/moment exerted in some directions, it is necessary to adopt the hybrid position/force control (HPFC) [2]. The main drawback of HPFC is that it requires force/moment sensors for any generalized coordinate which must be controlled in force/moment. Impedance control represents an intermediate approach between position control and HPFC and is widely adopted since it allows the control of the interaction forces with the environment without force/moment sensors, even if it has lower accuracy than HPFC [3,4].

The basic idea behind impedance control is to impose on the end-effector, subject to external forces, a trajectory which is followed with a programmed spatial compliant behavior. The end-effector compliance is defined through the stiffness and damping matrices, K and D, which correlate linearly with the force/moment exerted by the end-effector on the environment with the end-effector position/orientation error. Consequently, it is possible to achieve a compliant behavior with low stiffness in the directions which must be force-controlled, and a stiff behavior in the directions which must be position-controlled. Impedance control allows the elimination of the force/moment sensors on the end-effector through an indirect measurement based on the actuation force/moments; therefore, the accuracy of impedance control is remarkably lower in the presence of friction in the joints and gearboxes.

For translational robots, the end-effector external coordinates are always expressed in an orthogonal reference frame, and, in case of impedance control, K and D are 3×3 and represent, respectively, the zero-order (proportional) term and the first-order (derivative) term of a three-dimensional PD control in the external coordinates [5]. The matrices K and D can be expressed in a principal reference frame, selected on the basis of the specific task; when this principal reference frame is not parallel to the fixed reference frame, W, in which the end-effector coordinates are expressed, K and D are non-diagonal.

On the contrary, for impedance control of robots with rotational degrees of freedom of the end-effector, there are different possible approaches to define the stiffness and damping matrices, since the orientation of a rigid body in space can be described in many alternative ways [6]; this leads to different definitions of rotational stiffness and damping [7,8]. In general, for impedance control of robots with both translational and rotational degrees of freedom, the translational and rotational impedances are decoupled, and this results in block-diagonal matrices K and D, each one with two blocks: one block defines the translational stiffness (for K) or damping (for D), and the other block defines the rotational stiffness (for K) or damping (for D) [9].

The recalled basic concepts characterize classical impedance control, in the following referred to as KD, in which the end-effector stiffness and damping are linear; nonetheless, in the scientific literature several variants have been proposed. First of all, some researchers have conceived nonlinear impedance control algorithms, with nonlinear stiffness and damping of the end-effector, for instance for cooperative human–robot tasks, or to keep the end-effector within a restricted region in case of excessive contact forces [10–12].

Other alternative impedance control schemes are based on Fractional Calculus (FC), which deals with derivatives and integrals of non-integer order [13]. Using FC, the end-effector damping can be proportional not to the first-order derivative of the position error, but to a derivative with non-integer order μ [14,15], with possible benefits in terms of accuracy of the regulation of the contact forces [16]. This fractional-order impedance control, which is referred to as KD^μ, represents an n-dimensional version (where n is the number of external coordinates) of the fractional-order PD^μ control scheme for single-input single-output systems [17], while KD impedance control conceptually corresponds to the PD scheme.

In [18] a different application of FC to impedance control has been proposed, named KDHD: there is a damping term proportional to the first-order derivative of the position error according to matrix D, as in the KD scheme, but a second damping term is added, proportional to the fractional derivative of order 1/2 of the position error according to the matrix HD. Consequently, this algorithm generalizes to n-dimensional systems the $PDD^{1/2}$ scheme, whose benefits over the classical PD have been demonstrated in simulation [19] and experimentally for linear [20] and rotary axes [21].

In [18] the KDHD impedance control has been applied in simulation to a 3-PUU parallel robot, with three translational degrees of freedom of the end-effector. In the present paper the KD–KDHD comparison is extended considering the application to a six-degree-of-freedom manipulator, the Stewart platform, thus requiring the definition of both the translational and rotational impedances.

In the remainder of the paper: Section 2 discusses the theoretical definition and the digital implementation of the half-derivative of a time function; Section 3 proposes the KDHD impedance control for a six-degrees-of-freedom non-redundant parallel robot, highlighting the differences with respect to the classical KD impedance control; Section 4 deals with the Stewart platform geometry and its kinematics; Section 5 discusses the multibody modelling and the considered mechanical parameters; Section 6 presents the simulation results of the comparison between the KD and KDHD impedance controls; Section 7 outlines conclusions and future developments.

2. Half-Order Derivative: Definition and Digital Implementation Issues

As already stated, FC generalizes the concepts of derivative and integral to a generic non-integer order. In the scientific literature there are many alternative definitions of fractional derivatives and integrals, which are proven to be equivalent, but give rise to different numerical implementations. In the following, the Grünwald–Letnikov definition will be adopted since it leads to a robust discrete-time reduction [22]. Adopting this definition, the fractional operator for a continuous time function $x(t)$ is:

$$\frac{d^\alpha}{dt^\alpha} x(t) = x(t)^{(\alpha)} = \lim_{h \to 0} \frac{1}{h^\alpha} \sum_{k=0}^{\left[\frac{t-a}{h}\right]} (-1)^k \frac{\Gamma(\alpha+1)}{\Gamma(k+1)\Gamma(\alpha-k+1)} x(t - kh), \tag{1}$$

where: $\alpha \in \mathbb{R}^+$ is the differentiation order; a and t are the fixed and variable limits; Γ is the Gamma function; h is the time increment; $[y]$ is the integer part of y. To obtain a numerical implementation, Equation (1) must be rewritten using a small but finite sampling time T_s [23]:

$$x(t)^{(\alpha)} \cong x(kT_s)^{(\alpha)} \cong \frac{1}{T_s^\alpha}\left(\sum_{j=0}^{k} w_j^\alpha x((k-j)T_s)\right), k = [(t-a)/T_s], \tag{2}$$

where:

$$w_0^\alpha = 1, \ w_j^\alpha = \left(1 - \frac{\alpha+1}{j}\right) w_{j-1}^\alpha \ , j = 1, \ 2, \ \ldots . \tag{3}$$

In Equation (2) there is a summation of $k + 1$ sampled values of x, with k tending to infinity as time passes, which is computationally unfeasible for real-time control applications; therefore, only a fixed number of n previous steps is used in Equation (2), obtaining an nth order digital filter with memory length $L = nT_s$, which can be written in the z-transform notation:

$$D^{(\alpha)}(z) = \frac{1}{T_s^\alpha} \sum_{j=0}^{n} w_j^\alpha z^{-j}. \tag{4}$$

The application of digital filter (4) corresponds to considering only the recent past of the function, in the interval $[t–L, t]$. In general, this approximation is acceptable according to the so-called short-memory principle [23]. Nevertheless, the truncation of the summation of Equation (2) to $n + 1$ addends leads to an issue when filter (4) is applied to impedance control, as discussed in [18]. In reality, the summation of the filter coefficients tends to zero as k tends to infinity:

$$\lim_{k \to \infty} \sum_{j=0}^{k} w_j^\alpha = 0. \tag{5}$$

Consequently, any fractional-order derivative of a constant c is null. Unfortunately, considering only a finite number n of terms, this summation is non-zero, but it is a positive function of n, which tends to zero quite slowly, as discussed in [18]:

$$W_\alpha(n) = \sum_{j=0}^{n} w_j^\alpha > 0. \tag{6}$$

Therefore, the fractional derivative of a constant c evaluated in real time by means of the nth order digital filter (4) is non-zero, but equal to $cW_\alpha(n)/T_s^\alpha$, tending to zero as the filter order increases. This fact alters the behavior of the impedance control in a steady state: as a matter of fact, in a steady state only the stiffness term should be non-zero, with null damping terms, since the position error is constant, and, therefore, its time derivatives should be null. Nevertheless, if fractional-order derivatives are numerically evaluated with finite-order filters, the fractional-order damping term is also non-zero in the steady state, giving rise to a constant term which alters the relationship between position error and end-effector force, which should depend only on the stiffness matrix and not on the fractional-order damping matrix. To solve this issue, a proper compensation term is introduced, as discussed in Section 3.3.

3. The Proposed KDHD Impedance Control

3.1. KD Impedance Control of Six-Degree-of-Freedom Non-Redundant Parallel Robots

The KDHD impedance control has been proposed for the first time in [18], with application to a purely translating parallel robot. In the following, the KDHD impedance control is extended to a full-mobility, six-degree-of-freedom parallel robot. In particular, the considered case study is the Stewart platform, but the same approach can be used for any non-redundant six-degree-of-freedom parallel robot by only replacing the Jacobian matrix.

The KD impedance control with gravity compensation for a non-redundant six-degree-of-freedom parallel robot can be expressed by the following control law [24]:

$$
\begin{aligned}
\boldsymbol{\tau} &= \left(\mathbf{J}_p^T(\mathbf{q})\right)^{-1}\left(\mathbf{K}_{KD}\begin{bmatrix}\mathbf{x}_d-\mathbf{x}(\mathbf{q})\\ \mathbf{E}_d-\mathbf{E}(\mathbf{q})\end{bmatrix}+\mathbf{D}_{KD}\begin{bmatrix}(\mathbf{x}_d-\mathbf{x}(\mathbf{q}))^{(1)}\\ (\mathbf{E}_d-\mathbf{E}(\mathbf{q}))^{(1)}\end{bmatrix}\right)+\boldsymbol{\tau}_g(\mathbf{q})=\\
&= \left(\mathbf{J}_p^T(\mathbf{q})\right)^{-1}\left(\begin{bmatrix}\mathbf{K}_{KDt}&0\\0&\mathbf{K}_{KDr}\end{bmatrix}\begin{bmatrix}\mathbf{x}_d-\mathbf{x}(\mathbf{q})\\ \mathbf{E}_d-\mathbf{E}(\mathbf{q})\end{bmatrix}+\begin{bmatrix}\mathbf{D}_{KDt}&0\\0&\mathbf{D}_{KDr}\end{bmatrix}\begin{bmatrix}(\mathbf{x}_d-\mathbf{x}(\mathbf{q}))^{(1)}\\ (\mathbf{E}_d-\mathbf{E}(\mathbf{q}))^{(1)}\end{bmatrix}\right)+\boldsymbol{\tau}_g(\mathbf{q})
\end{aligned}
\tag{7}
$$

where $\boldsymbol{\tau}$ is the vector of the actuation forces, $\mathbf{q}$ is the vector of the internal coordinates, $\boldsymbol{\tau}_g(\mathbf{q})$ is the gravity compensation vector and $\mathbf{J}_p(\mathbf{q})$ is the Jacobian matrix of the parallel robot, which correlates to the time-derivative of the external and internal coordinates:

$$
\dot{\mathbf{q}}=\mathbf{J}_p\begin{bmatrix}\dot{\mathbf{x}}\\ \dot{\mathbf{E}}\end{bmatrix}.
\tag{8}
$$

In particular, for the Stewart platform the vector of the internal coordinates collects the six leg lengths ($\mathbf{q}=[q_1, q_2, q_3, q_4, q_5, q_6]^T$), while the vector of the external coordinates assembles the three end-effector coordinates in the fixed reference frame ($\mathbf{x}=[x, y, z]^T$) and the three components in the fixed reference frame of the angle-axis vector $\mathbf{E}=[E_x, E_y, E_z]^T$. The vector $\mathbf{E}=\theta\cdot\mathbf{e}$ represents a generic rotation with magnitude θ around an axis defined by the unit vector $\mathbf{e}$ according to the right-hand rule [6]. The use of the vector $\mathbf{E}$ to represent the end-effector orientation is the most suitable for impedance control due to its strict relationship with the angular velocity vector ω. As a matter of fact, it is possible to demonstrate that, if θ tends to zero, the angular velocity ω tends to be equal to the time-derivative of $\mathbf{E}$ [9]. This property is not very important for rotations with respect to a fixed frame, because in general θ is different from zero, but it is useful in impedance control, where the angle-axis representation is used to describe the relative rotation between the end effector frame, $\mathbf{E}(\mathbf{q})$, and the frame describing its reference orientation, $\mathbf{E}_d$, because θ is small. Consequently, if rotational damping is based on the time-derivative of $\mathbf{E}$, which tends to ω, it is based on an entity with a clear geometric meaning, realizing a strict analogy with the translational position and velocity vectors [9].

In Equation (7), $\mathbf{K}_{KD}$ and $\mathbf{D}_{KD}$ are the stiffness and damping matrices, and the vectors $\mathbf{x}_d$ and $\mathbf{E}_d$ describe the reference trajectory in the external coordinates. Let us note that $\mathbf{K}_{KD}$ and $\mathbf{D}_{KD}$ are 6×6 and block-diagonal, composed of 3×3 blocks representing the translational stiffness ($\mathbf{K}_{KDt}$), the rotational stiffness ($\mathbf{K}_{KDr}$), the translational damping ($\mathbf{D}_{KDt}$) and the rotational damping ($\mathbf{D}_{KDr}$). The remaining elements of $\mathbf{K}_{KD}$ and $\mathbf{D}_{KD}$ are null, since the translational and rotational behaviors are imposed to be decoupled.

In general, the stiffness and damping matrices are block-diagonal but non-diagonal in the fixed reference frame, W. Based on the task requirements, it is possible to choose a convenient principal reference frame in which stiffness and damping are decoupled for any direction. Moreover, it can be useful to have two distinct principal reference frames, one for the translations (PT) and one for the rotations (PR). Therefore, the 3×3 stiffness and damping submatrices are defined as diagonal in PT and PR and then transformed into frame W by means of the corresponding rotation matrices [24]:

$$
\mathbf{K}_{KDt}=\left(\mathbf{R}_W^{PT}\right)^T\mathbf{K}_{KDt}^{PT}\,\mathbf{R}_W^{PT},
\tag{9}
$$

$$K_{KDr} = \left(R_W^{PR} \right)^T K_{KDr}^{PR} \, R_W^{PR}, \tag{10}$$

$$D_{KDt} = \left(R_W^{PT} \right)^T D_{KDt}^{PT} \, R_W^{PT}, \tag{11}$$

$$D_{KDr} = \left(R_W^{PR} \right)^T D_{KDr}^{PR} \, R_W^{PR}. \tag{12}$$

3.2. KDHD Impedance Control of Six-Degree-of-Freedom Non-Redundant Parallel Robots

In the KDHD impedance control, damping is proportional not only to the first-order derivative of the external coordinates' error, according to matrix D_{KDHD}, but also to its half-order derivative (derivative of order 1/2), according to matrix HD_{KDHD}:

$$
\begin{aligned}
\tau =& \left(J_p^T(q) \right)^{-1} \left(K_{KDHD} \begin{bmatrix} x_d - x(q) \\ E_d - E(q) \end{bmatrix} + D_{KDHD} \begin{bmatrix} (x_d - x(q))^{(1)} \\ (E_d - E(q))^{(1)} \end{bmatrix} + \right. \\
&\left. + HD_{KDHD} \begin{bmatrix} (x_d - x(q))^{(1/2)} \\ (E_d - E(q))^{(1/2)} \end{bmatrix} \right) + \tau_g(q) = \\
=& \left(J_p^T(q) \right)^{-1} \left(\begin{bmatrix} K_{KDHDt} & 0 \\ 0 & K_{KDHDr} \end{bmatrix} \begin{bmatrix} x_d - x(q) \\ E_d - E(q) \end{bmatrix} + \begin{bmatrix} D_{KDHDt} & 0 \\ 0 & D_{KDHDr} \end{bmatrix} \begin{bmatrix} (x_d - x(q))^{(1)} \\ (E_d - E(q))^{(1)} \end{bmatrix} + \right. \\
&\left. + \begin{bmatrix} HD_{KDHDt} & 0 \\ 0 & HD_{KDHDr} \end{bmatrix} \begin{bmatrix} (x_d - x(q))^{(1/2)} \\ (E_d - E(q))^{(1/2)} \end{bmatrix} \right) + \tau_g(q)
\end{aligned}
\tag{13}
$$

Similarly to the matrices K_{KDHD} and D_{KDHD}, for which the Equations (9) to (12) are always valid, the two blocks of the matrix HD_{KDHD} can also be defined in the principal reference frames PT and PR, and then transformed to frame W:

$$HD_{KDt} = \left(R_W^{PT} \right)^T HD_{KDt}^{PT} \, R_W^{PT}, \tag{14}$$

$$HD_{KDr} = \left(R_W^{PR} \right)^T HD_{KDr}^{PR} \, R_W^{PR}. \tag{15}$$

3.3. KDHDc Impedance Control of Six-Degree-of-Freedom Non-Redundant Parallel Robots

As discussed in Section 2, the approximated real-time implementation of fractional-order derivatives by means of a digital filter with finite order n introduces an issue for impedance control, since it alters the relationship between position error and end-effector generalized force. To solve this problem, a stiffness-compensated KDHD impedance control, named KDHDc, is proposed, subtracting in the stiffness term the unwanted additional stiffness caused by the half-order damping and evaluated by means of Equation (6):

$$
\begin{aligned}
\tau =& \left(J_p^T(q) \right)^{-1} \left(\left(K_{KDHD} - \frac{W_{1/2}(n)}{T_s^{1/2}} HD_{KDHD} \right) \begin{bmatrix} x_d - x(q) \\ E_d - E(q) \end{bmatrix} + D_{KDHD} \begin{bmatrix} (x_d - x(q))^{(1)} \\ (E_d - E(q))^{(1)} \end{bmatrix} + \right. \\
&\left. + HD_{KDHD} \begin{bmatrix} (x_d - x(q))^{(1/2)} \\ (E_d - E(q))^{(1/2)} \end{bmatrix} \right) + \tau_g(q)
\end{aligned}
\tag{16}
$$

This expression extends the KDHDc impedance control proposed in [18] for purely translational robots to full-mobility, non-redundant manipulators.

4. Kinematic Model of the Stewart Platform

The Stewart platform, also known as the Gough–Stewart platform, has been selected as a case study since it is the probably the best-known parallel robot. From its introduction in 1956 by Gough as a tire testing machine and in 1965 by Stewart as an aircraft simulation machine, it has attracted considerable research interest in manufacturing and robotic applications. Several commercial machine tools based on this architecture have been proposed. Other application fields are animatronics, crane technology, positioning of satellite dishes, telescopes and surgery devices. Moreover, the Stewart platform is six-degree-of-freedom and non-redundant, which is necessary to apply KDHDc impedance control using Equations (7), (13) and (16). The

Stewart platform and its kinematics [25] have been exhaustively discussed and investigated in the scientific literature. In this Section, its geometry and the formulation of the Jacobian matrix, which is used in impedance algorithms, are briefly recalled.

The geometric scheme of a generic Stewart platform is represented in Figure 1. The base is fixed with respect to the fixed reference frame W. The reference frame P is fixed to the moving platform. The external coordinates are composed of the vector **x**, from the origin of W to the origin of P, and of the angle-axis vector **E**, which represents the orientation of P with respect to W. The internal coordinates of the robot are the lengths of the six variable-length legs, collected in the vector **q**. The legs are connected to the base and to the platform by universal joints, and the two parts of each extensible leg are connected by a cylindrical joint with idle rotation and actuated translation. The positions of the centers of the platform and base joints are respectively represented in the frame W by the vectors $\mathbf{P}_i$ and $\mathbf{B}_i$.

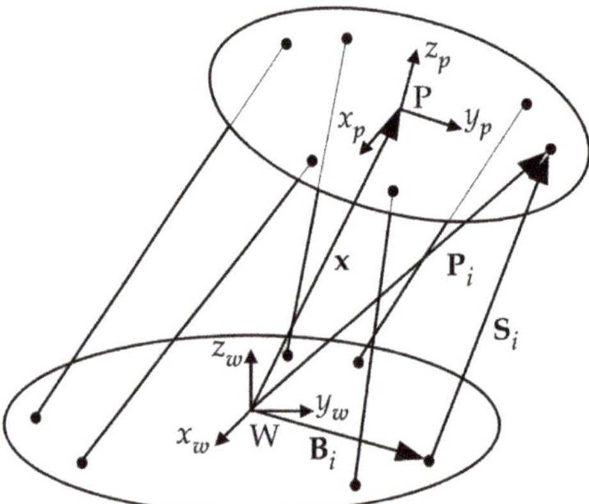

Figure 1. Geometric scheme of the Stewart platform; W(x_w, y_w,z_w): fixed reference frame; P(x_p, y_p,z_p): moving platform reference frame.

Let $\mathbf{S}_i$ be the *i*th leg vector in the base coordinate system, that is the difference between the position vectors $\mathbf{P}_i$ and $\mathbf{B}_i$; the vector $\mathbf{U}_i$ of the six Plücker coordinates of the *i*th leg is composed of the vector $\mathbf{S}_i$, normalized and expressed in the platform frame P, and by its moment about the platform reference frame origin, always expressed in frame P:

$$\mathbf{U}_i = \begin{bmatrix} \mathbf{R}_W^P \dfrac{\mathbf{S}_i}{|\mathbf{S}_i|} \\ \mathbf{R}_W^P \left[(\mathbf{B}_i - \mathbf{x}) \times \dfrac{\mathbf{S}_i}{|\mathbf{S}_i|} \right] \end{bmatrix}. \tag{17}$$

The static behavior of the Stewart Platform is described by the 6×6 matrix $\mathbf{U}(\mathbf{q})$, a function of the internal coordinates, which collects the six vectors of the Plücker coordinates [26]:

$$\boldsymbol{\tau} = (\mathbf{U}(\mathbf{q}))^{-1}\mathbf{F} = \begin{bmatrix} \mathbf{U}_1(\mathbf{q}) & \mathbf{U}_2(\mathbf{q}) & \mathbf{U}_3(\mathbf{q}) & \mathbf{U}_4(\mathbf{q}) & \mathbf{U}_5(\mathbf{q}) & \mathbf{U}_6(\mathbf{q}) \end{bmatrix}^{-1}\mathbf{F}, \tag{18}$$

where $\boldsymbol{\tau}$ is the vector of the forces exerted by the six legs and $\mathbf{F}$ is the six-dimensional vector of the generalized forces exerted by the robot on the environment, expressed on the platform reference frame. Consequently, it is evident that the transpose of $\mathbf{U}$ corresponds to the robot Jacobian matrix $\mathbf{J}_p$ and can be used in impedance control algorithms (7), (13) and (16). In order to work properly, the Stewart platform should avoid configurations in which the Jacobian matrix is singular or ill-conditioned; to obtain this, first of all the hexagons of the platform and base joints are never regular; moreover, it is convenient to place the joints of platform and base as close as possible two by two, giving rise to two hexagons similar to equilateral triangles, which are rotated $60°$ in the rest position [27], as in the model shown in Figure 2.

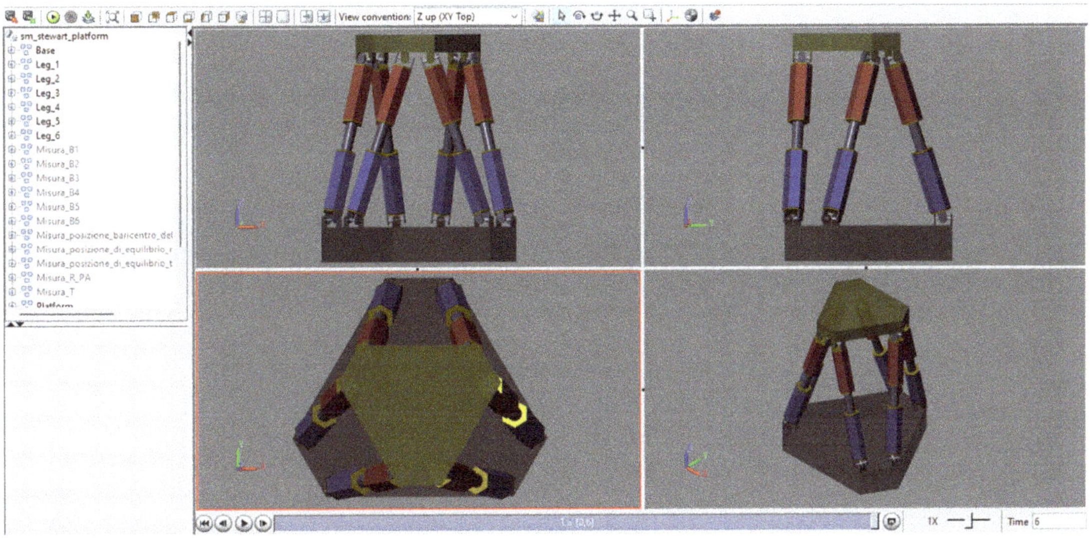

Figure 2. Multibody model of the Stewart platform.

5. Multibody Model of the Stewart Platform

The proposed KDHDc impedance control has been tested and compared to the classical KD control implementing a model of the Stewart platform in the multibody simulation environment Simscape Multibody™ by MathWorks (Figure 2). The main geometrical and inertial parameters considered in the simulations are collected in Table 1.

Table 1. Geometrical and inertial parameters of the 3-PUU parallel robot.

Symbol	Parameter	Value	Unit
$\vert \mathbf{B}_i \vert, i = 1\dots 6$	base platform radius	0.28	m
$\vert \mathbf{P}_i - \mathbf{x} \vert, i = 1\dots 6$	moving platform radius	0.3864	m
	angular distances of the platform joint centers	$20°/100°/20°/100°/20°/100°$	degrees
	angular distances of the base joint centers	$100°/20°/100°/20°/100°/20°$	degrees
m_{mp}	moving platform mass, with payload	63	kg
m_l	mass of the upper part of one leg	1	kg
m_s	mass of the lower part of one leg	1.5	kg
$[J_1, J_2, J_3]$	principal moments of inertia of the moving platform (frame P)	$[1.636, 1.636, 3.221]$	kg·m^2
$\mathbf{x}_{ref} = [0, 0, z_{ref}]$	reference workspace central position	$[0, 0, 0.69]$	m

6. Simulation Results

The KDHDc/KD comparison has been carried out in a wide variety of case studies; here, for reasons of space, the results of three case studies are reported:

(A) Approach/depart motion without contact with the environment: the end-effector follows a reference trajectory, defined by time-varying reference values of $\mathbf{x}_d$ and $\mathbf{E}_d$ (external coordinates), without external forces acting on the end-effector; for each external coordinate a trapezoidal speed law is imposed, and the end-effector compliance is isotropic: the stiffness and damping matrices are diagonal, with three equal elements for the translational and rotational submatrices.

(B)   Interaction with the environment: the end-effector reference external coordinates are constant, and an external generalized force is applied to the end-effector. The stiffness and damping matrices are diagonal in the world frame W, which coincides with the principal stiffness/damping reference frames PT and PR, but the end-effector behavior is not isotropic, since the three diagonal elements of each submatrix are not equal.

(C)   Interaction with the environment: similarly to case B, the end-effector reference external coordinates are constant, and an external generalized force is applied to the end-effector, but differently from case B the principal stiffness/damping frames PT and PR do not coincide with W; therefore, the stiffness and damping matrices are block-diagonal but not diagonal.

6.1. Case Study A: Approach/Depart Motion without Contact with the Environment

In this case study, the stiffness and damping matrices are diagonal and isotropic both for the KD and KDHD impedance controllers.

For the KD impedance algorithm:

- $\mathbf{K}_{KDt}$ is diagonal with diagonal values $k_{KDt,i}$, $i = 1 \ldots 3$;
- $\mathbf{K}_{KDr}$ is diagonal with diagonal values $k_{KDr,i}$, $i = 1 \ldots 3$;
- $\mathbf{D}_{KDt}$ is diagonal with diagonal values $d_{KDt,i}$, $i = 1 \ldots 3$;
- $\mathbf{D}_{KDr}$ is diagonal with diagonal values $d_{KDr,i}$, $i = 1 \ldots 3$.

After imposing the stiffness diagonal values, the diagonal values of the damping matrix are selected starting from the nondimensional damping coefficient, ζ_{KD}, according to the following expression [28]:

$$d_{KDt,i} = 2\zeta_{KDt}\sqrt{k_{KDt,i}m_{mp}}, i = 1\ldots3, \tag{19}$$

$$d_{KDr,i} = 2\zeta_{KDr}\sqrt{k_{KDr,i}J_i}, i = 1\ldots3, \tag{20}$$

in which the mass of the moving platform, m_{mp}, is considered for the translational damping, and the three moments of inertia of the moving platform, J_1, J_2 and J_3, are considered for the rotational damping. In general, the nondimensional damping coefficient can have different values for the translational impedance (ζ_{KDt}) and for the rotational impedance (ζ_{KDr}).

For the KDHDc impedance algorithm:

- $\mathbf{K}_{KDHDt}$ is diagonal with diagonal values $k_{KDHDt,i}$, $i = 1 \ldots 3$;
- $\mathbf{K}_{KDHDr}$ is diagonal with diagonal values $k_{KDHDr,i}$, $i = 1 \ldots 3$;
- $\mathbf{D}_{KDHDt}$ is diagonal with diagonal values $d_{KDHDt,i}$, $i = 1 \ldots 3$;
- $\mathbf{D}_{KDHDr}$ is diagonal with diagonal values $d_{KDHDr,i}$, $i = 1 \ldots 3$;
- $\mathbf{HD}_{KDHDt}$ is diagonal with diagonal values $hd_{KDHDt,i}$, $i = 1 \ldots 3$;
- $\mathbf{HD}_{KDHDr}$ is diagonal with diagonal values $hd_{KDHDr,i}$, $i = 1 \ldots 3$.

The diagonal values of the two damping matrices can be selected starting from the nondimensional coefficients ζ_{KDHD} and ψ_{KDHD}, according to the following expressions:

$$d_{KDHDt,i} = 2\zeta_{KDHDt}\sqrt{k_{KDHDt,i}m_{mp}}, \quad i = 1\ldots3, \tag{21}$$

$$d_{KDHDr,i} = 2\zeta_{KDHDr}\sqrt{k_{KDHDr,i}J_i}, \quad i = 1\ldots3, \tag{22}$$

$$hd_{KDHDt,i} = \psi_{KDHDt,i}^{3/4}m_{mp}^{1/4}, \quad i = 1\ldots3, \tag{23}$$

$$hd_{KDHDr,i} = \psi_{KDHDr,i}^{3/4}J_i^{1/4}, \quad i = 1\ldots3. \tag{24}$$

The coefficients ζ_{KDHD} and ψ_{KDHD} represent non-dimensionally the derivative and half-derivative damping terms [28]. In [20], three couples of PD and PDD$^{1/2}$ tunings, recalled in Table 2, have been compared in the control of a linear axis.

Table 2. Nondimensional parameters of the compared PD and PDD$^{1/2}$ tunings.

KD/KDHD Comparison	PD Control/ KD Impedance Control	PDD$^{1/2}$ Control/ KDHD Impedance Control	
	ζ_{KD}	ζ_{KDHD}	ψ_{KDHD}
I	0.8	0.46	0.7990
II	1	0.45	1.4266
III	1.2	0.48	2.1510

These couples of PD and PDD$^{1/2}$ tunings have been selected as the starting point of the study, since they have been obtained with reference to a nondimensional second-order purely inertial linear system, with transfer function $G(s) = 1/s^2$, according to the following method:

- A PD closed-loop control with a given ζ (reference PD) is applied to the position control of $G(s)$, applying a step input, and the settling energy of the step response is calculated.
- There are infinite combinations of ζ and ψ for a PDD$^{1/2}$ controller with the same proportional gain and the same settling energy of the reference PD; among these, the ζ–ψ combination which minimizes the settling time is selected.

Table 2 collects the PDD$^{1/2}$ ζ–ψ combinations associated to three different reference PD controllers, with $\zeta = 0.8$, 1, and 1.2. Simulations and experimental tests on a single mechatronic axis, as reported in [20], show that these PDD$^{1/2}$ tunings allow to improve the readiness with respect to PD, with a similar control effort. Moreover, even if this tuning is based on the step input, the benefits in terms of control readiness of the PDD$^{1/2}$ controller have been also demonstrated with different reference inputs [28]. In the following, for reasons of space, only the results for the comparison II (KD with $\zeta_{KD} = 1.0$ vs. KDHDc with $\zeta_{KDHD} = 0.45$ and $\psi_{KDHD} = 1.4266$) will be reported, the other two comparisons leading to qualitatively similar results.

The stiffness matrices for the KD and KDHDc impedance controls have been obtained imposing the following stiffness values:

- $k_{KDt,i} = k_{KDHDt,i} = 1 \times 10^3$ N/m, $i = 1 \ldots 3$,
- $k_{KDr,i} = k_{KDHDr,i} = 1 \times 10^2$ Nm/rad, $i = 1 \ldots 3$,

While the damping matrices have been calculated using Equations (19) to (24).

In the simulations, the half-derivative is calculated using the digital filter (4) with sampling time $T_s = 0.005$ s and filter order $n = 10$; these values are suitable for a real-time digital implementation on a commercial controller.

The two control laws have been compared in case of a trapezoidal reference trajectory, both translational and rotational, characterized by four phases. At the beginning of this trajectory, $\mathbf{x} = \mathbf{x}_d = \mathbf{x}_{ref}$ (Table 1, the reference position in the workspace center) and $\mathbf{E} = \mathbf{E}_d = 0$ (the platform is aligned to the base); then:

1. *Phase 1:* $\mathbf{x}_d$ varies with constant velocity from $\mathbf{x}_{ref}$ to $\mathbf{x}_{ref} + [0.05, 0.05, 0.05]^T$ [m], while $\mathbf{E}_d$ varies with constant velocity from 0 to $[0.2, -0.2, 0.2]^T$ [rad]. The duration of this phase is $t_{ramp} = 1$ s.
2. *Phase 2:* $\mathbf{x}_d$ remains constant in $\mathbf{x}_{ref} + [0.05, 0.05, 0.05]^T$ [m] and $\mathbf{E}_d$ remains constant in $[0.2, -0.2, 0.2]^T$ [rad] for $t_{stop} = 2$ s.
3. *Phase 3:* $\mathbf{x}_d$ and $\mathbf{E}_d$ return to the initial values ($\mathbf{x}_{ref}$ and 0) with constant velocity in t_{ramp}.
4. *Phase 4:* $\mathbf{x}_d$ and $\mathbf{E}_d$ remain constant in $\mathbf{x}_{ref}$ and 0 for t_{stop}.

The total simulation time is $T_{sim} = 2(t_{ramp} + t_{stop})$. Figures 3 and 4 show the simulation results of the KD–KDHDc comparison in terms of external coordinates and actuation forces.

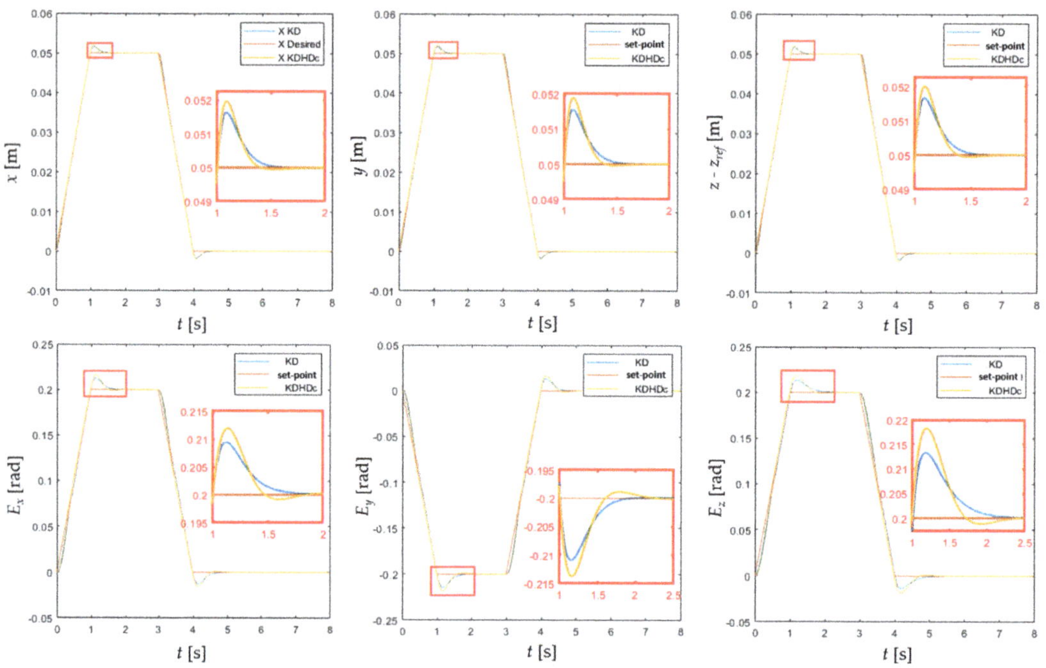

Figure 3. Case study A, KD–KDHDc comparison, external coordinates.

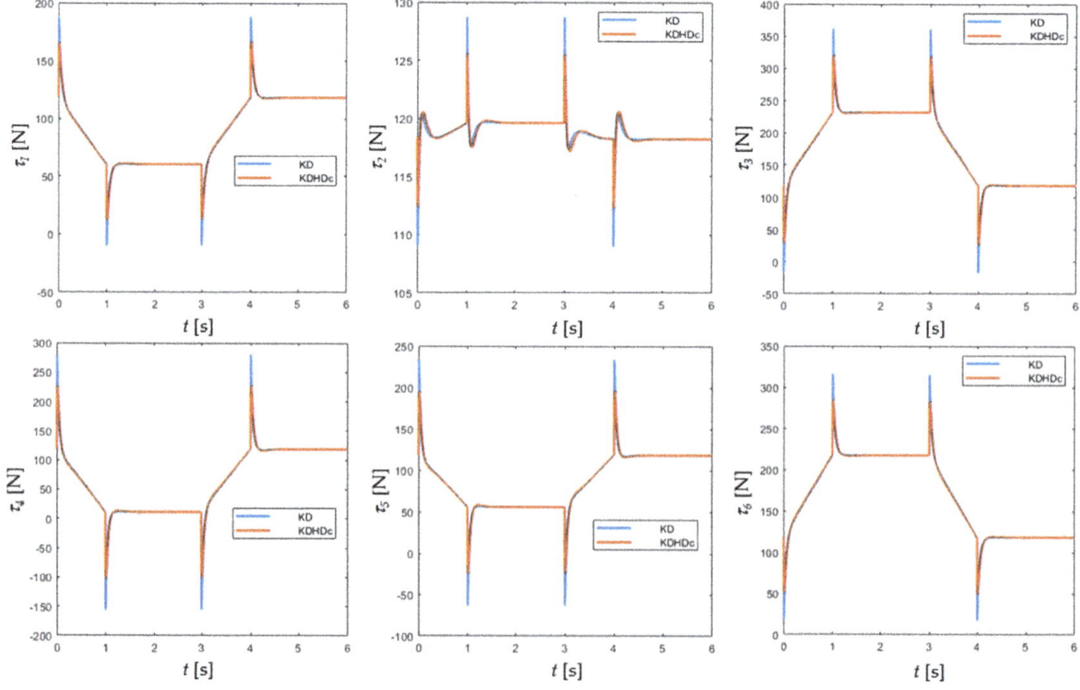

Figure 4. Case study A, KD–KDHDc comparison, actuation forces.

Adopting the KDHDc impedance control, the peaks of actuation forces are lower (average reduction over the six actuators: -13%, Figure 4), and the control effort, calculated according to the following equation,

$$E_{tot} = \sum_{i=1}^{6} \int_0^{T_{sim}} \tau_i^2 dt,$$ (25)

is slightly lower (-0.03%). On the other hand, the settling time to within 0.2% of the steady-state displacement for the KDHDc is slightly lower (-3.7% averaged over the six external coordinates, Figure 3), while the overshoot is higher (maximum overshoot for the translational coordinates: $+4.0\%$ for the KDHDc and $+3.4\%$ for the KD; maximum overshoot for the rotational coordinates: $+9.3\%$ for the KDHDc and $+7.4\%$ for the KD). Without discussing in detail the possible advantages of the KDHDc and its possible tuning criteria, it is interesting to observe that this results are qualitatively similar to the ones of the PDD$^{1/2}$ control scheme compared to the classical PD with same ζ and ψ (better readiness and slightly higher overshoot with lower maximum values of the control inputs [20]). This means that the tuning criteria developed for the PDD$^{1/2}$ control can provide useful indications for tuning the KDHDc impedance control.

6.2. Case Study B: Interaction with the Environment, Diagonal Stiffness and Damping Matrices

In this case study, the stiffness and damping matrices are diagonal in the world frame W, which coincides with PT and PR, but the imposed end-effector behavior is not isotropic, since the three diagonal elements of each 3×3 submatrix are not equal as in case study A:

- $k_{KDt,1} = k_{KDHDt,1} = 2 \times 10^3$ N/m; $k_{KDt,2} = k_{KDHDt,2} = k_{KDt,3} = k_{KDHDt,3} = 1 \times 10^4$ N/m,
- $k_{KDr,1} = k_{KDHDr,1} = 1 \times 10^2$ Nm/rad; $k_{KDr,2} = k_{KDHDr,2} = k_{KDr,3} = k_{KDHDr,3} = 1 \times 10^3$ Nm/rad,

Resulting in higher compliance along the x-axis both for the translations and for the rotations. As in case A, the diagonal values of the damping matrices are obtained using Equations (19) to (24) separately for each coordinate, and imposing KD with $\zeta = 1.0$ for the KD, and $\zeta = 0.45$ and $\psi = 1.4266$ for the KDHDc.

A force $\mathbf{F} = [100, 100, 100]^T$ N and a moment $\mathbf{M} = [20, -20, 20]^T$ Nm are applied to the end-effector at $t = 0$ s. The half-derivative is calculated adopting the same discrete-time implementation as in case A ($T_s = 0.005$ s, $n = 10$).

Figure 5 shows the simulation results of the KD–KDHDc comparison in terms of external coordinates. It is possible to note that the steady-state displacements of the external coordinates using the two algorithms coincide for both the algorithms with the expected values, $\mathbf{x}_{ss}$ and $\mathbf{E}_{ss}$, which can be obtained by inverting the translational and rotational diagonal submatrices:

$$\begin{aligned} \mathbf{x}_{ss} &= (\mathbf{K}_{KDt})^{-1}\mathbf{F} = (\mathbf{K}_{KDHDt})^{-1}\mathbf{F} = \begin{bmatrix} 0.05 & 0.01 & 0.01 \end{bmatrix}^T [\text{mm}] \\ \mathbf{E}_{ss} &= (\mathbf{K}_{KDr})^{-1}\mathbf{M} = (\mathbf{K}_{KDHDr})^{-1}\mathbf{M} = \begin{bmatrix} 0.200 & -0.020 & 0.020 \end{bmatrix}^T [\text{rad}] \end{aligned}.$$ (26)

It occurs thanks to the correctness of the stiffness compensation present in Equation (16) for the KDHDc and absent in Equation (13) for the KDHD; let us note that these correction terms are in percentage relevant with respect to the corresponding diagonal terms of the stiffness matrix: 158%, 105% 105%, 134%, 75% and 89%, respectively, for the six external coordinates. Adopting Equation (13), without this correction, the stiffness of the KDHD is much higher than the desired stiffness expressed by the matrix $\mathbf{K}_{KDHD}$, with diagonal values ranging from 175% to 258% of the corresponding desired values. Let us also note that for the first four external coordinates the negative correction is higher than the desired stiffness, giving rise to negative diagonal values of the proportional terms, without affecting the compensation effectiveness. In other words, the corrected stiffness term of Equation (16) is the sum of two terms, one positive, proportional to the matrix $\mathbf{K}_{KDHD}$ of the desired stiffness, and one negative, proportional to the matrix $W_{1/2}(n)/T_s^{1/2}\mathbf{HD}_{KDHD}$. The second negative term, which is the correction of the unwanted additional stiffness introduced by the half-derivative digital filter, depends on the filter order n and on the sampling time T_s, and can be even higher than the desired stiffness term. However, independently of which term is larger, the stiffness compensation acts correctly, and the KDHDc impedance control, Equation (16), exhibits in steady-state the desired stiffness for each external coordinate, as shown by the graphs of Figure 5.

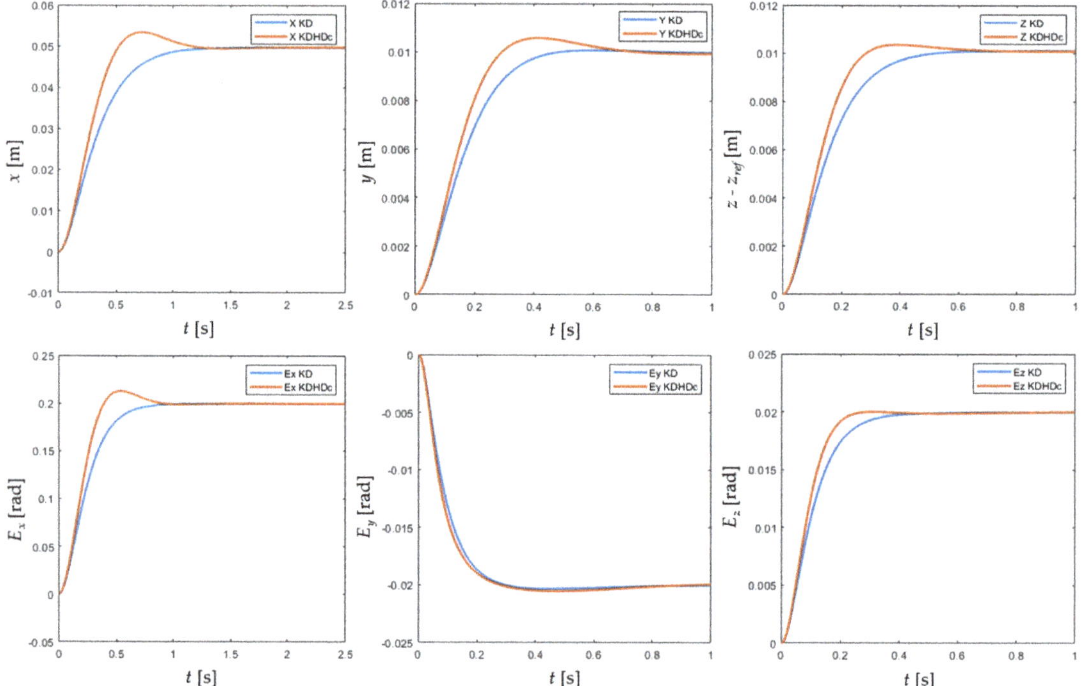

Figure 5. Case study B, KD–KDHDc comparison, external coordinates.

6.3. Case Study C: Interaction with the Environment, Non-Diagonal Stiffness and Damping Matrices

In this case study, the stiffness compensation of the impedance control law KDHDc is tested with the principal translational and rotational reference frames, PT and PR, rotated with respect to the fixed reference frame W: PT is rotated with respect to a W of $\theta_t = \pi/6$ rad about an axis represented by the unit vector $\mathbf{u}_t = [+1, -2, +4]^T/21^{1/2}$; PR is rotated with respect to a W of $\theta_t = \pi/3$ rad about an axis represented by the unit vector $\mathbf{u}_r = [+1, +2, +4]^T/21^{1/2}$. The rotation matrix between W and PT can be obtained from the axis-angle representation by the following formula [6]:

$$\mathbf{R}_W^{PT} = \begin{bmatrix} u_{tx}^2 + (1 - u_{tx}^2)c\theta_t & u_{tx}u_{ty}(1 - c\theta_t) - u_{tz}s\theta_t & u_{tx}u_{tz}(1 - c\theta_t) + u_{ty}s\theta_t \\ u_{tx}u_{ty}(1 - c\theta_t) + u_{tz}s\theta_t & u_{ty}^2 + \left(1 - u_{ty}^2\right)c\theta_t & u_{ty}u_{tz}(1 - \cos\theta_t) - u_{tx}s\theta_t \\ u_{tx}u_{tz}(1 - c\theta_t) - u_{ty}s\theta_t & u_{ty}u_{tz}(1 - c\theta_t) + u_{tx}s\theta_t & u_{tz}^2 + (1 - u_{tz}^2)c\theta_t \end{bmatrix}. \quad (27)$$

where $c\theta_t$ and $s\theta_t$ stand for $\cos\theta_t$ and $\sin\theta_t$. The rotation matrix between W and PR can be obtained from the same Equation (27), replacing θ_t and $\mathbf{u}_t$ with θ_r and $\mathbf{u}_r$.

The principal stiffness matrix and damping matrices have the same diagonal values of case study B, but the stiffness matrix and damping matrices expressed in W change due to the orientations of PT and PR; on the contrary, the external force and moment $\mathbf{F}$ and $\mathbf{M}$ are kept equal in frame W to case B. The half-derivative is calculated adopting the same discrete-time implementation as in cases A and B ($T_s = 0.005$ s, $n = 10$). Figure 6 shows the simulation results in terms of external coordinates for the KD and KDHDc control laws.

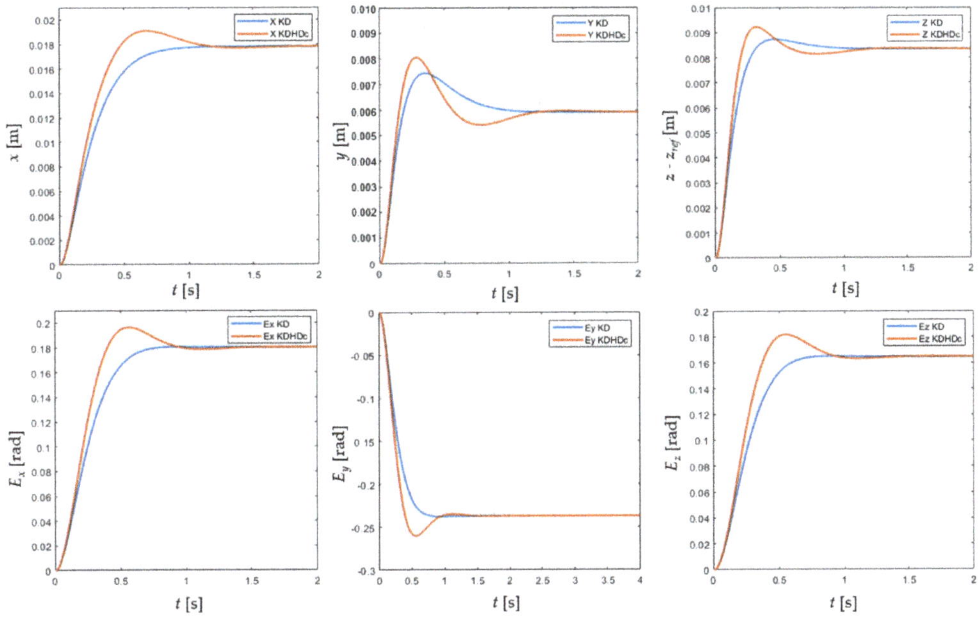

Figure 6. Case study C, KD–KDHDc comparison, external coordinates.

It is possible to note that also in this case the steady-state displacements of the external coordinates using the two algorithms coincide for both the algorithms with the expected values, $\mathbf{x}_{ss}$ and $\mathbf{E}_{ss}$, which can be obtained by inverting the translational and rotational diagonal submatrices, once transformed in the fixed reference frame by Equations (9) and (10), confirming the correctness of the KDHDc stiffness compensation:

$$\mathbf{x}_{ss} = \left(\mathbf{K}_{KDt}\right)^{-1}\mathbf{F} = \left(\left(\mathbf{R}_W^{PT}\right)^T\mathbf{K}_{KDt}^{PT}\,\mathbf{R}_W^{PT}\right)^{-1}\mathbf{F} = \begin{bmatrix} 0.0180 & 0.0059 & 0.0082 \end{bmatrix}^T [\text{mm}]$$
$$\mathbf{E}_{ss} = \left(\mathbf{K}_{KDr}\right)^{-1}\mathbf{M} = \left(\left(\mathbf{R}_W^{PR}\right)^T\mathbf{K}_{KDr}^{PR}\,\mathbf{R}_W^{PR}\right)^{-1}\mathbf{M} = \begin{bmatrix} 0.181 & -0.237 & 0.165 \end{bmatrix}^T [\text{rad}] \tag{28}$$

7. Conclusions and Future Research Directions

In this paper, the KDHDc fractional-order extension of the classical KD impedance control algorithm, already proposed in [18] for a purely translational parallel robot, has been applied to a six-degree-of-freedom parallel robot, the Stewart platform, requiring the definition of the rotational impedance. To this end, for the representation of the moving platform orientation, which influences the rotational impedance, the three components of the angle-axis vector $\mathbf{E}$ have been selected. Using these external coordinates for evaluating the orientation error and its derivatives has some theoretical advantages over other representations (for instance, Euler angles) due to the fact that $\mathbf{E}$ is a frame-independent natural invariant with a clear Euclidean-geometric meaning [6], and its first-order derivative has a close relationship with the angular velocity vector, which makes it particularly suitable for defining the damping term [9].

The conception of the KDHDc impedance control originates from the work on the $\text{PDD}^{1/2}$ control scheme, first developed for single-input single-output systems, and can be considered as its extension to impedance-controlled robots, which are a class of non-linear multi-input multi-output systems.

In this paper and in [18] non-redundant parallel robots are considered, but the proposed KDHDc impedance control can be applied also to non-redundant serial robots, only bearing in mind that the Jacobian matrix of serial robots usually transforms the internal coordinates into external coordinates and not vice versa, as for parallel robots. Consequently, the KDHDc impedance control equations proposed in [18] for translational parallel robots and in this paper for full-mobility parallel robots can be used, respectively, for translational and full-mobility serial robots by simply replacing $(\mathbf{J}^T(\mathbf{q}))^{-1}$ with $\mathbf{J}^T(\mathbf{q})$.

Even if the work is only a starting point, it is possible to summarize the following conclusions:

- The application of the half-derivative damping term allows for tuning of the impedance control with additional degrees of freedom, with potential benefits. For instance, the tuning criterion used in case study A, derived from the one proposed in [20] for a single-degree-of-freedom, an almost linear mechatronic axis, leads in free motion to lower actuation forces, almost equal control effort, lower settling time with slightly higher overshoot. These results are qualitatively similar to what happens for the single mechatronic axis, conforming the validity of the approach of deriving the KDHDc tuning form the $PDD^{1/2}$ tuning. However, the system can be tuned differently, focusing on other performance indices, exploiting the additional regulation opportunities provided by the half-derivative damping.
- The real-time digital implementation of the half-derivative term introduces a remarkable alteration to the stiffness of the impedance control in steady state, as discussed in Section 2, invalidating the capability of the KDHD impedance control to regulate the contact force between the end-effector and the environment through the measurement of the position error. Nevertheless, the proposed KDHDc compensation, Equation (16), has been proven to be effective also for the rotational behavior (case studies B and C).

Some of the possible future research directions are the following:

- a systematic investigation on possible tuning approaches for the KDHDc impedance control will be carried out, in order to maximize different performance indices; to this aim, both extension of $PDD^{1/2}$ tuning methods and numerical optimization techniques will be considered;
- the KDHDc impedance control will be applied to different serial and parallel architectures; by now, two kinds of mobilities have been taken into account: three translational degrees of freedom and full mobility, without redundancy (equal numbers of internal and external coordinates); in the following, also the cases of limited-degree-of-freedom robots, mixing translational and rotational motions, will be analyzed, both for serial [29,30] and parallel/hybrid manipulators [31,32];
- impedance control of redundant manipulators, which has interesting theoretical aspects [33], will also be considered;
- the problem of impedance control for manipulators equipped with flexure joints [34], requiring a proper stiffness compensation of the joint elastic return force, should be addressed;
- another possible application of the proposed fractional-order approach is impedance control of cable-driven parallel robots [35];
- as regards the validation methodology, only simulation results are available by now; in the future, besides performing experimental tests of the considered case studies (approach/depart motions without external forces, interactions with the environment with constant force/moment) the effectiveness of the KDHDc impedance control will be assessed in real working conditions, for example peg-in-hole or milling tasks.

Author Contributions: L.B. conceived the control algorithm and designed the simulation campaign; A.P. developed the multibody model and performed simulations; L.B. and A.P. prepared the manuscript. All authors have read and agreed to the published version of the manuscript.

Funding: This research received no external funding.

Institutional Review Board Statement: Not applicable.

Informed Consent Statement: Not applicable.

Data Availability Statement: Data is contained within the article. The simulation data presented in this study are available in figures and tables.

Conflicts of Interest: The authors declare no conflict of interest.

References

1. Craig, J. *Introduction to Robotics. Mechanics and Control*; Addison-Wesley: Boston, MA, USA, 1989.
2. Raibert, M.H.; Craig, J.J. Hybrid Position/Force Control of Manipulators. *ASME J. Dyn. Syst. Meas. Control* **1981**, *103*, 126–133. [CrossRef]
3. Caccavale, F.; Siciliano, B.; Villani, L. Robot Impedance Control with Nondiagonal Stiffness. *IEEE Trans. Autom. Control* **1999**, *44*, 1943–1946. [CrossRef]
4. Valency, T.; Zacksenhouse, M. Accuracy/Robustness Dilemma in Impedance Control. *J. Dyn. Syst. Meas. Control* **2003**, *125*, 310–319. [CrossRef]
5. Bruzzone, L.; Molfino, R.M.; Zoppi, M. An impedance-controlled parallel robot for high-speed assembly of white goods. *Ind. Robot Int. J.* **2005**, *32*, 226–233. [CrossRef]
6. Angeles, J. *Rational Kinematics*; Springer: New York, NY, USA, 1988.
7. Bonev, I.A.; Ryu, J. A new approach to orientation workspace analysis of 6-DOF parallel manipulators. *Mech. Mach. Theory* **2001**, *36*, 15–28. [CrossRef]
8. Caccavale, F.; Siciliano, B.; Villani, L. The role of Euler parameters in robot control. *Asian J. Control* **1999**, *1*, 25–34. [CrossRef]
9. Bruzzone, L.; Molfino, R.M. A geometric definition of rotational stiffness and damping applied to impedance control of parallel robots. *Int. J. Robot. Autom.* **2006**, *21*, 197–205. [CrossRef]
10. Ikeura, R.; Inooka, H. Variable impedance control of a robot for cooperation with a human. In Proceedings of the 1995 IEEE International Conference on Robotics and Automation, Nagoya, Japan, 21–27 May 1995; Volume 3, pp. 3097–3102.
11. Tsumugiwa, T.; Yokogawa, R.; Hara, K. Variable impedance control with virtual stiffness for human-robot cooperative peg-in-hole task. In Proceedings of the IEEE/RSJ International Conference on Intelligent Robots and Systems, Lausanne, Switzerland, 30 September–4 October 2002; Volume 2, pp. 1075–1081.
12. Shimizu, M. Nonlinear impedance control to maintain robot position within specified ranges. In Proceedings of the 2012 SICE Annual Conference (SICE), Akita, Japan, 20–23 August 2012; pp. 1287–1292.
13. Miller, K.S.; Ross, B. *An Introduction to the Fractional Calculus and Fractional Differential Equations*; John Wiley & Sons: New York, NY, USA, 1993.
14. Kizir, S.; Elşavi, A. Position-Based Fractional-Order Impedance Control of a 2 DOF Serial Manipulator. *Robotica* **2021**, *39*, 1560–1574. [CrossRef]
15. Liu, X.; Wang, S.; Luo, Y. Fractional-order impedance control design for robot manipulator. In Proceedings of the ASME 2021 International Design Engineering Technical Conferences and Computers and Information in Engineering Conference, Virtual, Online, 17–19 August 2021; Volume 7, p. V007T07A028.
16. Fotuhi, M.J.; Bingul, Z. Novel fractional hybrid impedance control of series elastic muscle-tendon actuator. *Ind. Robot Int. J. Robot. Res. Appl.* **2021**, *48*, 532–543. [CrossRef]
17. Podlubny, I. Fractional-order systems and $PI^{\lambda}D^{\mu}$ controllers. *IEEE Trans. Autom. Control* **1999**, *44*, 208–213. [CrossRef]
18. Bruzzone, L.; Fanghella, P.; Basso, D. Application of the Half-Order Derivative to Impedance Control of the 3-PUU Parallel Robot. *Actuators* **2022**, *11*, 45. [CrossRef]
19. Bruzzone, L.; Fanghella, P. Comparison of PDD1/2 and PDμ position controls of a second order linear system. In Proceedings of the IASTED International Conference on Modelling, Identification and Control, Innsbruck, Austria, 17–19 February 2014; pp. 182–188.
20. Bruzzone, L.; Fanghella, P. Fractional-order control of a micrometric linear axis. *J. Control Sci. Eng.* **2013**, *2013*, 947428. [CrossRef]
21. Bruzzone, L.; Fanghella, P.; Baggetta, M. Experimental assessment of fractional-order $PDD^{1/2}$ control of a brushless DC motor with inertial load. *Actuators* **2020**, *9*, 13. [CrossRef]
22. Machado, J.T. Fractional-order derivative approximations in discrete-time control systems. *J. Syst. Anal. Model. Simul.* **1999**, *34*, 419–434.
23. Das, S. *Functional Fractional Calculus*; Springer: Berlin/Heidelberg, Germany, 2011.
24. Bruzzone, L.; Callegari, M. Application of the rotation matrix natural invariants to impedance control of rotational parallel robots. *Adv. Mech. Eng.* **2010**, *2010*, 284976. [CrossRef]
25. Stewart, D. A Platform with Six Degrees of Freedom. *Proc. Inst. Mech. Eng.* **1965**, *180*, 371–386. [CrossRef]
26. Fichter, E.F. A Stewart Platform-based Manipulator: General Theory and Practical Construction. *Int. J. Robot. Res.* **1986**, *5*, 157–182. [CrossRef]
27. Ma, O.; Angeles, J. Optimum Architecture Design of Platform Manipulators. In Proceedings of the 5th International Conference on Advanced Robotics, Pisa, Italy, 19–22 June 1991; Volume 2, pp. 1130–1135.

28. Bruzzone, L.; Bozzini, G. PDD1/2 control of purely inertial systems: Nondimensional analysis of the ramp response. In Proceedings of the IASTED International Conference on Modelling, Identification and Control, Innsbruck, Austria, 14–16 February 2011; pp. 308–315.
29. Makino, H.; Furuya, N. Selective compliance assembly robot arm. In Proceedings of the First International Conference on Assembly Automation (ICAA), Brighton, UK, 25–27 March 1980; pp. 77–86.
30. Bruzzone, L.; Bozzini, G. A statically balanced SCARA-like industrial manipulator with high energetic efficiency. *Meccanica* **2011**, *46*, 771–784. [CrossRef]
31. Fang, Y.; Tsai, L.-W. Structure synthesis of a class of 4-DoF and 5-DoF parallel manipulators with identical limb structures. *Int. J. Robot. Res.* **2002**, *21*, 799–810. [CrossRef]
32. Dong, C.; Liu, H.; Xiao, J.; Huang, T. Dynamic Modeling and Design of a 5-DOF Hybrid Robot for Machining. *Mech. Mach. Theory* **2021**, *165*, 2021165. [CrossRef]
33. Xiong, G.; Zhou, Y.; Yao, J. Null-Space Impedance Control of 7-Degree-of-Freedom Redundant Manipulators Based on the Arm Angles. *Int. J. Adv. Robot. Syst.* **2020**, *17*, 1729881420925297. [CrossRef]
34. Bruzzone, L.; Molfino, R.M. A novel parallel robot for current microassembly applications. *Assem. Autom.* **2006**, *26*, 299–306. [CrossRef]
35. Reichert, C.; Müller, K.; Bruckmann, T. Robust Internal Force-Based Impedance Control for Cable-Driven Parallel Robots. *Mech. Mach. Sci.* **2015**, *32*, 131–143.

Article

Dynamic Modeling, Workspace Analysis and Multi-Objective Structural Optimization of the Large-Span High-Speed Cable-Driven Parallel Camera Robot

Yu Su [1],*, Yuanying Qiu [2], Peng Liu [3], Junwei Tian [1], Qin Wang [1] and Xingang Wang [1]

[1] Department of Intelligent Manufacturing Engineering, School of Mechatronic Engineering, Xi'an Technological University, Xi'an 710021, China; tianjunwei@xatu.edu.cn (J.T.); wangqin@xatu.edu.cn (Q.W.); wangxingang@st.xatu.edu.cn (X.W.)

[2] Key Laboratory of Electronic, Equipment Structure Design, Xidian University, Xi'an 710071, China; yyqiu@mail.xidian.edu.cn

[3] School of Mechanical Engineering, Xi'an University of Science and Technology, Xi'an 710054, China; liupeng@xust.edu.cn

* Correspondence: suyu@xatu.edu.cn

Abstract: Since most of the cable-driven parallel manipulators (CDPMs) are small in dimension or low in speed, the self-weight or inertia of the cable is neglected when dealing with the problems of kinematics, dynamics and workspace. The cable is treated as a massless straight line, and the inertia of the cable is not discussed. However, the camera robot is a large-span high-speed CDPM. Thus, the self-weight and inertia of the cable cannot be negligible. The curved cable due to the self-weight is modeled as a catenary to accurately account for its sagging effect. Moreover, the dynamic model of the camera robot is derived by decomposing the motion of the cable into an in-plane motion and an out-plane motion, based on which an iterative-based tension distribution algorithm and a workspace generation algorithm are presented. An optimization model is presented to simultaneously improve the workspace volume, anti-wind disturbance ability and impulse of tensions on the camera and pan–tilt device system (CPTDS) by selecting the proper optimal variables under the linear and nonlinear constraints. An improved genetic algorithm (GA) is proposed, and the simulation results demonstrate that the improved GA offers a stronger ability in global optimization compared to the standard genetic algorithm (SGA). The ideal-point method is employed to avoid the subjective influence of the designer when performing the multi-objective optimization, and a remarkable improvement of the performance is obtained through the optimization. Furthermore, the distribution characteristics of the optimization objects are studied, and some valuable conclusions are summarized, which will provide some valuable references in designing large-span high-speed CDPMs.

Keywords: cable-driven parallel manipulator (CDPM); high-speed manipulator; large-span structure; structural optimization; genetic algorithm (GA); ideal-point method

Citation: Su, Y.; Qiu, Y.; Liu, P.; Tian, J.; Wang, Q.; Wang, X. Dynamic Modeling, Workspace Analysis and Multi-Objective Structural Optimization of the Large-Span High-Speed Cable-Driven Parallel Camera Robot. *Machines* **2022**, *10*, 565. https://doi.org/10.3390/machines10070565

Academic Editors: Zhufeng Shao, Dan Zhang and Stéphane Caro

Received: 6 June 2022
Accepted: 11 July 2022
Published: 14 July 2022

1. Introduction

The cable-driven parallel camera robot (camera robot for short, see Figure 1) is a type of CDPM (cable-driven parallel manipulator) consisting of four computer-driven servo motors that enable the controlled release of four cables that act in parallel on an end-effector, i.e., the CPTDS (camera and pan–tilt device system). The camera robot can take pictures and videos with an aerial view that conventional cameras would have difficulty in realizing [1], making it the best candidate for aerial panoramic photographing in big venues, such as stadiums, football fields, theaters, etc.

There have existed some commercial camera robots, e.g., Skycam, Spidiercam and Cablecam [2], which can also be applied in agricultural remote sensing [3] and traffic monitoring [4] and film production [5]. A camera robot is a typical of large-span high-speed

CDPM with redundant actuation [6], whose performance is largely determined by its structural parameters [7,8].

The essence of structural optimization is to establish the mapping relationships between optimal variables and optimization objects. Determining an optimal structure of a CDPM meeting a set of optimization objectives is generally challenging, which can be solved by formulating a constrained optimization problem. There has been plenty of prior work in the structural optimization of CDPMs, which employ various optimization objects, such as workspace size [9], condition number of Jacobian matrix [10], tension factor [11,12], stiffness [13–15], avoiding collision of cables [16,17], manipulability [18] and the maximum acceptable horizontal distance [19].

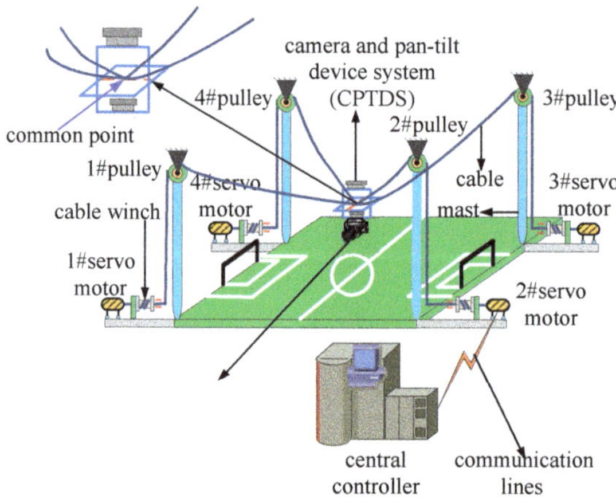

Figure 1. Schematic of a camera robot configuration.

However, most of the above optimization methods belong to traditional optimization methods, such as the simplex method, interval analysis, Dynamic-Q, grouped coordinate descent and enumeration method, which are sensitive to initial values and easily fall into local optimums, leading to a failure to obtain a global optimal solution. Until now, many optimal methods have been developed to overcome this difficulty. Genetic algorithms (GA) are an intelligent optimization method offering a powerful global search ability and can effectively escape from local optimums compared to the traditional optimization method, causing it to be a mainstream structural optimization approach of CDPMs.

Li et al. proposed a GA-based multi-objective optimization method to obtain the best global dexterity index and overall stiffness index of a planar three degrees of freedom (DOF) CDPM [20]. Jamwal et al. developed a GA-based multi-objective optimization method to conduct the optimal design of a cable-driven ankle rehabilitation robot using the minimum global condition number, the maximum workspace utilization index and the minimum cable tension norm as optimal objectives [21]. Arsenault et al. used GA to optimize the geometry of planar CDPMs with four cables with the objective of reaching a desired pre-stress stable wrench-closure workspace [22].

Bahrami et al. optimized the workspace volume, kinematic performance indices and actuating energy of a spatial CDPM by means of GA, [23]. Amine and Hamida investigated the structural optimization of a cable-driven upper limb rehabilitation robot (LAWEX) based on GA, where the objectives were the simultaneous minimization of the robot size and the tensions in the cables [24,25]. Nevertheless, cables in these studies were all treated as massless straight lines without considering the self-weights, which are only suited to small-dimensional CDPMs.

There are two aspects that require attention in the structural optimization of the camera robots: large-span and high-speed motion. For the large-span CDPMs, the sagging effect due to the self-weight of the cable cannot be neglected [26,27], as the profile of the cable is no longer a straight line but a curve. Generally, the deformed curve under self-weight can be described as a catenary or a parabola [28,29].

Jiang et al. optimized the dimensional parameters of a under-restrained six-DOF CDPM (URPM4-3R3T) with workspace size as the objective function [30]. Yao et al. obtained the better dimension parameters for a CDPM with four cables used for the feed supporting system of the five-hundred-meter aperture spherical radio telescope (FAST) to meet the workspace requirements [31]. Tang et al. optimized dimensional parameters for constructing a CDPM with six cables for FAST based on the sensitivity design method and three tension performance evaluating functions [32].

Du determined the structural parameters to simultaneously improve the stiffness and dexterity of large workspace CDPMs [33]. Wei presented an approach on the stability analysis of cable-driven parallel robots considering cable mass based on cable tensions and stiffness matrix condition numbers [34]. However the motions of manipulators are low-speed or quasi-static, and thus the dynamics of the manipulators were not considered.

The highest speed of the camera robot that is known currently is up to 9 m/s [35]. Hence, the maneuverability of the camera robot due to the high-speed motions must be considered. Barrette et al. proposed a notion of a dynamic workspace of CDPMs on the conditions of the dynamic and tension for a planar CDPM [36]. Kawamura et al. developed a high-speed manipulation (FALCON-7) by using a CDPM and studied its dynamics based on the vector closure condition [37].

Gagliardini et al. proposed an improved dynamic feasible workspace considering the inertia of a moving platform, external wrenches applied on the moving platform and the Coriolis forces corresponding to a constant moving platform twist concurrently [38]. Yu et al. determined the dynamic workspace of a camera robot by taking the intersection of workspaces with accelerations at different directions [35] and studied the relationships between the reachable area of the workspace bottom and heights of masts. Kieu et al. developed a modified kinematic equation considering cable nonlinear tension and analyzed the wrench-feasible workspace at various motion accelerations [39]. However, the self-weights of the cables in these studies were all negligible. In our preliminary work, the dynamic modeling and cable tension distribution considering the self-weight and inertia of the cable were simultaneously investigated [1,40], which could provide a theoretical basis for the structural optimization of the camera robot. However, the essence of the tension distribution algorithm in our preliminary work is a multi-dimensional parameter optimization problem, which is time-consuming and has the risk of failure in determining the optimization parameters.

Thus, the algorithm is not fit for CDPMs whose number of cables is much greater than the DOF of the end-effector, which limits the scope of applications. In this paper, the dynamical model of the camera robot is established considering the self-weight and inertia of the cable simultaneously. Furthermore, the tension distribution algorithm is simplified based on the iterative idea to reduce the computing time and computational complexity. Based on the dynamical model, a dynamic workspace generating approach is presented. On the basis of the dynamic workspace, the structural optimization of the camera robot is studied applying a GA, and the ideal-point method is used to deal with multi-objective problems.

The organization of the rest of the paper is as follows: Section 2 establishes the dynamic model of the camera robot by employing the catenary model of the cable and considering the inertia of the cable. Section 3 proposes an iterative-based optimization algorithm based on the dynamic model to determine the tension distribution. Section 4 develops a workspace generation algorithm based on the judging conditions of the dynamic force-feasible workspace. Section 5 presents an optimization model aiming at achieving the maximum workspace volume, anti-wind disturbance and impulse of tensions on CPTDS. Section 6 employs the ideal point method to deal with the multi-objective optimization based on an improved genetic algorithm and studies the distribution characteristics of the optimization objects. Section 7 summarizes our conclusions.

2. Dynamic Model of the Camera Robot

2.1. Description of the Camera Robot

The structure of the camera robot is showed in Figure 1, consisting of a CPTDS driven by four cables in parallel. One end of the cable is connected to the CPTDS, and the other end is wound on the pulley and extends to connect to the cable winch and servo motor. The CPTDS can fly freely in every direction because the cables can be shortened and lengthened through winding driven by four servo motors mounted on the ground, which receive control commands from the central controller.

As shown in Figure 1, the four cables intersect at a common point, and therefore the cables are only responsible for translation. The camera is mounted on a pan–tilt device, which resembles a composite hinge structure. By means of the pan–tilt device, the camera realizes pan (yawing) and tilt (pitching) motion. As a result, the CPTDS can be looked as an ideal mass point P with three translational DOFs, at which the four cables meet. Thus, the position vector $P = \begin{bmatrix} x & y & z \end{bmatrix}^\mathrm{T}$ of the point P is expressed in the global fixed frame $\{oxyz\}$, and it also denotes the end point of the cable. The camera robot can be categorized to completely restrained positioning mechanisms (CRPMs) [41]—namely, the number of the cables m is equal to DOF of the CPTDS n plus one ($m = n + 1$).

2.2. Catenary Equation of the Cable

In order to guarantee the stability of the camera robot, the cables must be inextensible and offer high strength. A previous study indicated that a cable can be seen as a catenary under the sag influence. As a consequence, the profile of the cable i ($\{i = 1,2,3,4\}$) is a catenary within a vertical plane, noted $o_i{}^s x_i{}^s z_i{}^s$, is shown in Figure 2. For analysis, the symbols used in Figure 2 are defined as follows: $B_i = \begin{bmatrix} B_{i,x} & B_{i,y} & B_{i,z} \end{bmatrix}^\mathrm{T}$ denotes the position vector of cable drawing point B_i on the pulley in $\{oxyz\}$. $\{o_i{}^s x_i{}^s z_i{}^s\}$ is the local moving frame with the origin $o_i{}^s$ attached to B_i, and the direction of $z_i{}^s$ is straight down.

t_i is the tension in the cable i at the end point P, h_i is the horizontal component of t_i, and v_i is the vertical component of t_i. L_i is the horizontal length of the span of cable i, and C_i is the vertical length of the span of cable i. ρ is linear density of the inextensible cable, and m_p is the mass of the CPTDS. S_i is the length of the cable i, and f_i is the sag of cable i at the center of the horizontal span. The catenary equation can be written as follows [42]:

$$z_i^s = \frac{h_i}{\rho g}\left[\cosh \alpha_i - \cosh\left(\frac{2\beta_i x_i^s}{l_i} - \alpha_i\right)\right] \tag{1}$$

where g is the gravitational acceleration, and

$$\alpha_i = \sinh^{-1}\left[\frac{\beta_i(C_i/L_i)}{\sinh \beta_i}\right] + \beta_i \tag{2a}$$

$$\beta_i = \frac{\rho L_i}{2h_i} \tag{2b}$$

Equation (1) represents mathematically a family of catenaries. The whole catenary can be determined being given coordinates of arbitrary points on the catenary. Consequently, the length S_i of the cable i can be calculated as follows:

$$S_i = L_i - \frac{h_i \beta_i}{\rho l_i}\left[\frac{L_i}{16\beta_i}\left(e^{4\beta_i - 2\alpha_i} - e^{-4\beta_i + 2\alpha_i}\right) + \frac{1}{2}\right] \tag{3}$$

According to Equation (3), the length of the cable is closely interrelated with the tension. When $x_i^s = L_i/2$, the sag f_i can be obtained as follows:

$$f_i = \frac{8h_i \sinh\beta_i \sinh^{-1}\left(\frac{\rho g C_i/2h_i}{\sinh\beta_i}\right) - \rho C_i g}{2\rho g} \tag{4}$$

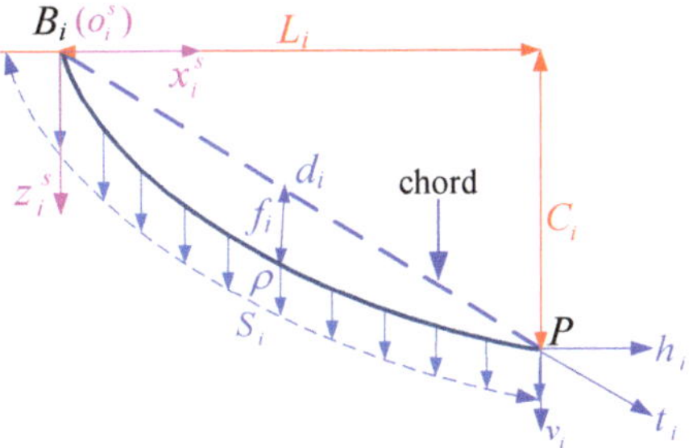

Figure 2. Catenary model of the cable in the vertical plane.

2.3. Inertia of the Rapidly Varying-Length Cable

Due to the high-speed motions of the camera robot, the length of cable changes rapidly, leading to the non-negligible inertia of the cable. As shown in Figure 3, q_i is a node on cable i with the time-varying curve length s_i away from B_i, whose position vector can be denoted by $q_i(s_i, t) = [x_i \quad y_i \quad z_i]^\mathrm{T}$ in the global frame $\{oxyz\}$. The relationship q_i and q_i^s can be calculated as follows:

$$q_i = Qq_i^s + B_i \tag{5}$$

where $q_i^s = [x_i^s \quad z_i^s]^\mathrm{T}$ in the local moving frame $\{o_i{}^s x_i{}^s z_i{}^s\}$, and $Q = \begin{bmatrix} \cos\theta_i & 0 \\ \sin\theta_i & 0 \\ 0 & -1 \end{bmatrix}$ is the rotation matrix from frame $\{o_i{}^s x_i{}^s z_i{}^s\}$ to frame $\{oxyz\}$, and θ_i is the angle between x_i^s and x.

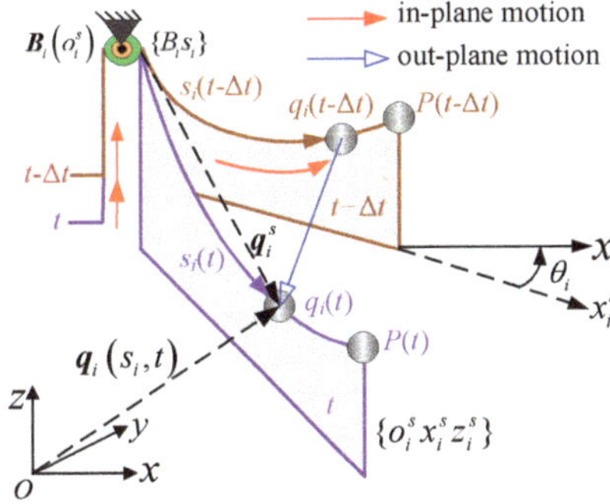

Figure 3. In-plane and out-plane motions of the cable.

Since the motion of q_i is decomposed to an out-plane motion between the vertical planes and an in-plane motion along the cable as shown in Figure 3, the velocity and acceleration of q_i can be calculated by taking the first and second-derivative of $q_i(s_i, t)$ with respect to s_i and time t, respectively, in the following expression [1]:

$$\begin{cases} \dfrac{dq_i}{dt} = \dfrac{\partial q_i}{\partial t} + \dfrac{\partial q_i}{\partial s_i}\dot{s}_i \\ \dfrac{d^2q_i}{dt^2} = \dfrac{\partial^2 q_i}{\partial t^2} + 2\dfrac{\partial^2 q_i}{\partial s_i \partial t}\dot{s}_i + \dfrac{\partial^2 q_i}{\partial s_i^2}\dot{s}_i + \dfrac{\partial q_i}{\partial s_i}\ddot{s}_i \end{cases} \tag{6}$$

where $\frac{\partial^2 q_i}{\partial t^2}$ is yielded by the cable's out-plane motion and $\frac{\partial q_i}{\partial s_i}\ddot{s}_i$ by the the cable's in-plane motion; $\frac{\partial^2 q_i}{\partial s_i \partial t}$ and $\frac{\partial^2 q_i}{\partial s_i^2}\dot{s}_i$ are yielded by the combination of cable's in-plane motion and out-plane motion. $\dot{s}_i = ds_i/dt$, and $\ddot{s}_i = d\dot{s}_i/dt$.

Since the cable drawing point B_i on the pulley i is attached to the mast, $\dot{B}_i = 0$. Noting that $\partial q_i^s = \frac{d_i^s}{S_i}\partial s_i$ and $\frac{\partial q_i}{\partial t} = \dot{P} = [\dot{x} \quad \dot{y} \quad \dot{z}]^T$. Thus, the derivatives of $q_i(s_i, t)$ with respect to the curve length coordinate s_i can be obtained as follows:

$$\begin{cases} \frac{\partial q_i}{\partial s_i} & = \frac{\partial}{\partial s_i}\left(q_i^s\right) = Q\frac{\partial q_i^s}{\partial s_i} = Q\frac{d_i^s}{S_i} = \frac{d_i}{S_i} \\ \frac{\partial q_i}{\partial s_i \partial t} & = \frac{\partial}{\partial s_i}\left(\frac{\partial q_i}{\partial t}\right) = \frac{\partial}{\partial s_i}\dot{P} \\ \frac{\partial^2 q_i}{\partial s_i^2} & = \frac{\partial}{\partial s_i}\left(\frac{\partial q_i}{\partial s_i}\right) = Q\frac{\partial}{\partial s_i}\left(\frac{d_i^s}{S_i}\right) = \frac{\partial}{\partial s_i}\left(\frac{d_i}{S_i}\right) \end{cases} \tag{7}$$

Moreover, $\frac{\partial^2 q_i}{\partial t^2} = \ddot{P} = [\ddot{x} \quad \ddot{y} \quad \ddot{z}]^T$, which is the acceleration of the CPTDS. Substituting Equation (7) into (6), the inertia of the cable i can be found by integrating the whole cable:

$$I_i = \rho \int_0^{S_i} \frac{d^2 q_i}{dt^2}\partial s_i = \rho\left(\ddot{P}S_i + 2\dot{P}\dot{s}_i + \frac{d_i}{S_i}\dot{s}_i^2 + d_i\ddot{s}_i\right) \tag{8}$$

2.4. Dynamic Equation of the Camera Robot

As illustrated in Figure 4, the cable tension t_i of the cable i is along the tangential direction of the catenary at the cable end node P. As a consequence, t_i can be decomposed into the components that are along the directions of the x, y and z axes, respectively, which can be written in terms of $[h_i \cos\theta_i \quad h_i \sin\theta_i \quad h_i \tan\gamma_i]^T$, where h_i is the horizontal component of t_i as described in Equation (1) and γ_i is the angle between h_i and t_i. The inertia $I_i = [I_{i,x} \quad I_{i,x} \quad I_{i,x}]^T$ is a 3×1 vector. $W = [W_x \quad W_y \quad W_z]^T$ is the external force on the CPTDS. The tangent $\tan\gamma_i$ at P can be computed as the following equation:

$$\tan\gamma_i = \left.\frac{dz_i^s}{dx_i^s}\right|_{x_i^s = L_i} = -\sinh(2\beta_i - \alpha_i) \tag{9}$$

Thus, the dynamic equilibrium equation can be written as follows:

$$\begin{cases} m_P\ddot{x} + \sum_{i=1}^{4} I_{i,x} & = \sum_{i=1}^{4} h_i \cos\theta_i + W_x \\ m_P\ddot{y} + \sum_{i=1}^{4} I_{i,y} & = \sum_{i=1}^{4} h_i \sin\theta_i + W_y \\ m_P\ddot{z} + \sum_{i=1}^{4} I_{i,z} & = \sum_{i=1}^{4} h_i \tan\gamma_i - \sum_{i=1}^{4} \rho g S_i - mg + W_z \end{cases} \tag{10}$$

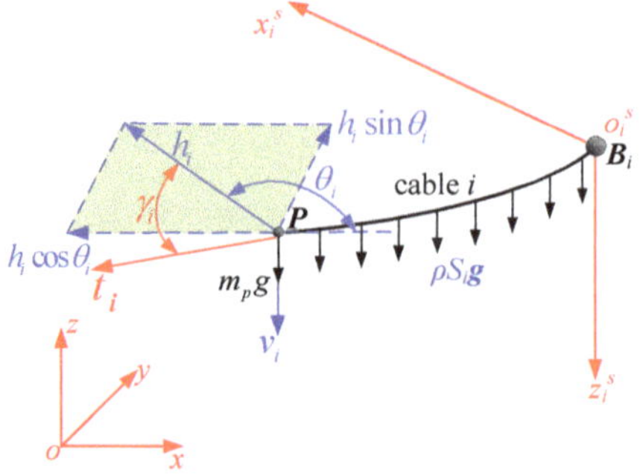

Figure 4. Catenary model of the cable in a vertical plane.

Equation (10) can be further transferred to the following matrix form:

$$M\ddot{P} + \ddot{I}_{cab} = JH + W + G + G_{cab} \tag{11}$$

where $M = \begin{bmatrix} m_p & 0 & 0 \\ 0 & m_p & 0 \\ 0 & 0 & m_p \end{bmatrix}$ and m_p is the mass of the CPTDS. $I_{cab} = \begin{bmatrix} \sum\limits_{i=1}^{4} I_{i,x} & \sum\limits_{i=1}^{4} I_{i,y} & \sum\limits_{i=1}^{4} I_{i,z} \end{bmatrix}^{\mathrm{T}}$

and $H = \begin{bmatrix} h_1 & h_2 & h_3 & h_4 \end{bmatrix}^{\mathrm{T}}$. $J = \begin{bmatrix} \cos\theta_1 & \cos\theta_2 & \cos\theta_3 & \cos\theta_4 \\ \sin\theta_1 & \sin\theta_2 & \sin\theta_3 & \sin\theta_4 \\ \tan\gamma_1 & \tan\gamma_2 & \tan\gamma_3 & \tan\gamma_4 \end{bmatrix}$ is the Jacobian ma-

trix of the camera robot. $G = \begin{bmatrix} 0 & 0 & -m_p g \end{bmatrix}^{\mathrm{T}}$ denotes the gravity vector of the CPTDS,

$G_{cab} = \begin{bmatrix} 0 & 0 & \sum\limits_{i=1}^{4} \rho g S_i \end{bmatrix}^{\mathrm{T}}$ is the total gravity vector of the four cables.

3. Iterative-Based Tension Distribution Algorithm

Equation (11) can be further simplified to the following formulation:

$$JH = F \tag{12}$$

where $F = M\ddot{P} + I_{cab} - W - G - G_{cab}$ is the generalized external force. As Figure 4 shows, the vertical component of the tension $v_i = h_i \tan\gamma_i$. Thus, the cable tension t_i of cable i can be calculated through the following equation:

$$t_i = h_i \sqrt{1 + \tan^2\gamma_i} \tag{13}$$

The horizontal component h_i of t_i should subject to the following restraint condition:

$$h_{i,\min} \leq h_i \leq h_{i,\max} \tag{14}$$

where the lower bound of the cable tension $h_{i,\min}$ is required to keep cable i tight; the upper bound of the cable tension $h_{i,\max}$ is defined to account for the the output torque of the servo motor i and the maximum tension that the cable i can withstand without breaking. In this paper, $h_{i,\min} = h_{\min}$ and $h_{i,\max} = h_{\max}$.

According to Equation (12), it is a non-linear transcendental equation because J and F are associated with H. In this paper, we propose an iterative-based algorithm to determine H, which is different from the algorithm in our previous paper. The algorithm termination condition is when the difference of two sags obtained from adjacent steps is small enough, which could meet after several iterative steps. As a result, although the algorithm in this manuscript does not meet an optimization goal, and the computing time and computational complexity are greatly reduced.

After obtaining H, the cable tension $T = \begin{bmatrix} t_1 & t_2 & t_3 & t_4 \end{bmatrix}^{\mathrm{T}}$ can be calculated according to Equation (13). It is well-known that the initial values have a significant impact on the iterative method. After several trails, the tensions and lengths of the cables obtained by the massless straight line model of the cable are used as the initial values. Since the profile of the cable is a straight line, the initial sag $f_0 = \begin{bmatrix} 0 & 0 & 0 & 0 \end{bmatrix}^{\mathrm{T}}$. ε is a small value and to be used as the threshold of the iterative-based algorithm. Thus, the iterative-based tension distribution algorithm can be summarized as Algorithm 1:

Algorithm 1: Distribution of tensions

Input: $B_i, P, \dot{P}, \ddot{P}, \rho, \Delta t$

Output: H

1   $flag = 1$;

2   $S_{0,i} = \|d_i\|_2$ %Initial cable length;

3   $\tan\gamma_{0,i} = C_i / L_i$ %Initial tangent;

4   $\hat{J}_{0,i} = [\cos\theta_{0,i} \quad \sin\theta_{0,i} \quad \tan\gamma_{0,i}]^{\mathrm{T}}$;

5   $J_0 = [\hat{J}_{0,1} \quad \hat{J}_{0,2} \quad \hat{J}_{0,3} \quad \hat{J}_{0,4}]^{\mathrm{T}}$ %Initial Jacobian matrix;

6   $I_{0,cab} = \sum\limits_{i=1}^{4} \rho\left(\ddot{P}S_{0,i} + 2\dot{P}\dot{s}_i + d_i\frac{\dot{s}_i^2}{S_{0,i}} + d_i\ddot{s}_i \right)$;

7   $G_{0,cab} = \rho \sum\limits_{i=1}^{4} S_{0,i}$ %Initial cable mass;

8   $F_0 = M\ddot{P} + I_{0,cab} - W - G - G_{0,cab}$;

9   $H_0 = J_0 F_0$ %Initial H;

10   **while** $flag \neq 0$ **do**

11     $\alpha_i = \sinh^{-1}\left[\frac{\beta_i(C_i/L_i)}{\sinh\beta_i}\right] + \beta_i, \ \beta_i = \frac{\rho g L_i}{2h_{0,i}}$;

12     $S_i = L_i - \frac{h_{0,i}\beta_i}{qL_i}\left[\frac{L_i}{16\beta_i}\left(e^{4\beta_i - 2\alpha_i} - e^{-4\beta_i + 2\alpha_i}\right) + \frac{1}{2} \right]$;

13     $\tan\gamma_i = -\sinh(2\beta_i - \alpha_i)$ % Compute tangent ;

14     $H = J^+ F$ % Compute H;

15     $f_i = \dfrac{8h_i \sinh\beta_i\sinh^{-1}\left(\frac{\rho g C_i/2h_i}{\sinh\beta_i}\right) - \rho g C_i}{2\rho g}$;

16   **if** all $(|f_i - f_{0,i}| > \varepsilon)$ **then**

17       $J_0 = J$ %Update J;

18       $f_0 = f$ %Update f;

19

20   **else**

21       $flag = 0$;

22

23   **end**

24   **end**

25   **if** all $(h_{\min} \leq h_i \leq h_{\max})$ **then**

26     $H = H_0$ %Final output;

27   **else**

28   H=Null % H is empty;

29   **end**

---

4. Workspace Analysis

4.1. Dynamic Force-Feasible Workspace

There are many workspace criteria proposed to tackle with the influence on the workspace of the unilateral nature of the wrenches (combination of force and moment) applied on the end-effector by cables, among which the wrench-closure workspace (WCW) [43] and wrench-feasible workspace (WFW) [44] are of particular interest. In fact, the WFW is of more practical significance because the cable tensions must be limited within a reasonable range.

There is no moment applied on the CPTDS because this is a mass point. Thus, the camera robot only has a force-feasible workspace (FFW). A position of the end-effector of the CDPM is said to be force-feasible in a particular structure and for a specified set of forces, if the tensions in the cables can counteract any external force of the specified set applied to CPTDS. Liu proposed the generalized determining conditions of WFW/FFW [45] as follows:

(1)   The Jacobian matrix *J* is full rank.

(2)   The tensions in the cables are all positive and in a definite range.

(3)   The magnitude and direction of the projections of column vectors arbitrarily chosen from the Jacobian matrix *J* and the external force *J* are both balanced on a normal vector of the hyperplane determined by *J*.

In this paper, the dynamic force feasible workspace (DFFW) was applied based on the FFW. For the camera robot, the Jacobian matrix *J* and generalized external force *F* have some unique features. On the one hand, the Jacobian matrix *J* depends not only upon the geometry configuration of the manipulator but also on the tensions and self-weights of cables. On the other hand, the generalized external force *F* should consist of the inertias of the cables and CPTDS.

Based on the proposed generalized determining conditions of FFW, we can summarize the judging conditions of DFFW, including the direction balance condition (DBC) and magnitude balance condition (MBC). The DBC, which is also the force-closure condition, is given as follows:

$$
\begin{cases}
q^{\mathrm{T}}F > 0 \exists k : q^{\mathrm{T}}\hat{J}_k < 0 \\
q^{\mathrm{T}}F < 0 \exists j : q^{\mathrm{T}}\hat{J}_j > 0
\end{cases}
\tag{15}
$$

where $q = \hat{J}_a \times \hat{J}_b$ is a unit normal vector of a hyperplane determined by the column vectors of *J* and pointing towards the exterior of the zonotope as shown in Figure 5. $\hat{J}_i = [\cos\theta_i \quad \sin\theta_i \quad \tan\gamma_i]^{\mathrm{T}}$ is the column vector of *J*. *a* and *b* are the subscripts of two linearly independent column vectors arbitrarily chosen from *J* with $a, b \in \{1, 2, 3, 4\}$; *j* and *k* are the subscripts of the rest column vectors with $j, k \in \{1, 2, 3, 4\} - \{a, b\}$.

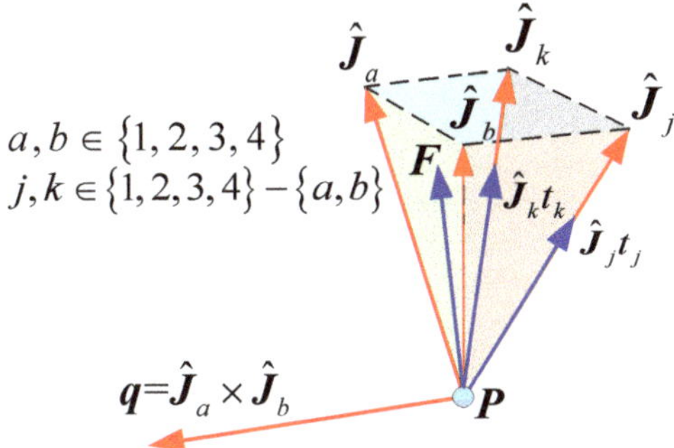

Figure 5. Normal vector of the hyperplane and the generalized external force.

As displayed in Figure 6a, the projections of the tension $\hat{J}_j h_j$ (or $\hat{J}_k h_k$) and the generalized external force *F* on *q* are both scalars. Consequently, there always exists a tension $\hat{J}_j h_j$ (or $\hat{J}_k h_k$) to resist the the generalized external force *F* as long as the signs of $q^{\mathrm{T}}\hat{J}_j$ (or $q^{\mathrm{T}} \hat{bm}J_k$) and $q^{\mathrm{T}}F$ are different. However, *H* ranges from $h_{\min}$ to $h_{\max}$. Hence, the components of *H* $\hat{J}_j h_j$ and $\hat{J}_k h_k$ are bound with $[\hat{J}_j h_{\min}, \hat{J}_j h_{\max}]$ and $[\hat{J}_k h_{\min}, \hat{J}_k h_{\max}]$ as shown in Figure 6b. Therefore, the MBC is given as follows:

$$
\Gamma^-_{\min} \leq \Gamma^+_{\max} + \left|q^{\mathrm{T}}F\right| \bigcap \Gamma^+_{\min} + \left|q^{\mathrm{T}}F\right| \leq \Gamma^-_{\max}
\tag{16}
$$

where $\Gamma^+_{\min}$ and $\Gamma^+_{\max}$ are the lower and upper limit of the sum of projections of the rest column vectors on the positive direction of *q*, respectively; $\Gamma^-_{\min}$ and $\Gamma^-_{\max}$ are the lower and upper limit of the sum of projections of the rest column vectors on the negative direction of *q*, respectively. They can be computed as follows:

$$\Gamma_{\min}^{+} = \sum_{q^{\mathrm{T}}\hat{j}_{j}>0} \left(q^{\mathrm{T}}\hat{j}_{j}\right)h_{\min},\ \Gamma_{\max}^{+} = \sum_{q^{\mathrm{T}}\hat{j}_{j}>0} \left(q^{\mathrm{T}}\hat{j}_{j}\right)h_{\max} \tag{17a}$$

$$\Gamma_{\min}^{-} = \sum_{q^{\mathrm{T}}\hat{j}_{k}<0} \left(q^{\mathrm{T}}\hat{j}_{k}\right)h_{\min},\ \Gamma_{\max}^{-} = \sum_{q^{\mathrm{T}}\hat{j}_{k}<0} \left(q^{\mathrm{T}}\hat{j}_{k}\right)h_{\max} \tag{17b}$$

The pre-condition to determine DBC or MBC is the Jacobian matrix J is full rank. Thus, the DFFW can be described as the following set:

$$\begin{aligned}&\{P:\ (rank(J)=3)\cap\forall(q\in\mathbf{R}^3,\ q^{\mathrm{T}}F>0),\ \exists k:q^{\mathrm{T}}\hat{j}_k<0\cap|\Gamma_{\min}^-|\leq\Gamma_{\max}^++q^{\mathrm{T}}F\cap\Gamma_{\min}^++q^{\mathrm{T}}F\leq|\Gamma_{\max}^-|\}\\&\bigcup\{P:\ (rank(J)=3)\cap\forall(q\in\mathbf{R}^3,\ q^{\mathrm{T}}F<0),\ \exists j:q^{\mathrm{T}}\hat{j}_j>0\cap|\Gamma_{\min}^-|\leq\Gamma_{\max}^++|q^{\mathrm{T}}F|\cap\Gamma_{\min}^++|q^{\mathrm{T}}F|\leq\Gamma_{\max}^-\}\end{aligned} \tag{18}$$

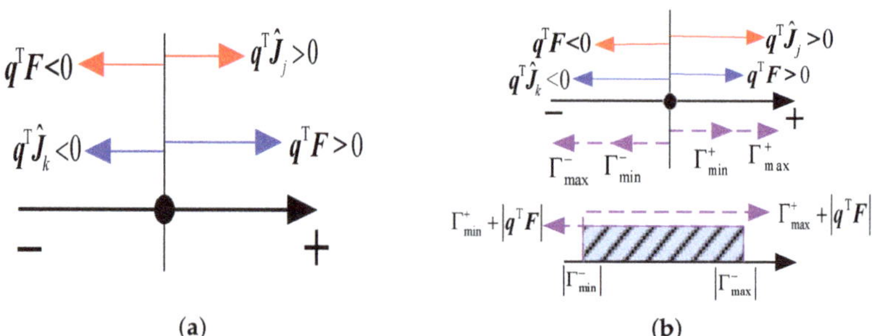

Figure 6. The projections of the tension and generalized external force in DBC and MBC. (**a**) The projections in DBC. (**b**) The projections in MBC.

4.2. Procedure of Generating DFFW

The DFFW can be generated by judging whether a position X meets the conditions demonstrating in Equation (18) or not. The search space of X is decided by the position of the pulleys and the ground, which is a cuboid for the camera robot. Before generating DFFW, the generalized external force F and the Jacobian matrix J must be calculated according to Algorithm 1. To calculate the generalized external force F, it is required to determine the CPTDS's velocity $\dot{X}$ and acceleration $\ddot{X}$.

$\mathbf{V} = \left\{v_x^+, v_x^-, v_y^+, v_y^-, v_z^+, v_z^-\right\}$ is defined as the maximal allowable velocity set of the CPTDS, in which v_x^+, v_x^-, v_y^+, v_y^-, v_z^+ and v_z^- denote the maximum translation speeds along the positive and negative directions of the x-axis, y-axis and z-axis, respectively. $A = \left\{a_x^+, a_x^-, a_y^+, a_y^-, a_z^+, a_z^-\right\}$ is defined as the maximal allowable acceleration set of the CPTDS, in which a_x^+, a_x^-, a_y^+, a_y^-, a_z^+ and a_z^- denote the maximum translation accelerations along the positive and negative directions of the x-axis, y-axis and z-axis, respectively.

Thus, a set \$, named a velocity–acceleration pair is constituted by choosing an element from the set V and A separately while setting the other elements of V and A equal to zero. For example, the velocity–acceleration pair $\$ = \{v_x^+, a_x^+\}$ represents the CPTDS's velocity and acceleration along the x-axis positive direction.

Since there are six possibilities for selecting two column vectors randomly from four column vectors of the Jacobian matrix J, the inner for-loop of Algorithm 2 will execute six times. The search space refers to SearchCube, which is a cube. After every inner for-loop, the position should move to the new position $P + \Delta P$. Consequently, the algorithm of generating the subspace of DFFW can be summarized as Algorithm 2.

For a selected velocity–acceleration pair $\$_l$ ($l = 1, 2, \cdots 6$), a subspace of DFFW referred to as DSS_l will be generated through implementing Algorithm 2 correspondingly. By repeatedly performing Algorithm 2 six times and taking the intersection of the subspaces

DSS$_l$, the DFFW DS will be generated. Mathematically, the relationship between DSS$_l$ and DS can be described as follows:

$$\$_l \rightarrow \text{DSS}_l \subseteq \text{DS}$$

$$\text{DS} = \bigcap_{l=1,2,\cdots 6} \text{DSS}_l \tag{19}$$

Algorithm 2: Generate the subspace of DFFW

    Input: input parameters P, $\$_l$, J, F
    Output: DSS$_l$, $\{l = 1, 2, \cdots 6\}$

```
 1  a, b ∈ {1, 2, 3, 4} ;
 2  j, k ∈ {1, 2, 3, 4} − {a, b} ;
 3  while P ∈ SearchCube do
 4  │   if rank(J) = 3 then
 5  │   │   for i ← 1 to 6 do
 6  │   │   │   if qᵀF > 0 & ∃k : qᵀĴₖ < 0 then
 7  │   │   │   │   Judgeᵢ = 1 %Direction balance;
 8  │   │   │   else if qᵀF < 0 & ∃j : qᵀĴⱼ > 0 then
 9  │   │   │   │   Judgeᵢ = 1 %Direction balance;
10  │   │   │   else
11  │   │   │   │   Judgeᵢ = 0 %Direction imbalance;
12  │   │   │   end
13  │   │   │   if ∏_{i=1,2···6} Judgeᵢ = 1 & |Γ⁻_min| ≤ Γ⁺_max + |qᵀF| ∩ Γ⁺_min + |qᵀF| ≤ |Γ⁻_max|
    │   │   │   then
14  │   │   │   │   P ∈ DSS_l %Magnitude balance;
15  │   │   │   else
16  │   │   │   │   P ∉ DSS_l
17  │   │   │   end
18  │   │   end
19  │   else
20  │   │   P ∉ DSS_l;
21  │   end
22  │   P = P + ΔP % Move to the new position;
23  end
```

---

5. Optimization Model Establishment

5.1. Optimization Variables

The pulleys of the camera robot can be easily mounted on top of masts or the surface of buildings according to the characteristics of the shooting place, whose positions have a strong influence on the workspace, kinematics and dynamics. Therefore, it is reasonable to employ the positions of the pulleys as the optimization variables for structural optimization. Since the horizontal projection of the four pulleys on the ground is a rectangle, the global frame $\{oxyz\}$ can be established with the origin fixed on the projection point of the #1 pulley. As Figure 7 displayed, we can use three dimensional parameters to describe the positions of four pulleys in $\{oxyz\}$, i.e., the length of the rectangle *len*, width of the rectangle *wid* and the pulley height *hei*. Thus, the position of #1 pulley is $B_1 = [0, 0, hei]^T$, #2 pulley $B_2 = [len, 0, hei]^T$, #3 pulley $B_3 = [len, wid, hei]^T$ and #4 pulley $B_4 = [0, wid, hei]^T$.

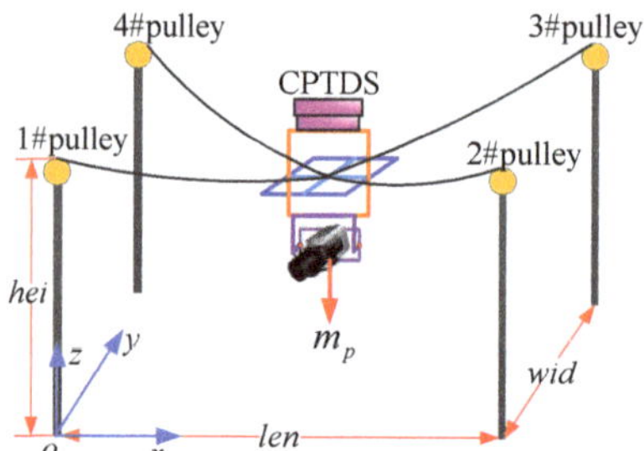

Figure 7. Optimization variables of the camera robot.

In addition to the positions of the pulleys, the mass of the CPTDS also can be adjusted. Underweight or overweight CPTDS may affect the performance of the camera robot. Thus, the performance of the camera robot can be improved by choosing a suitable mass of the CPTDS. Hence, it is necessary to use the mass of the CPTDS m_p as the optimization variables for structural optimization.

The optimization variables of structural optimization of the camera robot can be written in the following vector form:

$$\boldsymbol{D} = \left[len, wid, hei, m_p\right]^T = [d_1, d_2, d_3, d_4]^T, \tag{20}$$

and the physical significance and unit of each optimization variable are shown in Table 1.

Table 1. The physical significance and unit of each optimization variable.

Element	Symbol	Physical Significance	Unit
d_1	len	length of the rectangle	m
d_2	wid	width of the rectangle	m
d_3	hei	height of the pulley	m
d_4	m_p	mass of the CPTDS	kg

5.2. Optimization Objects

The camera robot is a high-speed and high-maneuverable manipulator, which will bring great challenges to take real-time video pictures. In order to find the video images with a sufficient scope and high-resolution ratio, a sufficiently large workspace and a stable enough camera are needed. In this paper, we attempt to enlarge the workspace and enhance the shooting stability through the structural optimization.

5.2.1. Workspace Volume

In order to satisfy the shooting requirements, it is desired to enable the camera robot to achieve as high ability of tracking photography as possible. The volume of the workspace is directly related to the tracking photography ability of the camera robot. The larger the workspace, the more areas the camera robot can capture. Naturally, the workspace volume can be integrated into the optimization model of the camera robot as the optimization object. Given the large-span and high-speed characteristics of the camera robot, the DFFW presented in the previous section is employed in this paper. As shown in Figure 8a, the rectangle determined by four pulleys forms the upper surface of the cube, which is the theoretical maximum workspace, referred to as the maximum shooting workspace (MSW). However, the shape of the actual workspace (DFFW) is similar to a reversed quadrilateral prism with a smaller bottom and a larger top as a result of restrictions on the tensions in cables.

As illustrated in Figure 8b, the horizontal section of the DFFW is a rectangle, which is analogous to the rectangle formed by four pulleys. In order to guarantee the shooting effect, a cube named the minimum shooting task-space (MSTS) is defined within MSW and is required to be included within the DFFW.

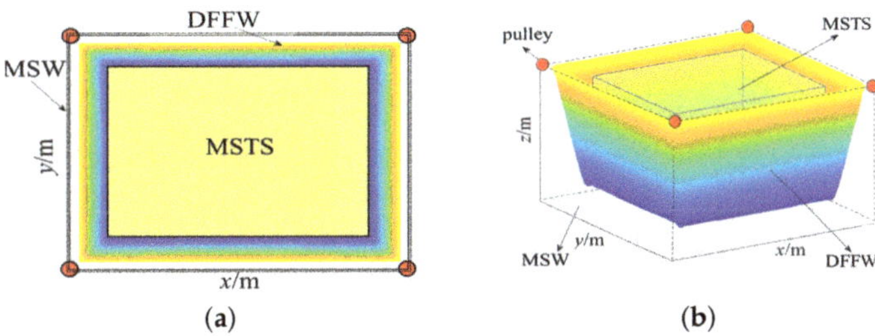

Figure 8. The DFFW, MSTS and MSW of the camera robot. (**a**) Top view. (**b**) Main view.

In this paper, we discretize the MSW, yielding the total number of point within the MSW N_{total}. Then, we scan the every point within the MSW and record the number of points meeting the dynamic force-feasible condition N_{DFFW} described in Equations (19) and (20). Finally, the volume of DFFW *vol* can be obtained by computing the ratio of N_{DFFW} to N_{total}. The larger volume of DFFW, the stronger the ability of the tracking photography. Thus, the optimization objective for the workspace volume is as follows:

$$f_1(\boldsymbol{D}) = \max \quad vol(d_1, d_2, d_3, d_4) \tag{21}$$

5.2.2. Anti-Wind Disturbance Ability

Camera robots often function in high-rise cable support structures, which are inevitably disturbed by wind due to the frequent outdoor operations. Thus, enhancing the ability of anti-wind disturbance is crucial for maintaining the stability of the camera when taking videos and pictures. There are two kinds of wind forces acting on buildings, i.e., steady wind pressure and fluctuating wind pressure [46]. Since the period of the stable wind pressure is much larger than the natural vibration period of the general structure, its force can be regarded as a static force [47]. However, the high-frequency pulsation components of the fluctuating wind pressure lead to vibration responses of the camera robot, which have an influence on the camera robot's normal operations.

The frequency characteristics of fluctuating wind can be expressed by its power spectrum, which is related to the surface roughness and geomorphic conditions. Considering that the camera robot mainly works in the open areas of urban spaces, the correction coefficient to the basic wind pressure w_0 is caused by the surface roughness $\mu_{w_0} = 0.731$. Given the air density $\rho_a = 1.225 \text{ kg/m}^3$ and the basic wind pressure $w_0 = 0.35 \text{ KN/m}^2$, the maximum 10-minute average wind speed $\bar{V}_{\max}$ with a 30-year return period can be calculated as follows [47]:

$$\bar{V}_{\max} = \sqrt{\frac{2\mu_{w0}k_0w_0}{\rho_a}} \tag{22}$$

where k_0 is the return period coefficient and $k_0 = 1$ is when the return period is 30 years. Since the frequently used Davenport spectrum overestimates the turbulence energy at high frequencies, these frequencies are of great significance for flexible tall structures. Therefore, this paper employs the modified Davenport wind speed spectrum—namely, the Maier spectrum, which can be written as follows [47]:

$$\begin{cases} S_v(z,f) = \dfrac{2x^2}{(1+3x^2)^{\frac{4}{3}}} \cdot \dfrac{\sigma_V^2}{f} \\[2mm] x = \dfrac{L_v^* f}{\bar{V}(z)} \\[2mm] \sigma_V = \sqrt{6k}\,\bar{V}_{10} \\[2mm] \bar{V}(z) = \bar{V}_{10}\left(\dfrac{z}{10}\right)^{\alpha} \end{cases} \tag{23}$$

where z is the height of the CPTDS above the ground, $L_v^* = 1200$ m is the overall scale of the turbulence, and $\bar{V}_{10}$ is average wind speed at 10 m. As the camera robot works under the common wind speed, the power spectrum of fluctuating wind can be obtained, which is shown in Figure 9.

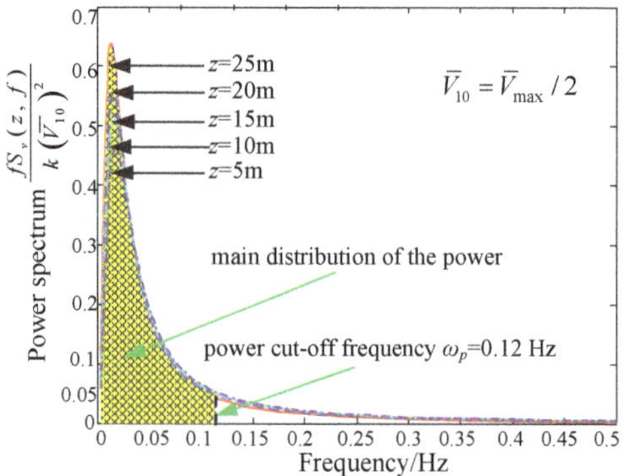

Figure 9. Power spectrum of the pulsating wind (wind speed is the half of the maximum 30-year wind speed).

On the one hand, as the height z changes from 5 to 25 m (the main working range of the camera robot), the power spectrum varies accordingly. It can be seen from Figure 10 that the maximal value of the power spectrum increases with the increasing pulley height. However, the values of five curves mainly exist in the range of 0.01–0.12 Hz, which is also the most-concentrated region of the energy of the pulsating wind. The maximum values of the five power spectrum curves both emerge at 0.02 Hz and then fall quickly; when the frequency is more than 0.12 Hz, the power spectrum is less than 0.05. Hence, we define 0.12 Hz as the energy cut-off frequency. In the study of anti-wind disturbance, we focused on the energy concentrated region, i.e., the frequency region of the pulsating wind between 0.01 and 0.12 Hz. On the other hand, we obtained the minimum first-order natural frequencies on the horizontal sections of the dynamic feasible workspace at different heights. As shown in Figure 10, the minimum first-order nature frequency is 0.13 Hz when $z = 10$ m. Therefore, the CPTDS will produce a vibration as the result of the wind-excitation at certain areas within the workspace (such as the workspace boundaries). Therefore, in order to improve the ability of the camera robot to resist wind disturbance, it is necessary to increase its first-order natural frequency to keep away from the range of 0.01–0.12 Hz.

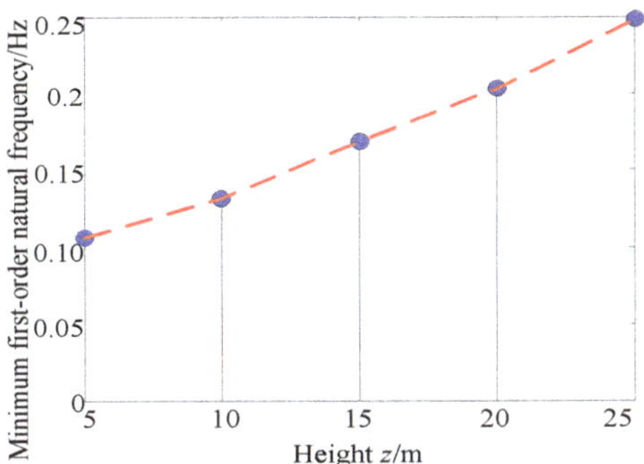

Figure 10. The minimum first-order natural frequency on the horizontal section of the workspace at different heights (Design variables: $len = 90$ m, $wid = 80$ m, $hei = 25$ m and $m_p = 50$ kg).

The transverse and longitudinal vibrations of the cable may occur in the vertical plane when the cable is disturbed by the wind. For the cable, the influence of longitudinal vibration on the end-effector is much greater than that of transverse vibration. Diao et al. indicated that the influence of transverse vibration on the end-effector is only 1.4%, while that of longitudinal vibration on the end-effector is 98.4% [48]. Liu et al. suggested that the longitudinal vibration of the cable has a greater influence on the feed cabin supported in parallel by six cables [49]. As seen in Figures 9 and 10, the lower the frequency of the camera robot, the more likely the camera robot was excited. As the first-order vibration has the most important effects in structural response, it is necessary to calculate the first-order natural frequency of the camera robot in the following form:

$$\omega_i = \frac{1}{\sqrt{\mathrm{eig}\left\{\left[(J^+)^{\mathrm{T}}MJ^+\right]\mathrm{diag}(S_1, S_2, S_3, S_4)/EA\right\}}} \quad i = 1, 2, 3 \tag{24}$$

where J^+ is the Moore–Penrose generalized inverse of the Jacobin matrix J. eig is an eigenfunction to solve the eigenvalues. For the camera robot, there are three eigenvalues as the result of its three translational DOF. $\mathrm{diag}(S_1, S_2, S_3, S_4) = \begin{bmatrix} S_1 & 0 & 0 & 0 \\ 0 & S_2 & 0 & 0 \\ 0 & 0 & S_3 & 0 \\ 0 & 0 & 0 & S_4 \end{bmatrix}$ is a diagonal matrix. E and A are the elastic modulus and cross-section area of the cable separately. The first-order natural frequency ω_1 is the minimum of Equation (24):

$$\omega_1 = \min\{\omega_i\} \tag{25}$$

$$\mathrm{GFNF} = \frac{1}{N_{total}} \sum_{i=1}^{N_{total}} \omega_1(D) \tag{26}$$

The GFNF should be as large as possible to improve the ability to resist the wind disturbance of the camera robot in the whole workspace. Therefore, the optimization objective of anti-wind disturbance performance is as follows:

$$f_2(D) = \max \frac{1}{N_{total}} \sum_{i=1}^{N_{total}} \omega_1(d_1, d_2, d_3, d_4) \tag{27}$$

5.2.3. Impulse of Tensions on CPTDS

When the camera robot works normally, the cables remain tight, and hence the cable tension will exert a tractive force on the CPTDS. Due to the high maneuverable, the tension changes quickly in a tiny period of time, causing an impulse on the CPTDS exerted by the cables, which will clearly have a great influence on the camera robot, such as vibration of the camera or blurred shooting video pictures. The impact should be reduced as much as possible to ensure the normal operation of the camera robot. If the external force applied on the CPTDS and locus of the CPTDS are known, the equivalent force F_{eq} of the tensions in the four cables can be computed according to Algorithm 1.

$$F_{eq} = \sum_{i=1}^{4} t_i \tag{28}$$

As shown in Figure 11, four sampling points, P_1, P_2, P_3 and P_4, were selected. P_1, P_2, P_3 and P_4 are located in the upper, left, lower and right surfaces of the MSTS, respectively. Thus, the sampling locus of the cable tension impact function is $P_1 \rightarrow P_2 \rightarrow P_3 \rightarrow P_4 \rightarrow P_1$. As the four sampling points are evenly distributed in MSTS and the acceleration varies dramatically at the points that the direction of locus changes, the impulse of tensions on the CPTDS will be great. Therefore, the selection of such a sampling locus can truly reflect the impact of the tensions on the CPTDS. The sum of impulse imp along the sampling locus can be calculated by computing the impulse at every time point, which can be used as the evaluating function of impulse of tensions on the CPTDS.

$$imp = \left\| \int_0^T F_{eq} dt \right\|_2 \tag{29}$$

where T is duration of the locus. In order to ensure that the impulse of CPTDS is as slight as possible, the impulse sum imp should be kept as small as possible:

$$f_3(\boldsymbol{D}) = \min \quad imp(d_1, d_2, d_3, d_4) \tag{30}$$

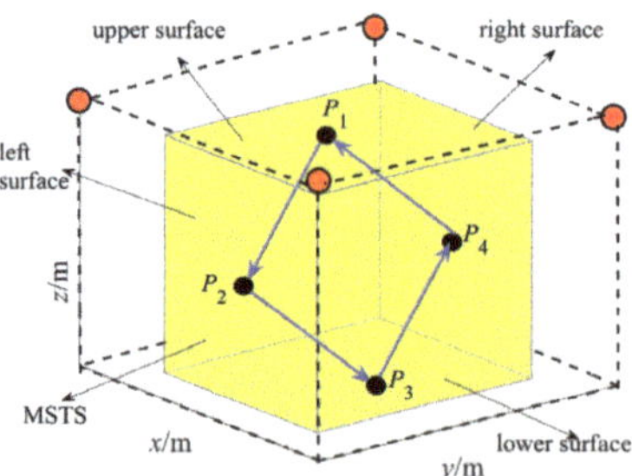

Figure 11. Sampling locus of cable tension impulse function.

5.3. Constraints

5.3.1. Linear Constraints

Considering the installation and site conditions, the optimization variable vector $\boldsymbol{D}$ should meet the following linear constraints to ensure that the camera robot has feasibility of structure.

$$\boldsymbol{lb} \leq \boldsymbol{D} \leq \boldsymbol{ub} \tag{31}$$

where $\boldsymbol{lb} = [lb_1 \quad lb_2 \quad lb_3 \quad lb_4]^T$ is the lower bound; $\boldsymbol{ub} = [ub_1 \quad ub_2 \quad ub_3 \quad ub_4]^T$ is the upper bound.

5.3.2. Nonlinear Constraints

As described in the previous sections, the horizontal components of tensions have upper bound $h_{\max}$ and lower bound $h_{\min}$. Furthermore, in order to ensure that the CPTDS has sufficient tracking and shooting abilities in the workspace, the workspace volume $vol(D)$ determined by the optimization variable vector D must be larger than the volume of MSTS Ω. To ensure that the camera robot has a strong ability of anti-wind disturbance, it is necessary to guarantee that the GFNF determined by D must be larger than the energy cut-off frequency $\omega_p = 0.12$ Hz. To sum up, the nonlinear constraints of camera robot structural optimization can be summarized as follows:

$$\begin{cases} g_1(D) = h_{\min}(D) - h_{\max} \leq 0 \\ g_2(D) = -h_{\max}(D) + h_{\min} \leq 0 \\ g_3(D) = -V(D) + \Omega \leq 0 \\ g_4(D) = -\omega(D) + \omega_p \leq 0 \end{cases} \tag{32}$$

where $h_{\min}$ and $h_{\max}$ are the minimum and maximum values of the horizontal components of tensions with regard to the optimization variable vector D, respectively.

6. GA-Based Structural Optimization

6.1. Information Entropy-Based Adaptive Multi-Island GA

Structure optimization of a camera robot is intrinsically a highly nonlinear optimization problem. Moreover, it is difficult to obtain a continuous and derivable analytical expression of the workspace of the camera robot. Therefore, it is difficult to deal with this problem for the traditional optimization methods. However, SGA is subject to premature convergence, thereby, falling into local minimum easily. GA is an intelligent optimization method that simulates the evolution of organisms in nature and genetic laws. It was first proposed by Professor Holland in 1975 in the book *Adaptation in Natural and Artistic Systems*. Multi-island GA divides each population into several sub populations—named islands—which can be viewed as a niche. As only the excellent individuals migrate between the islands, it can mean that excellent individuals spread to the whole population and improve the evolution level of the whole population [50]. In this paper, we propose a improved genetic algorithm incorporating into three improvements, namely fitness calibration, sub-population division and adaptive changes of crossover probability and mutation probability.

6.1.1. Fitness Degree Calibration

There may be special individuals with abnormal fitness in the initial population, leading to the dominance of the whole population possibly. At the end of the genetic process, the fitness of individuals tends to be consistent and the solution will swing around the optimal solution so that the optimal solution can not be searched. The selection ability of the population can be improved through enlarging fitness degree appropriately, which is the principle of fitness degree calibration and can be expressed as follows:

$$\widetilde{fit} = \frac{fit_0 + |fit_{\min}|}{|fit_{\min}| + fit_{\max} + \delta} \tag{33}$$

where fit_0 is the original fitness degree; $fit_{\min}$ and $fit_{\max}$ are the lower and upper bound of fitness degree, respectively; and $\delta \in (0, 1)$ is a positive real number to avoid the zero denominator. The purpose of $|fit_{\min}|$ is to guarantee the fitness degree is always positive after calibration.

6.1.2. Sub-Population Division

The mechanism of sub-population division and migration arises from the idea of multi-island GA, whose group is divided into several sub-populations referred to as islands. The selection, crossover and mutation operations of GA are performed on each island, and the migration operation is performed between different islands periodically. The steps of multi-island GA are shown in Figure 12, where t is the current generation, n_f is the number of sub-population, q is positive integer, and m_f is the migration interval.

GOs represents a genetic operation, such as selection, crossover, or mutation. Only when the number of generations reaches the integral multiple of the migration interval m_f, will the individuals be transferred according to the migration rate i_f between the islands; otherwise, the migrations will not occur but instead GOs. Through the sub-population division and migration, the diversity of solution to GA is improved to thus improve the ability to search the global optimal solution [51].

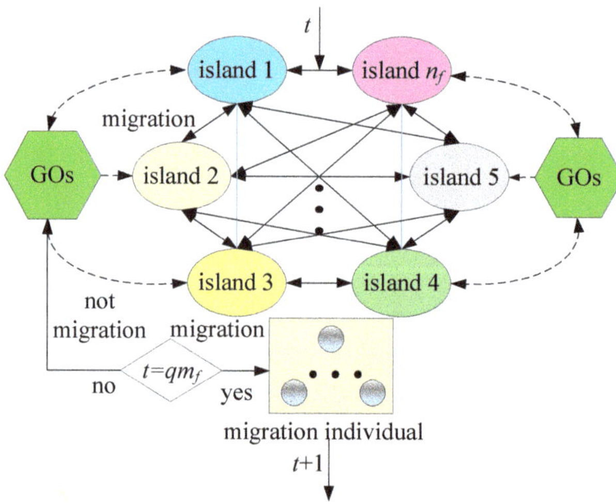

Figure 12. Algorithm steps of the multi-island GA.

6.1.3. Adaptive Changes of Crossover Probability and Mutation Probability

When GA evolves to a certain generation, the fitness of the population will converge, and the population diversity will decline, allowing the algorithm to easily fall into the local optimal solution. In this paper, we regulate the crossover probability p_c and mutation probability p_m adaptively based on the information entropy of the population. Information entropy is a concept proposed by Shannon indicating the disorder degree of the system [52]. At the early stage of evolution, the diversity of the population and the disorder degree is high, and thus the information entropy is high. As the evolution process of the superior winning and the bad eliminated, the disorder degree of the population decreases, and then the information entropy decreases, which conforms to the law of biological development. The information entropy of the population can be defined as follows:

$$I(t) = -\sum_{i=1}^{n_p} \left(p(t)_i \cdot \log_2 p(t)_i \right)$$

$$p(t)_i = \frac{fit(t)_i}{\sum\limits_{i=1}^{n_p} fit(t)_i}$$

(34)

where n_p is the total number of individuals in the population, and $fit(t)_i$ is the fitness degree of the ith individual in the population of the tth generation. Thus, the adaptive operators of crossover rate $P_c(t)$ and mutation $P_m(t)$ rate of the population of the tth generation are as follows:

$$P_c(t) = \frac{1 + \left(e^{I(t)}/e^{I_{\max}}\right)}{2} P_{c0} \tag{35a}$$

$$P_m(t) = \frac{1 + \left(e^{I(t)}/e^{I_{\max}}\right)}{2} P_{m0} \tag{35b}$$

$I_{\max}$ is the maximum information entropy and $I_{\max} = \log_2 n_p$. P_{c0} and P_{m0} are the initial values of crossover rate and mutation rate. According to GA theory, crossover rate ranges from 0.4 to 0.9, and the mutation rate ranges from 0.01 to 0.1 [51]. Therefore, P_{c0} and P_{m0} should be calculated several times according to the value range to obtain the appropriate value.

6.1.4. Population Information Entropy-Based Adaptive Multi-Island GA

Based on the three improvements, we propose an improved GA, i.e., the population information entropy-based adaptive multi-island Genetic Algorithm (PIEAMIGA) as illustrated in Figure 13. Compared to the standard genetic algorithm (SGA), PIEAMIGA adds three modules, namely sub-groups division, fitness calibration and migration, whose crossover rate and mutation rate vary according to the information entropy of the population. The optimization variable vector $\boldsymbol{D} = [d_1, d_2, d_3, d_4]^\mathrm{T}$ is the chromosome of PIEAMIGA that is coded by real values, and the selection is based on stochastic universal sampling. The entire process will stop when $t = G$, and the current optimization parameters will be exported; otherwise, it will re-generate a population and proceed to the next iteration.

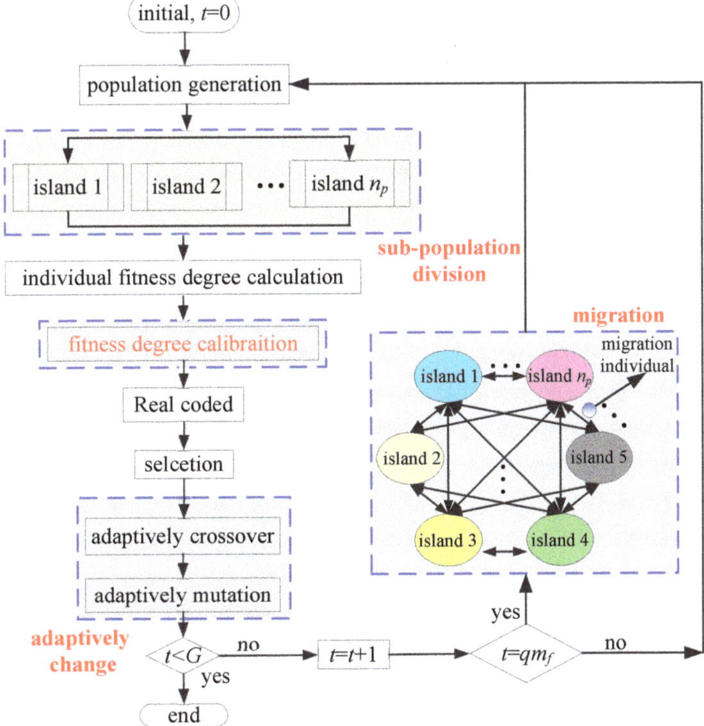

Figure 13. Flowchart of PIEAMIGA.

6.2. Multi-Objective Structural Optimization

The main method to deal with the multi-objective structural optimization problem of CDPM is the weighted coefficient method [33,53,54]. However, the weighted coefficient method has three shortcomings: first, the weight coefficients are determined according to the designer's subjective intention; second, the weight coefficients are difficult to quantify accurately; third, a single weight coefficient is difficult to represent all design intents due to the infinite selections of weight coefficients. In addition to the weighted coefficient method, there are some other multi-objective structural optimization methods for CDPMs, such as the enumeration method [55] and tabular method [56]. However, these methods are not objective, because the optimal criteria are determined by designers subjectively. Accordingly, we apply the ideal-point method to tackle the multi-objective structural optimization problem.

Process of the ideal-point method: first, three optimization objectives are conducted to obtain the optimal solution with regard to each objective; secondly, the solution spaces of the three objectives are normalized to the interval [0, 1], and the optimal value of each objective is normalized to "0"; finally, the minimum sum of the distances between the three optimization objectives and "0" is taken as the optimization objective, namely the "ideal point". The PIEAMIGA was implemented with MATLAB programming language and executed on a notebook computer with a 2.2 GHz Intel Core I5-5200U CPU and a 12 G RAM.

6.2.1. The Optimal Value and Worst Value of Single-Objective Optimization

Since the GA itself does not have the ability to deal with constraints, it is necessary to transform the constrained optimization problem into unconstrained optimization problem, which can be realized through the penalty function method. Thus, the fitness function of the optimization object $f_i(\boldsymbol{D})$ with regard to design variable vector $\boldsymbol{D}$ is as follows:

$$\tilde{F}_i(\boldsymbol{D}) = fit\{\text{sgn}(i) \cdot Tra[f_i(\boldsymbol{D})] + Pun \sum_{j=1}^{4} [g_j(\boldsymbol{D}_j) + |g_j(\boldsymbol{D})|]\} \tag{36}$$

where $fit()$ is the fitness degree function, and $\{i = 1, 2, 3\}$. *Tra* is a transfer function. *Pun* is a large positive number and is used as the penalty factor.

The parameters of SGA and PIEAMIGA are listed in Table 2. The simulation parameters of the optimization model are shown in Table 3. According to the site and installation conditions, the range of the length *len* and width *wid* of the horizontal projection rectangle of pulleys are [80, 90] m and [70, 80] m, respectively; the range of the height of the pulleys' points is [25, 30] m. The range of the mass of the CPTDS is [20, 50] kg. The MSTS is a 70 m × 60 m × 22 m cuboid, whose volume Ω is 92,400 m³. Based on the parameters listed in Tables 2 and 3, PIEAMIGA and SGA are programmed by MATLAB to find the optimal solution and value with regard to three optimization objects separately, which are listed in Table 4. The worst values with regard to the single optimization object are also listed in Table 4.

It can be seen from Table 4 that the optimal solution obtained by PIEAMIGA is better than that obtained by SGA. Workspace volume increases by 8.52%, the first-order natural frequency increases by 9.73%, and the impulse exerted by tensions on the CPTDS to CPTDS is reduced by 10.08%. Although three modules are added into PIEAMIGA based on SGA, the computing time of PIEAMIGA increases by 4.48% compared with that of SGA because the multi-island GA adopts the parallel mechanism, and operations in each island are performed in parallel. It also can be seen from Table 4 that PIEAMIGA has a stronger global search ability as the result of the adaptive crossover and mutation rate.

Table 2. Parameters of PIEAMIGA and SGA.

PIEAMIGA		SGA	
total evolutionary generations	200	total evolutionary generations	200
population size n_p	40	population size n_p	40
penalty factor Pun	10^7	penalty factor Pun	10^7
initial of crossover rate P_{c0}	0.01	fixed crossover rate P_c	0.6
initial of mutation rate P_{m0}	0.85	fixed mutation rate P_m	0.008
correction for fitness degree δ	3		
number of sub-population n_f	8		
migration interval m_f	5		
migration rate i_f	0.5		

Table 3. Simulation parameters of the optimization.

Parameter	Value	Parameter	Value
lower bound of design variable L	$\{80, 70, 25, 20\}^{\mathrm{T}}$	gravitational acceleration g	$9.8\ \mathrm{m/s^2}$
upper bound of design variable U	$\{90, 80, 30, 50\}^{\mathrm{T}}$	impulse sampling time $\Delta t/\mathrm{s}$	0.1
volume of MSTS $\Omega/\mathrm{m^3}$	92400	position of sampling point P_2/m	$[45, 40, 24]^{\mathrm{T}}$
maximum allowable velocity set $V/\mathrm{m/s}$	$\{9, -9, 9, -9, 9, -9\}$	position of sampling point P_2/m	$[15, 40, 15]^{\mathrm{T}}$
maximum allowable acceleration set $A/\mathrm{m/s^2}$	$\{3.5, -3.5, 3.5, -3.5, 3.5, -3.5\}$	position of sampling point P_3/m	$[45, 40, 6]^{\mathrm{T}}$
lower limit of horizontal component h_{min}/N	55	position of sampling point P_4/m	$[75, 40, 15]^{\mathrm{T}}$
upper limit of horizontal component h_{max}/N	4000	time of $P_1 \rightarrow P_2/\mathrm{s}$	7.5
elastic modulus of the cables E/Gpa	28	time of $P_2 \rightarrow P_3/\mathrm{s}$	7.5
cross sectional area of the cable $A/\mathrm{mm^2}$	20.34	time of $P_3 \rightarrow P_4/\mathrm{s}$	7.5
linear density of the cable $\rho/\mathrm{Kg/m}$	0.188	time of $P_4 \rightarrow P_1/\mathrm{s}$	7.5

Table 4. Comparison between PIEAMIGA and SGA.

Algorithm		PIEAMIGA	SGA
optimal solution	D_1^*	[89.89, 79.92, 27, 99, 49.97]	[90.00, 80.00, 29.89, 40.03]
	D_2^*	[80.00, 70.00, 30.00, 20.00]	[80.00, 70.00, 29.50, 24.85]
	D_3^*	[83.77, 80.00, 26.57, 20.00]	[85.74, 79.25, 25.72, 20.35]
optimal value	$f_1^*/\mathrm{m^3}$	183,750	169,325
	f_2^*/Hz	0.3100	0.2875
	$f_3^*/\mathrm{N{\cdot}s}$	24,907	27,699
computing time	t/min	70	67
the worst value of IEPAMIGA			
$f'_1 = 84,875\ \mathrm{m^3}$		$f'_2 = 0.1042\ \mathrm{Hz}$	$f'_3 = 63,027\ \mathrm{N{\cdot}s}$

The convergence processes of each objective by the PIEAMIGA and SGA are shown in Figure 14. In the early stage of evaluation, the optimal solution changes rapidly due to the randomness of population and great differences between individuals, which is a sharply changed line in Figure 14a. With the progress of evolution, the number of excellent individuals in the population gradually increases; thus, the change of optimal solution slows down, and the local convergence appears, which is a horizontal line in Figure 14. Furthermore, the process of local convergence is prolonged gradually as the result of the increasing proportion of excellent individuals in the population.

Thus, the change of the optimal solution is not as violent as in the initial stage, and the length of the horizontal line is gradually longer with the increase in generation of evolution. There exist a jump after the local convergence process on the curve because the optimal solution jumps out of a local optimal solution and evolves towards a better optimal solution. It is inevitable that the performance of the offspring will be degraded compared with the parent because the new individuals are not always better than the parent individuals, which is a fluctuating line in Figure 14a. It can also be seen from Figure 14 that PIEAMIGA shows the better "climbing" (downhill) ability with the progress of evolution, and thus the gap between the two curves becomes increasingly larger. Finally, PIEAMIGA converges to the global optimal solution while SGA can only converge to the sub-optimal solution.

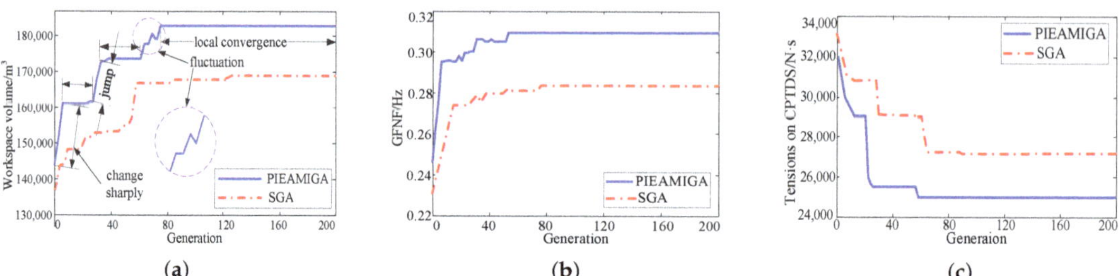

Figure 14. Comparison of the convergence process between IEPAMIGA and SGA. (**a**) Convergence of $f_1(\boldsymbol{D})$; (**b**) convergence of $f_2(\boldsymbol{D})$; and (**c**) convergence of $f_3(\boldsymbol{D})$.

Figure 15a shows that the workspace boundary extends outwards after using PIEAMIGA and that the volume increases by 14,425 m^3 compared to SGA. Figure 15b shows that the first-order natural frequency on the horizontal section of workspace that the height z is 6 m has significantly increased after using PIEAMIGA, and the GFNF is increased by 0.0452 Hz compared to SGA. Figure 15c shows that the impulse sum to the CPTDS has decreased significantly after using PIEAMIGA, and the impulse sum is decreased by 2792 N·s compared to SGA.

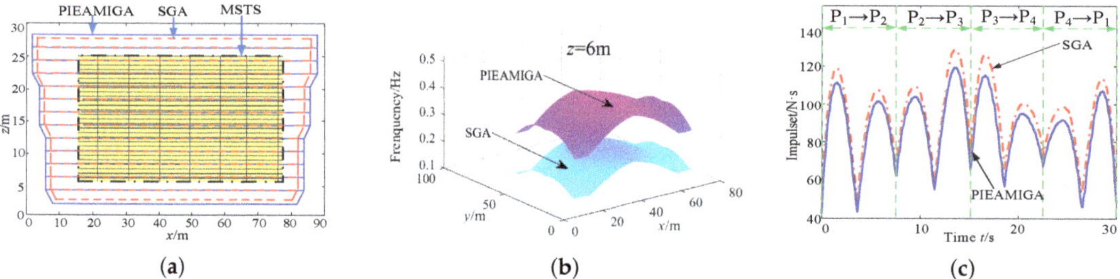

Figure 15. Comparison of the workspace, the first-order natural frequency and the impulse on CPTDS between PIEAMIGA and SGA. (**a**) Comparison of workspace; (**b**) comparison of the first-order natural frequency; and (**c**) comparison of the impulse on the CPTDS.

6.2.2. Multi-Objective Optimization Based on the Ideal Point Approach

In the previous sections, we performed structural optimization with regard to the three optimization objectives, respectively. However, we want to optimize the three optimization objectives simultaneously, i.e., the largest workspace, the highest first-order nature frequency and the smallest impulse on CPTDS exerted by tensions. Mathematically, it is a multi-objective optimization problem and will be solved by using the ideal-point method.

Since the physical meaning of the three optimization objectives is different, it is necessary to normalize the three optimization objectives $f_1(\boldsymbol{D})$, $f_2(\boldsymbol{D})$ and $f_3(\boldsymbol{D})$, respectively. Thus, the three optimization objects become continuous functions in the [0,1] interval, and the multi-objective function is uniform in dimension. The maximum and minimum values of each optimization objective need to be normalized.

As shown in Table 4, for $f_1(\boldsymbol{D})$, $f_{1,\min} = f'_1 = 84{,}875$ m^3, $f_{1,\max} = f^*_1 = 183{,}750$ m^3; for $f_2(\boldsymbol{D})$, $f_{2,\min} = f'_2 = 0.1042$ Hz, $f_{2,\max} = f^*_2 = 0.3100$ Hz; for $f_3(\boldsymbol{D})$, $f_{3,\min} = f^*_3 = 24{,}907$ N·s, $f_{3,\max} = f'_3 = 63{,}207$ N·s. Thus, we can find three objectives after normalization:

$$\begin{cases} \bar{f}_1(\boldsymbol{D}) = \dfrac{f_{1,\max}-f_1(\boldsymbol{D})}{f_{1,\max}-f_{1,\min}} \\ \bar{f}_2(\boldsymbol{D}) = \dfrac{f_{2,\max}-f_2(\boldsymbol{D})}{f_{2,\max}-f_{2,\min}} \\ \bar{f}_3(\boldsymbol{D}) = \dfrac{f_3(\boldsymbol{D})-f_{3,\min}}{f_{3,\max}-f_{3,\min}} \end{cases} \tag{37}$$

According to the definition of the ideal-point method, the minimum sum of the distances between the three objective function and the three ideal points is taken as the optimization objective, and the linear and nonlinear constraints proposed in previous are taken as the optimization constraints:

$$\begin{cases} \min \ \ W(\boldsymbol{D}) = \sum_{i=1}^{3} \left| \bar{f}_i(D) \right| \\ \text{s.t.} \ \ \ \boldsymbol{L} \leq \boldsymbol{D} \leq \boldsymbol{U} \\ g_1(\boldsymbol{D}) = H_{\min}(\boldsymbol{D}) - h_{\max} \leq 0 \\ g_2(\boldsymbol{D}) = -H_{\max}(\boldsymbol{D}) + h_{\min} \leq 0 \\ g_3(\boldsymbol{D}) = -V(\boldsymbol{D}) + \Omega \leq 0 \\ g_4(\boldsymbol{D}) = -\omega(\boldsymbol{D}) + \omega_p \leq 0 \end{cases} \tag{38}$$

6.3. Results and Discussion of Multi-Objective Structural Optimization

According to the simulation parameters listed in Tables 3 and 4, the multi-objective structural optimization is conducted by real-coded PIEAMIGA in the MATLAB environment. The results are listed in Table 5. It can be seen that the three optimization results are balanced, none of which is dominant, while others perform poorly.

Table 5. The results of multi-objective structural optimization.

Optimal Solution $\boldsymbol{D}^*$	$[83.27, 77.93, 27.46, 25.39]^{\mathrm{T}}$	
optimization objective	ideal point	optimization result
$f_1(\boldsymbol{D})/\mathrm{m}^3$	183,750	140,750
$f_2(\boldsymbol{D})/\mathrm{Hz}$	0.3100	0.2371
$f_3(\boldsymbol{D})/\mathrm{N\cdot s}$	24,907	30,653

Figure 16 shows the DFFW corresponding to the optimal solution D^*, which completely contains the MSTS, and thus the filming work can be successfully completed.

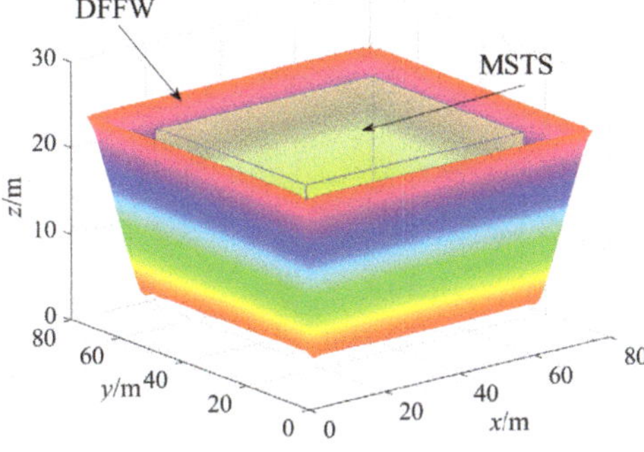

Figure 16. DFFW and MSTS with regard to the optimal solution D^*, i.e., *len* = 83.27 m, *wid* = 77.93 m, *hei* = 27.46 m and m_P = 25.39 kg.

Figure 17 and Table 6 illustrate the relationship of the workspace volume *vol* with the pulley point height *hei* and the mass of CPTDS m_p when the optimization variables d_1 and d_2 are fixed. In this case, *len* = d_1^* and *wid* = d_2^*. The pulley height *hei* is not positively correlated with volume of workspace *vol*. It can be known from the curved surface that the volume is small at both ends and large in the middle. It can be seen from Table 6 that the workspace volumes are about the same when the pulley height is equal to 27 and 28 m, and thus the maximum value is near $hei = 27.5$ m.

Generally speaking, the higher the pulley height hei is, the larger the volume is. However, the tension range is also a decisive factor of workspace. With the increase in height hei, the angle between the tangent line at the cable end and the direction of gravity decreases; thus, the cable tension must be increased to balance the gravity. The increase in height hei will cause an increase in the length and weight of the cable, and thus an increase in the cable tension. Thus, it is possible that the tensions exceed the upper limit h_{max}, leading to a decrease in the workspace that is mainly located in the area near the pulley point close to the upper surface of the workspace.

Relatively speaking, the relationship between the mass of CPTDS and the workspace volume is obvious, which has two characteristics: (1) Positive correlation. The light mass of CPTDS may give rise to the decrease in cable tension, or even less than the lower limit h_{min}. The reduced volume of workspace is mainly located at the boundaries of the workspace, where the area of the cable tension is relatively small. (2) At the beginning of the curve, the increase is obvious; however, at the end of the curve, the change is small.

When the mass m_p is small, increasing the mass of CPTDS will significantly enhance the cable tension so that many position points that originally do not meet the lower limit h_{min} satisfy the lower limit h_{min} again. Therefore, the volume of workspace increases. However, when the mass m_p increases to a certain extent, the minimum tension at most of the m_p points have been greater than h_{min}. Therefore, the increase in the workspace volume vol is limited.

Figure 18 shows the spatial distribution of the first-order natural frequency corresponding to the "ideal point" on three horizontal sections of the workspaces at different heights, i.e., $z = 5$ m, $z = 15$ m and $z = 25$ m. Clearly, the higher the height along the z axis is, the higher the natural frequency is. Since the camera robot belongs to the suspended CDPM, the four pulley points are all above the CPTDS. Therefore, the higher the height along z axis is, the shorter the cable length is. According to Equation (25), the shorter the cable length, the higher the natural frequency of the camera robot. Similarly, there is always a long cable when the CPTDS at one of the four corners of the workspace; hence, the first-order natural frequency of the camera robot will be smaller, leading to a downward sharp corner.

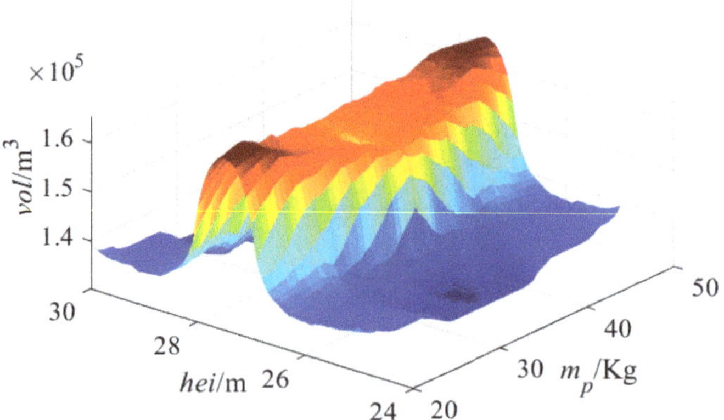

Figure 17. Relationship of the workspace volume vol with the pulley point height hei and the mass of CPTDS m_p when $len = d_1^* = 83.27$ m and $wid = d_2^* = 77.93$ m.

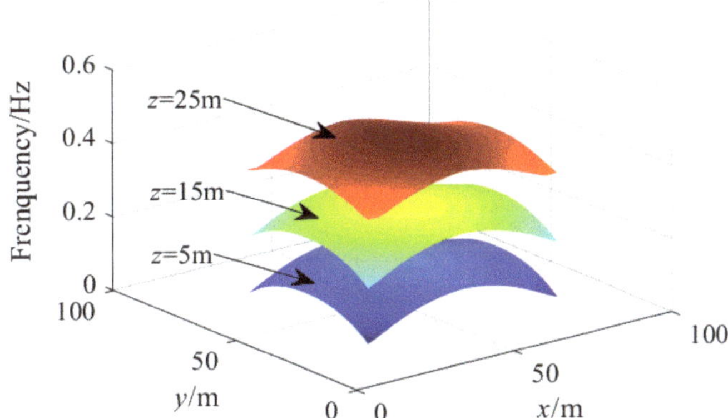

Figure 18. Distribution of first-order natural frequency on the horizontal sections of the workspaces at different heights *hei* for the optimal solution D^*, i.e., *len* = 83.27 m, *wid* = 77.93 m, *hei* = 27.46 m and m_p = 25.39 kg.

Table 6. The workspace volume when the pulley point height *hei* and the mass of CPTDS m_p vary (unit m^3).

Height \ Mass	20/kg	30/kg	40/kg	50/kg
25/m	138,570	132,200	132,000	139,300
26/m	136,754	134,200	134,200	137,000
27/m	165,300	159,700	160,000	153,700
28/m	159,700	160,000	159,800	158,000
29/m	137,000	133,660	132,800	137,000
30/m	135,300	132,200	132,200	139,300

Figure 19 shows the minimum value of the first-order natural frequency on the different horizontal sections of the workspace with different heights, which are significantly improved compared with those shown in Figure 10. The minimum first-order natural frequency on each horizontal section all exceeds 0.12 Hz. At z = 5 m, the minimum first-order natural frequency is equal to 0.13 Hz. Although it has been improved, it is still close to the energy truncation frequency ω_p = 0.12 Hz. Therefore, it is still vulnerable to wind disturbance. Thus, some measures are essential to enhance the first-order natural frequency of the camera robot, such as the stiffness improvement and tension regulation.

Figure 20 and Table 7 show the relationship of GFNF with the pulley point height *hei* and the mass of CPTDS m_P when the optimization variables d_1 and d_2 are fixed. In this case, *len* = d_1^* and *wid* = d_2^*. The relationship between GFNF and *hei* is not proportional, but exhibits the characteristics of "high at both ends, low in the middle". It can be concluded from Equation (21) that the critical factor of determining the GFNF is the length of the cable. On the one hand, the smaller the height *hei* is, the smaller the length of the cable is. Thus, the frequency is larger when *hei* = 25 m; However, on the other hand, the decrease in height *hei* will lead to the decrease in the height of workspace, lead to losing some points in workspace.

Since the cable length at these "losing point" is shorter, the first-order natural frequency is larger. Therefore, the whole level of the natural frequency in the workspace will decrease after losing these points. Hence, the contradiction between them determines that the value of GFNF is larger when *hei* is equal to the maximum or minimum height and smaller when *hei* is in the middle height. The relationship between the mass of CPTDS m_p and GFNF is very obvious. The greater the mass is, the lower the frequency is. We can observed that the GFNFs are the same when the pulley height at 27 m and 28 m. We can also draw the conclusion that the GFNF is more sensitive to the pulley height than the mass of the CPTDS.

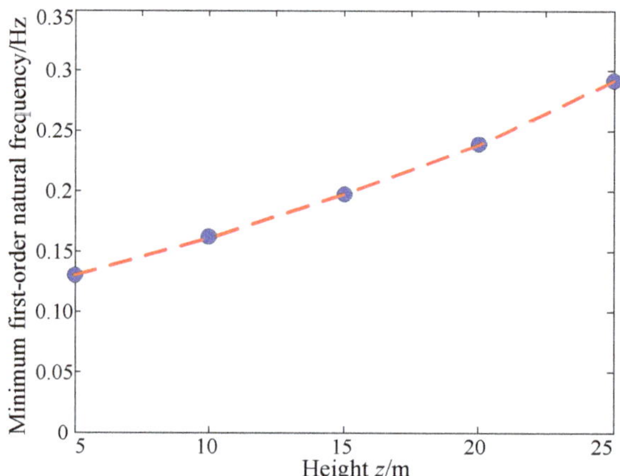

Figure 19. The minimum first-order natural frequency on the horizontal planes of workspace at different heights with regard to the optimal solution $\boldsymbol{D}^*$, i.e., $len = d_1^* = 83.27$ m, $wid = d_2^* = 77.93$ m, $hei = d_3^* = 27.46$ m and $m_p = d_1^* = 25.39$ kg.

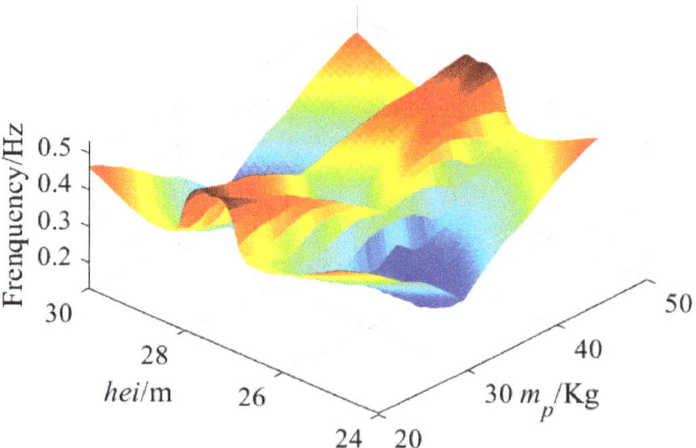

Figure 20. Relationship of GFNF with the pulley height hei and the mass of CPTDS m_p when $len = d_1^* = 83.27$ m and $wid = d_2^* = 77.93$ m.

Table 7. The frequency when the pulley point height hei and the mass of CPTDS m_p vary (unit Hz).

Height \ Mass	20/kg	30/kg	40/kg	50/kg
25/m	0.4461	0.4313	0.4313	0.4461
26/m	0.4398	0.4241	0.4241	0.4398
27/m	0.4479	0.4422	0.4422	0.4479
28/m	0.4479	0.4422	0.4422	0.4479
29/m	0.4398	0.4241	0.4241	0.4398
30/m	0.4455	0.4313	0.4313	0.4461

Figure 21 shows the impulse applied on the CPTDS and the accelerations of the CPTDS corresponding to the optimization solution following the sampling locus shown in Figure 11. Since every straight line segment of the locus is planned by a quintic polynomial, the acceleration curve is a cubic curve, and the value of acceleration at each point change motion direction of the locus is 0. It can be seen that the impulse is larger in the time point with larger acceleration, reflecting the tensions exert a large impulse on the CPTDS at this time point. The maximum value appears at t = 15 s, when the CPTDS moves near the point P_3 and is close to the lower surface of the workspace.

Therefore, the cable length and tension are larger than those at the rest points and thus the impulse. It can be observed that the curve is not completely symmetrical on both sides with P_3 as the center. However, the left side is larger and the right side is smaller. As the segment $P_2 \rightarrow P_3$ is on the left side of P_3 and the motion direction is consistent with the direction of gravity, the tension is larger; the motion direction of the segment $P_3 \rightarrow P_4$ on the right side of P_3 is contrary to the gravity direction, and therefore the tension is smaller. Thus, the impulse on the segment $P_2 \rightarrow P_3$ is larger than that on the segment $P_3 \rightarrow P_4$. Similarly, the impulse applied on the CPTDS on the segment $P_1 \rightarrow P_2$ is larger than that on the segment $P_4 \rightarrow P_1$.

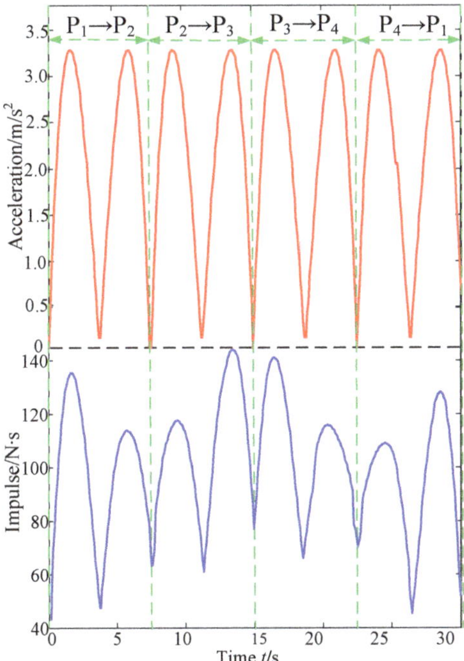

Figure 21. The impulse applied on the CPTDS and acceleration of the CPTDS corresponding to the optimization solution following the sampling locus with regard to the optimal solution D^*, i.e., $len = d_1^* = 83.27$ m, $wid = d_2^* = 77.93$ m, $hei = d_3^* = 27.46$ m, $m_P = d_1^* = 25.39$ kg.

Figure 22 and Table 8 show the relationship of the impulse sum on the CPTDS along the sampling locus with the pulley point height *hei* and the mass of CPTDS m_p when the optimization variables d_1 and d_2 are fixed. In this case, $len = d_1^*$ and $wid = d_2^*$. There is a positive correlation of the impulse sum with the pulley point height *hei* and the mass of CPTDS m_p. The higher the pulley point height *hei* and the greater the mass m_p of the CPTDS are, the greater the impulse sum will be. In addition, it can be observed from Table 8 that the impulse sum linearly increases with increasing of the pulley point height and the mass of the CPTDS.

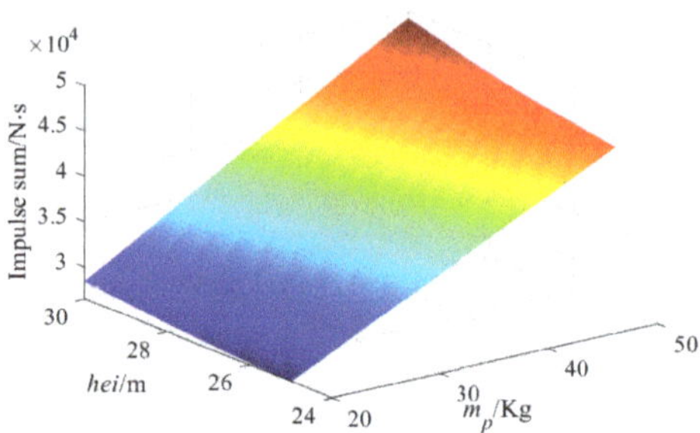

Figure 22. Relationship of impulse sum on the CPTDS along the sampling locus with the pulley point height *hei* and the mass of CPTDS m_p when $len = d_1^* = 83.27$ m and $wid = d_2^* = 77.93$ m.

Table 8. The impulse sum on the CPTDS when the pulley point height *hei* and the mass of CPTDS m_p vary (unit N·s).

Height \ Mass	20/kg	30/kg	40/kg	50/kg
25/m	26,450	32,330	38,220	44,110
26/m	26,740	32,780	38,820	44,860
27/m	27,080	33,300	39,520	45,750
28/m	27,460	33,880	40,310	46,750
29/m	27,930	34,590	41,270	47,940
30/m	28,430	35,370	42,220	49,260

7. Conclusions

In this paper, the dynamic modeling of the camera robot was presented considering the self-weight and inertia of the cable simultaneously, which was also fit to other large-span and high-speed manipulators, and a tension distribution algorithm was developed based on the iteration method. According to the dynamical model, an approach of generating a dynamic feasible workspace was proposed. In order to improve the shooting ability of the camera robot, the structure of the camera was optimized by optimizing the three objectives separately and then mixed together using the ideal point method through a modified GA. Furthermore, the characteristics of the objectives were analyzed by varying the design parameters. The main contributions of this study include the following:

First, the dynamic model of the large-span high-speed camera robot with redundant actuation combining with the cable mass and inertia is established. Based on this model, an iterative-based tension distribution algorithm (Algorithm 1) was proposed to determine the tensions in cables given the position of the CPTDS. In addition, a dynamic force-feasible workspace (DFFW) generation algorithm (Algorithm 2) was proposed according to the characteristics of the camera robot based on the judging conditions of DFFW that contain the direction balance condition (DBC) and the magnitude balance condition (MBC) simultaneously. In this study, the three-DOF four-cable-driven camera robot was considered for illustrative case-studies. The presented algorithms were applicable to any cable-driven parallel manipulators while modifying the Jacobian matrix J, the generalized external force M, the unit normal vector q and the velocity–acceleration pair $.

Secondly, an optimization model of the camera robot was set up aiming to achieve the best workspace volume of DFFW, anti-wind disturbance ability and impulse exerted by tensions on CPTDS, where the anti-wind disturbance ability can be evaluated through the GFNF (global first-order nature frequency).

Thirdly, the multi-island and information entropy ideas were used to improve the SGA. Thus, an improved genetic algorithm was created in order to optimize the structure of the camera robot, namely the population information entropy-based adaptive genetic algorithm (PIEAMIGA). It can be seen from Figures 14 and 15 that PIEAMIGA offered the stronger global search capability and the better optimization results compared with SGA with regard to each optimization objective while the computing time increased slightly as shown in Table 4.

Fourthly, the ideal-point method was employed to deal with the multi-objective structural optimization to avoid the influence of the subjective intention of the designer. The ideal point of each optimization objective was obtained by conducting single-objective structural optimization through PIEAMIGA. The optimization results with regard to the optimal solution D^* by using PIEAMIGA are shown in Table 5 and demonstrate that the three optimization results were balanced. Figures 16, 18, 19 and 21 illustrate the workspace shape, spatial distribution of first-order natural frequency, the minimum first-order natural frequency on different horizontal planes of workspace and the impulse on the CPTDS when the optimization variables are equal to the optimal solution D^*.

Moreover, the relationship of the three optimization objectives with the pulley height *hei* and the mass of CPTDS m_p were studied as illustrated in Figures 17, 20 and 22 as well as Tables 6–8. Moreover, some laws were summarized, and these will be valuable for the design of camera robots. Furthermore, the design methodology presented in this paper can be extended to other classes of large-span high-speed CDPMs as well. Apart from the camera robot, the dynamic workspace generation approach and design methodology presented in this paper can be extended to other classes of large-span high-speed CDPMs as well. The next steps involve dynamic modeling for the complex spatial robot architectures and finally practical testing. Our future research will be focused on the motion control issue based on the dynamical model in this paper.

Author Contributions: Conceptualization, Y.S. and Y.Q.; methodology, Y.S.; software, P.L.; validation, Y.S. and P.L.; formal analysis, Y.Q.; investigation, P.L.; resources, Q.W.; data curation, X.W.; writing—original draft preparation, Y.S.; writing—review and editing, P.L.; supervision, Y.Q.; project administration, Y.S.; and funding acquisition, Y.S. and J.T. All authors have read and agreed to the published version of the manuscript.

Funding: This research was funded by Science and Technology Project of Shaanxi Province of China (Grant numbers: 2020JM-558, 2020GY-188 and 2015KTZDGY-02-01).

Institutional Review Board Statement: Not applicable.

Informed Consent Statement: Not applicable.

Data Availability Statement: Not applicable.

Conflicts of Interest: The authors declare no conflict of interest.

References

1. Su, Y.; Qiu, Y.Y.; Sheng, Y. Optimal cable tension distribution of the high-speed redundant driven camera robots considering cable sag and inertia Effects. *Adv. Mech. Eng.* **2014**, *2014*, 729020. [CrossRef]
2. Su, Y.; Qiu, Y.Y.; Liu, P. The continuity and real-time performance of the cable tension determining for a suspend cable-driven parallel camera robot. *Adv. Robot.* **2015**, *29*, 743–752. [CrossRef]
3. Bai, G.; Ge, Y.; Scoby, D.; Leavitt, B.; Stoerger, V.; Kirchgessner, N.; Irmak, S.; Graef, G.; Schnable, J.; Awada, T. NU-Spidercam: A large-scale, cable-driven, integrated sensing and robotic system for advanced phenotyping, remote sensing, and agronomic research. *Comput. Electron. Agric.* **2019**, *160*, 71–81. [CrossRef]
4. Yu, X.D.; Duan, L.; Tian, Q. Highway traffic information extraction from Skycam MPEG video. In Proceedings of the 2002 IEEE fifth International Conference on Intelligent Transportation Systems, Singapore, 3–6 September 2002; pp. 37–42. [CrossRef]
5. Uva, M. *The Grip Book*, 4th ed.; Elsevier: Burlington, MA, USA, 2010; Chapter Cablecam, pp. 295–296. [CrossRef]
6. Wei, H.L.; Qiu, Y.Y.; Sheng, Y. On the Cable Pseudo-Drag Problem of Cable-Driven Parallel Camera Robots at High Speeds. *Robotica* **2019**, *37*, 1695–1709. [CrossRef]
7. Wei, H.L.; Qiu, Y.Y.; Sheng, Y. An approach to evaluate stability for cable-driven parallel camera robots with hybrid tension-stiffness properties. *Int. J. Adv. Robot. Syst.* **2015**, *12*, 1–12. [CrossRef]
8. Liu, P.; Qiu, Y.Y.; Sheng, Y. A new hybrid force-position measure approach on the stability for a camera robot. *J. Mech. Eng. Sci.* **2016**, *230*, 2508–2516. [CrossRef]
9. Ouyang, B.; Shang, W.W. Wrench-feasible workspace based optimization of the fixed and moving platforms for cable-driven parallel manipulators. *Robot. Comput.-Integr. Manuf.* **2014**, *30*, 629–635. [CrossRef]

10. Fattah, A.; Agrawal, S.K. On the design of cable-suspended planar parallel robots. *J. Mech. Des.* **2005**, *127*, 1021–1028. [CrossRef]
11. Zhang, Z.K.; Shao, Z.F.; Wang, L.P. Optimization and implementation of a high-speed 3-DOFs translational cable-driven parallel robot. *Mech. Mach. Theory* **2020**, *145*, 103693. [CrossRef]
12. Hussein, H.; Santo, J.C.; Izard, J.B.; Gouttefarde, M. Smallest maximum cable tension determination for cable-driven parallel robots. *IEEE Trans. Robot.* **2021**, *34*, 1186–1205. [CrossRef]
13. Hamed, J.; Khajepour, A.; Fidan, B.; Rushton, M. Kinematically-constrained redundant cable-driven parallel robots: Modeling, redundancy analysis, and stiffness optimization. *IEEE/ASME Trans. Mechatron.* **2017**, *22*, 921–930. [CrossRef]
14. Anson, M.; Alamdari, A.; Venkat, K. Orientation workspace and stiffness optimization of cable-driven parallel manipulators with base mobility. *J. Mech. Robot.* **2017**, *9*, 031011. [CrossRef]
15. Cui, Z.W.; Tang, X.Q.; Hou, S.H.; Sun, H.N.; Wang, D.J. Optimization design of redundant cable driven parallel robots based on constant stiffness space. In Proceedings of the 2019 IEEE International Conference on Robotics and Biomimetics (ROBIO), Dali, China, 6–8 December 2019; pp. 1041–1046. [CrossRef]
16. Otis, M.J.; Perreault, S.; Nguyen-Dang, T.; Lambert, P.; Gouttefarde, M.; Laurendeau, D.; Gosselin, C. Determination and management of cable interferences between two 6-DOF foot platforms in a cable-driven locomotion interface. *IEEE Trans. Syst. Man, Cybern. Part A Syst. Humans* **2009**, *39*, 528–544. [CrossRef]
17. Otis, J.D.; Youssef, K. Reconfigurable fully constrained cable driven parallel mechanism for avoiding interference between cables. *Mech. Mach. Theory* **2020**, *148*, 103781. [CrossRef]
18. Gallina, P.; Rosati, G. Manipulability of a planar wire driven haptic device. *Mech. Mach. Theory* **2002**, *37*, 215–228. [CrossRef]
19. Gouttefarde, M.; Collard, J.F.; Nicolas, R.; Baradat, C. Geometry selection of a redundantly actuated cable-suspended parallel robot. *IEEE Trans. Robot.* **2017**, *31*, 501–510. [CrossRef]
20. Li, Y.; Xu, Q. GA-based multi-objective optimal design of a planar 3-DOF cable-dDriven parallel manipulator. In Proceedings of the 2006 IEEE International Conference on Robotics and Biomimetics, Kunming, China, 17–20 December 2006; pp. 1360–1365. [CrossRef]
21. Jamwal, P.K.; Xie, S.Q.; Aw, K.C. Multi criteria optimal design of cable driven ankle rehabilitation robot. In *Mobile Robots-State of the Art in Land, Sea, Air, and Collaborative Missions*; InTech: Rijeka, Croatia, 2009; Volume 1, Chapter 14, pp. 303–336. [CrossRef]
22. Arsenault, M. Optimization of the prestress stable wrench closure workspace of planar parallel three-degree-of-freedom cable-driven mechanisms with four cables. In Proceedings of the 2010 IEEE International Conference on Robotics and Automation, Anchorage, AK, USA, 3–7 May 2010; pp. 1182–1187. [CrossRef]
23. Bahrami, A.; Bahrami, M.N. Optimal design of a spatial four cable driven parallel manipulator. In Proceedings of the 2011 IEEE International Conference on Robotics and Biomimetics, Karon Beach, Thailand, 7–11 December 2011; pp. 2143–2149. [CrossRef]
24. Amine, L.M.; Giuseppe, C.; Said, Z. On the optimal design of cable driven parallel robot with a prescribed workspace for upper limb rehabilitation tasks. *J. Bionic Eng.* **2019**, *16*, 503–513. [CrossRef]
25. Hamida, I.B.; Laribi, M.A.; Mlika, A.; Romdhane, L.; Carbone, G. Multi-objective optimal design of a cable driven parallel robot for rehabilitation tasks. *Mech. Mach. Theory* **2021**, *156*, 104141. [CrossRef]
26. Riehl, N.; Gouttefarde, M.; Krut, S.; Baradat, C.; Pierrot, F. Effects of non-negligible cable mass on the static behavior of large workspace cable-driven parallel mechanisms. In Proceedings of the 2011 IEEE International Conference on Robotics & Automation, Karon Beach, Thailand, 7–11 December 2011; pp. 2143–2149. [CrossRef]
27. Dallej, T.; Gouttefarde, M.; Andreff, N.; Hervé, P.; Martinet, P. Modeling and vision-based control of large-dimension cable-driven parallel robots using a multiple-camera setup. *Mechatronics* **2019**, *61*, 20–36. [CrossRef]
28. Gouttefarde, M.; Collard, J.; Riehl, N.; Baradat, C. Simplified static analysis of large-dimension parallel cable-driven robots. In Proceedings of the 2012 IEEE International Conference on Robotics and Automation, Saint Paul, MN, USA, 14–18 May 2012; pp. 2299–2305. [CrossRef]
29. Nguyen, D.Q.; Gouttefarde, M.; Company, O.; Pierrot, F. On the simplifications of cable model in static analysis of large-dimension cable-driven parallel robots. In Proceedings of the 2013 IEEE/RSJ International Conference on Intelligent Robots and Systems, Tokyo, Japan, 3–7 November 2013; pp. 928–934. [CrossRef]
30. Jiang, X.L.; Huang, Y.J.; Zheng, Y.Q. Dimension optimization design of an under-restrained 6-DOF four-cable-driven parallel manipulator based on least square-support vector regression. In Proceedings of the 2011 IEEE International Conference on Mechatronics and Automation, Beijing, China, 7–10 August 2011; pp. 458–463. [CrossRef]
31. Yao, R.; Tang, X.; Wang, J.; Huang, P. Dimensional optimization design of the four-cable-driven parallel manipulator in FAST. *IEEE/ASME Trans. Mechatron.* **2010**, *15*, 932–941. [CrossRef]
32. Tang, X.Q.; Yao, R. Dimensional design on the six-cable driven parallel manipulator of FAST. *J. Mech. Des.* **2011**, *133*, 111012. [CrossRef]
33. Du, J.; Bao, H.; Cui, C. Stiffness and dexterous performances optimization of large workspace cable-driven parallel manipulators. *Adv. Robot.* **2014**, *28*, 187–196. [CrossRef]
34. Wei, H.L.; Qiu, Y.Y.; Luo, L.F.; Lu, Q.H. An approach on stability analysis of cable-driven parallel robots considering cable mass. *AIP Adv.* **2021**, *11*, 055014. [CrossRef]
35. Yu, L.L.; Qiu, Y.Y.; Su, Y. Dynamic workspace of a high-speed cable-driven camera robot. *Eng. Mech.* **2013**, *30*, 245–250.
36. Barrette, G.; Gosselin, C.M. Determination of the dynamic workspace of cable-driven planar parallel mechanisms. *J. Mech. Des.* **2005**, *127*, 242–248. [CrossRef]

37. Kawamura, S.; Kino, H.; Won, C. High-speed manipulation by using parallel wire-driven robots. *Robotica* **2000**, *18*, 13–21. [CrossRef]

38. Lenarčič, J.; Merlet, J.P. (Eds.) Determination of a dynamic feasible workspace for cable-driven parallel robots. In *Advances in Robot Kinematics 2016*; Springer Nature: Cham, Switzerland, 2009; Volume 4, Chapter 35, pp. 361–370. [CrossRef]

39. Kieu, V.N.D.; Huang, S.C. Dynamic and Wrench-Feasible Workspace Analysis of a Cable-Driven Parallel Robot Considering a Nonlinear Cable Tension Model. *Appl. Sci.* **2022**, *12*, 244. [CrossRef]

40. Su, Y.; Qiu, Y.Y.; Wei, H.L. Dynamic modeling and tension optimal distribution of cable-driven parallel robots considering cable mass and inertia force effects. *Eng. Mech.* **2016**, *33*, 231–239. [CrossRef]

41. Verhoeven, R. Analysis of the Workspace of Tendon-Based Stewart Platforms. Ph.D. Thesis, University of Duisburg-Essen, Duisburg, Germany, 2004.

42. Irivne, H.M. *Cable Structures*, 1st ed.; The MIT Series in Structural Mechanics; MIT Press: Cambridge, MA, USA, 1981. [CrossRef]

43. Diao, X.; Ou, M. A method of verifying force-closure condition for general cable manipulators with seven cables. *Mech. Mach. Theory* **2007**, *42*, 1563–1576. [CrossRef]

44. Gouttefarde, M.; Daney, D.; Merlet, J.P. Interval-analysis-based determination of the wrench-feasible workspace of parallel cable-driven robots. *IEEE Trans. Robot.* **2011**, *27*, 1–13. [CrossRef]

45. Liu, X.; Qiu, Y.Y.; Sheng, Y. Proofs of Existence Conditions for Workspaces of Wire-driven parallel robots and a uniform solution strategy for the workspaces. *J. Mech. Eng.* **2010**, *46*, 27–34. [CrossRef]

46. Qiu, Y.Y.; Duan, B.Y.; Wei, Q.; Nan, R.D.; Peng, B. Optimal distribution of the cable tensions and structural vibration control of the cable-cabin flexible structure. *Struct. Eng. Mech.* **2002**, *14*, 39–56. [CrossRef]

47. Qiu, Y.Y.; Chen, J.; Duan, B.Y. Simulation of the Wind Induced Vibration of the Suspended Feed Cabin in the largest Radio Telescope under random force. *Chin. J. A Mech.* **2000**, *17*, 91–95. [CrossRef]

48. Diao, X.M.; Ma, O. Vibration analysis of cable-driven parallel manipulators. *Multibody Syst. Dyn.* **2009**, *21*, 347–360. [CrossRef]

49. Liu, Z.H.; Tang, X.Q.; Shao, Z. Research on longitudinal vibration characteristic of the six-cable-driven parallel manipulator in FAST. *Adv. Mech. Eng.* **2013**, *2013*, 547416. [CrossRef]

50. Zhang, W.; Qi, H.; Yu, Z.Q.; He, M.J.; Ren, Y.T.; Li, Y. Optimization configuration of selective solar absorber using multi-island genetic algorithm. *Sol. Energy* **2021**, *224*, 947–955. [CrossRef]

51. Srinivas, M.; Patnaik, L.M. Adaptive probabilities of crossover and mutation in genetic algorithms. *IEEE Trans. Syst. Man Cybern.* **2002**, *24*, 656–667. [CrossRef]

52. Shannon, C.E. A mathematical theory of communication. *Bell Syst. Tech. J.* **1948**, *27*, 379–423. [CrossRef]

53. Khakpour, H.; Birglen, L.; Tahan, S.A. Analysis and optimization of a new differentially driven cable parallel robot. *J. Mech. Robot.* **2015**, *7*, 034503. [CrossRef]

54. Sun, H.; Hou, S.; Li, Q.; Tang, X. Research on the configuration of cable-driven parallel robots for vibration suppression of spatial flexible structures. *Aerosp. Sci. Technol.* **2021**, *109*, 106434. [CrossRef]

55. Abdolshah, S.; Zanotto, D.; Rosati, G.; Agrawal, S.K. Optimizing Stiffness and Dexterity of Planar Adaptive Cable-Driven Parallel Robots. *J. Mech. Robot.* **2017**, *9*, 031004. [CrossRef]

56. Zhang, F.; Shang, W.; Zhang, B.; Cong, S. Design optimization of redundantly actuated cable-driven parallel robots for automated Warehouse System. *IEEE Access* **2020**, *8*, 56867–56879. [CrossRef]

Article

Kinematic and Dynamic Modeling and Workspace Analysis of a Suspended Cable-Driven Parallel Robot for Schönflies Motions

Ruobing Wang , Yanlin Xie, Xigang Chenand Yangmin Li *

Department of Industrial and Systems Engineering, The Hong Kong Polytechnic University, Kowloon, Hong Kong SAR, China; ruobing.wang@connect.polyu.hk (R.W.); yanlin.xie@connect.polyu.hk (Y.X.); xigang.chen@connect.polyu.hk (X.C.)
* Correspondence: yangmin.li@polyu.edu.hk

Abstract: In recent years, cable-driven parallel robots (CDPRs) have drawn more and more attention due to the properties of large workspace, large payload capacity, and ease of reconfiguration. In this paper, we present a kinematic and dynamic modeling and workspace analysis for a novel suspended CDPR which generates Schönflies motions. Firstly, the architecture of the robot is introduced, and the inverse and forward kinematic problems of the robot are solved through a geometrical approach. Then, the dynamic equation of the robot is derived by separately considering the moving platform and the drive trains. Based on the dynamic equation, the dynamic feasible workspace of the robot is determined under different values of accelerations. Finally, experiments are performed on a prototype of the robot to demonstrate the correctness of the derived models and workspace.

Keywords: cable-driven parallel robot; kinematics; dynamics; workspace

Citation: Wang, R.; Xie, Y.; Chen, X.; Li, Y. Kinematic and Dynamic Modeling and Workspace Analysis of a Suspended Cable-Driven Parallel Robot for Schönflies Motions. *Machines* **2022**, *10*, 451. https://doi.org/10.3390/machines10060451

Academic Editors: Dan Zhang, Zhufeng Shao and Stéphane Caro

Received: 23 May 2022
Accepted: 2 June 2022
Published: 6 June 2022

Publisher's Note: MDPI stays neutral with regard to jurisdictional claims in published maps and institutional affiliations.

1. Introduction

Cable-driven parallel robots (CDPRs) are particular types of parallel robots in which the rigid kinematic chains are replaced by flexible cables [1]. In recent years, CDPRs have drawn more and more attention due to the properties of large workspace, large payload capacity, and ease of reconfiguration [2,3]. CDPRs are divided into the suspended type and the fully constrained type according to the tensioning methods. For the suspended CDPRs, the cables are all located over the moving platform, and the gravity of the moving platform is essential to keep the cables in tension. Since the space below the moving platform is free of cables, the suspended CDPRs are generally used to obtain a large workspace. The suspended CDPRs usually work under static conditions because of the limited gravity [4]. For the fully constrained CDPRs, the cables are located on both sides of the moving platform and pull against each other to ensure positive cable tensions. The fully constrained CDPRs are suitable to generate high-speed motions with large accelerations [5].

The existence of the flexible cables greatly complicates the modeling and analysis of CDPRs. On the one hand, the elasticity of the cables leads to low stiffness and deteriorates the positioning accuracy of CDPRs. On the other hand, resulting from the unilateral property, the cables can only apply pull forces and cannot provide push forces. Thus, a greater number of cables than the degrees of freedom (DOFs) are generally required to fully control CDPRs [6]. Many recent works have contributed to the modeling and analysis of CDPRs. In [7], the authors present the kinetostatic model of a 3-DOF CDPR involving pulley kinematics and cable elasticity. Meanwhile, a novel pulley structure is proposed to improve the positioning accuracy of CDPRs. The kinematic model of CDPRs considering pulley mechanisms is presented in [8]. Based on the kinematic model, a kinematic calibration method is further developed for CDPRs. In [9], the forward kinetostatic problem of a 6-DOF underactuated CDPR is solved using an unsupervised neural network approach. This novel approach is computationally efficient and has an accuracy similar to that of other optimization methods. In [10], the authors provide a kinematic and dynamic analysis of a

CDPR composed of multiple cranes by considering the hydraulic actuation system. A 3-DOF CDPR driven by five-bar winding mechanisms for workspace expansion is presented in [11]. The kinematic and dynamic analysis of the CDPR is performed based on the finite element method and the Lagrange formulation. In [12], the dynamic model of CDPRs is derived by combining the finite element model of the cables with the rigid-body dynamics of the moving platform.

Another important issue is determining the workspace of CDPRs, as the workspace is one of the decisive properties for the potential applications of CDPRs. The workspace of CDPRs is strongly coupled to the cable tensions because of the unilateral property of the cables [13]. A generalized ray-based method is proposed in [14] to solve the wrench-closure workspace of CDPRs. The authors further introduce a graph representation to visualize the high-dimensional workspace. In [15], the wrench-feasible workspace of CDPRs is computed through interval analysis, in which the sets fully inside or outside the workspace can be efficiently determined. The dynamic feasible workspace of a 6-DOF CDPR is solved by investigating the dynamic equilibrium of the moving platform in [16]. The authors in [17,18] use a geometric method to calculate the cylindrical operation workspace of a 3-DOF CDPR tensioned by passive springs. The relation between the number of springs and the shape of the workspace is investigated. In [19], a geometric approach based on convex analysis is proposed to compute the workspace of CDPRs subject to the constraints of the cable tensions. In [20], the author proposes the differential workspace hull method to calculate the workspace of CDPRs. This method adopts a triangulation representation to approximate the boundary of the workspace and can handle various criteria for the workspace.

Although much effort has been devoted to the modeling and analysis of CDPRs, the existing works mainly focus on the purely translational 3-DOF CDPRs [7,11,17,18] and the spatial 6-DOF CDPRs [9,12,16,20]. The CDPRs with other types of motions have seldom been reported, and there is no systematic approach for the modeling and analysis of such kinds of CDPRs. In many working scenarios, not all directions of motions are essential for the application requirements, and the redundant DOFs may increase the cost and complicate the robot systems [21]. Therefore, there is an urgent need to investigate the lower-mobility CDPRs with sub-spatial motions. Among the lower-mobility motion types, Schönflies motion, which contains three-dimensional translation and one-dimensional rotation about the vertical axis, is the most widely used kind of sub-spatial motion in robotics [22]. Aiming to narrow the research gap and complement the field of lower-mobility CDPRs, in this paper, we present a kinematic and dynamic modeling and workspace analysis for a novel suspended CDPR which generates Schönflies motions. In comparison with the existing works, the main contributions of this paper are summarized as follows:

1. The structure of the novel CDPR which generates Schönflies motions is introduced to complement the field of lower-mobility CDPRs.
2. The closed-form solutions for the inverse and forward kinematics of the robot are derived based on a geometrical approach.
3. The dynamic model of the robot is formulated based on the virtual power principle, which lays the foundation for the workspace determination and the model-based control of the robot.
4. The dynamic feasible workspace of the robot is determined under different values of accelerations, which facilitates the motion planning and control of the robot.

The remaining parts of this paper are arranged as follows. The architecture of the novel CDPR is introduced in Section 2. Then, the inverse and forward kinematic problems of the robot are solved in Section 3. Section 4 presents the dynamic modeling of the robot by separately considering the moving platform and the drive trains. Section 5 determines the dynamic feasible workspace of the robot under different values of accelerations. The prototype and experiments of the robot are presented in Section 6. Finally, Section 7 concludes this paper and discusses suggestions for future work.

2. Architecture Description

The robot studied in this paper is a novel suspended CDPR which generates Schönflies motions. A prototype of the robot is shown in Figure 1. In this section, we briefly introduce the architecture of the robot and define some notations used in later sections.

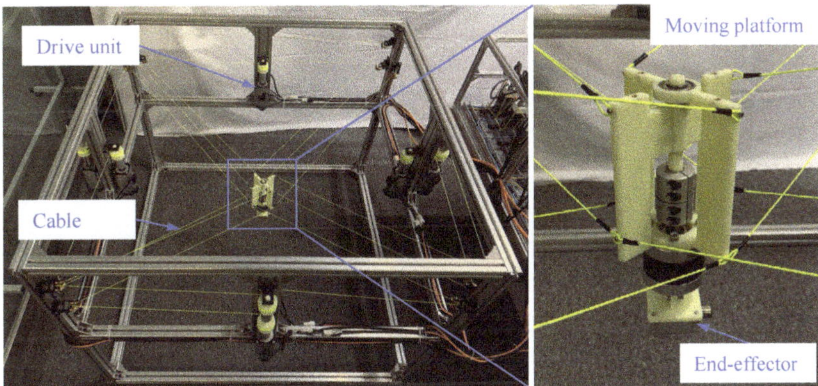

Figure 1. Prototype of the novel cable-driven parallel robot (CDPR).

2.1. Cable Arrangement

Figure 2 shows the kinematic diagram of the novel CDPR studied in this paper. The robot has twelve driving cables linking the base and the moving platform. The cable attachment points of the robot are uniformly arranged at the vertices of the base and the moving platform. The base of the robot is cuboid-shaped, and its size is determined by three parameters: L_{A1}, L_{A2}, and L_{A3}. Let $\mathcal{T}_A$ be the inertial frame located at the geometric center point O of the base. The cable attachment points on the base are named A_{ij}, where $i \in \{1, 2, 3, 4\}$ and $j \in \{1, 2, 3\}$. The locations from point O to points A_{ij} are represented by vectors $\boldsymbol{a}_{ij}$. The moving platform of the robot is composed of three parts which are articulated together. The size of the moving platform is depicted by three parameters: L_{B1}, L_{B2}, and ΔL_B. Let $\mathcal{T}_B$ be the moving frame attached at the geometric center point P of the moving platform. The vector linking point O and point P are denoted as $\boldsymbol{p} = (x \ y \ z)^T$. The cable attachment points on the moving platform are named B_{ij}. The locations from point P to points B_{ij} are represented by vectors $\boldsymbol{b}_{ij}$. The cables linking points A_{ij} and points B_{ij} have the lengths l_{ij}, and their directions are represented by unit length vectors $\boldsymbol{s}_{ij}$. Based on the above defined parameters, the positions of the cable attachment points of the robot are given as

$$
\begin{aligned}
&\boldsymbol{a}_{11} = \left(-\tfrac{1}{2}L_{A1} \quad -\tfrac{1}{2}L_{A2} \quad \tfrac{1}{2}L_{A3}\right)^T, \quad \boldsymbol{a}_{12} = \boldsymbol{a}_{33} = \left(-\tfrac{1}{2}L_{A1} \quad -\tfrac{1}{2}L_{A2} \quad \tfrac{1}{2}L_{A3} - L_{B2}\right)^T, \\
&\boldsymbol{a}_{21} = \left(\tfrac{1}{2}L_{A1} \quad -\tfrac{1}{2}L_{A2} \quad \tfrac{1}{2}L_{A3}\right)^T, \quad \boldsymbol{a}_{22} = \boldsymbol{a}_{43} = \left(\tfrac{1}{2}L_{A1} \quad -\tfrac{1}{2}L_{A2} \quad \tfrac{1}{2}L_{A3} - L_{B2} + \Delta L_B\right)^T, \\
&\boldsymbol{a}_{31} = \left(\tfrac{1}{2}L_{A1} \quad \tfrac{1}{2}L_{A2} \quad \tfrac{1}{2}L_{A3}\right)^T, \quad \boldsymbol{a}_{32} = \boldsymbol{a}_{13} = \left(\tfrac{1}{2}L_{A1} \quad \tfrac{1}{2}L_{A2} \quad \tfrac{1}{2}L_{A3} - L_{B2}\right)^T, \\
&\boldsymbol{a}_{41} = \left(-\tfrac{1}{2}L_{A1} \quad \tfrac{1}{2}L_{A2} \quad \tfrac{1}{2}L_{A3}\right)^T, \quad \boldsymbol{a}_{42} = \boldsymbol{a}_{23} = \left(-\tfrac{1}{2}L_{A1} \quad \tfrac{1}{2}L_{A2} \quad \tfrac{1}{2}L_{A3} - L_{B2} + \Delta L_B\right)^T,
\end{aligned} \tag{1}
$$

$$
\begin{aligned}
&{}^0\boldsymbol{b}_{11} = \left(\tfrac{1}{2}L_{B1} \quad 0 \quad \tfrac{1}{2}L_{B2}\right)^T, \quad {}^0\boldsymbol{b}_{12} = {}^0\boldsymbol{b}_{13} = \left(\tfrac{1}{2}L_{B1} \quad 0 \quad -\tfrac{1}{2}L_{B2}\right)^T, \\
&{}^0\boldsymbol{b}_{21} = \left(\tfrac{1}{2}L_{B1} \quad 0 \quad \tfrac{1}{2}L_{B2}\right)^T, \quad {}^0\boldsymbol{b}_{22} = {}^0\boldsymbol{b}_{23} = \left(\tfrac{1}{2}L_{B1} \quad 0 \quad -\tfrac{1}{2}L_{B2} + \Delta L_B\right)^T, \\
&{}^0\boldsymbol{b}_{31} = \left(-\tfrac{1}{2}L_{B1} \quad 0 \quad \tfrac{1}{2}L_{B2}\right)^T, \quad {}^0\boldsymbol{b}_{32} = {}^0\boldsymbol{b}_{33} = \left(-\tfrac{1}{2}L_{B1} \quad 0 \quad -\tfrac{1}{2}L_{B2}\right)^T, \\
&{}^0\boldsymbol{b}_{41} = \left(-\tfrac{1}{2}L_{B1} \quad 0 \quad \tfrac{1}{2}L_{B2}\right)^T, \quad {}^0\boldsymbol{b}_{42} = {}^0\boldsymbol{b}_{43} = \left(-\tfrac{1}{2}L_{B1} \quad 0 \quad -\tfrac{1}{2}L_{B2} + \Delta L_B\right)^T,
\end{aligned} \tag{2}
$$

$$b_{ij} = \begin{cases} R_z(\theta_1)\,{}^0b_{ij} & i \in \{1,3\},\, j \in \{1,2,3\} \\ R_z(\theta_2)\,{}^0b_{ij} & i \in \{2,4\},\, j \in \{1,2,3\} \end{cases}' \tag{3}$$

where $R_z(\cdot)$ denotes the rotation matrix about the z axis, and θ_1 and θ_2 are two angles representing the orientation of the moving platform. The detailed definitions of θ_1 and θ_2 will be given in Section 2.2.

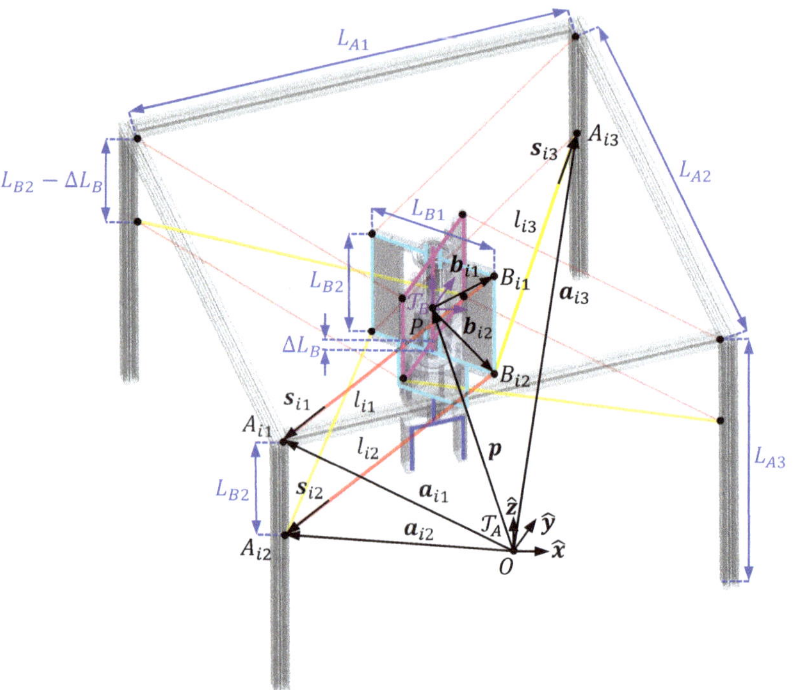

Figure 2. Kinematic diagram of the novel CDPR.

The twelve driving cables of the robot consist of four pairs of parallel cables (red lines in Figure 2) and four independent cables (yellow lines in Figure 2). During the modeling process, we use $j \in \{1,2\}$ to represent the parallel cables and use $j \in \{3\}$ to indicate the independent cables. The working principle of the parallel cables is demonstrated in Figure 3. Each pair of parallel cables is driven by two identical winches which are connected in a series and directly coupled to the same motor. The parallel cables are then passed through two identical guiding pulleys and connected to the moving platform. The arrangement of the winches and the pulleys ensures that the two cables are parallel, so they always have the same length. Based on Equations (1) and (2), we have $||A_{i1}A_{i2}|| = ||B_{i1}B_{i2}||$ and $A_{i1}A_{i2}//B_{i1}B_{i2}$, and thus the quadrilateral $A_{i1}A_{i2}B_{i2}B_{i1}$ forms a parallelogram. According to the proprieties of the parallelogram, we define $l_i = l_{i1} = l_{i2}$ and $s_i = s_{i1} = s_{i2}$. The parallel cables are used to constrain the rotational motion of the moving platform. When a pair of parallel cables are in tension, the moving platform cannot rotate about the normal direction of the parallelogram formed by the parallel cables. The four pairs of parallel cables are specially arranged so that the moving platform of the robot can only rotate about the vertical axis. Thus, the robot is ensured to perform Schönflies motions.

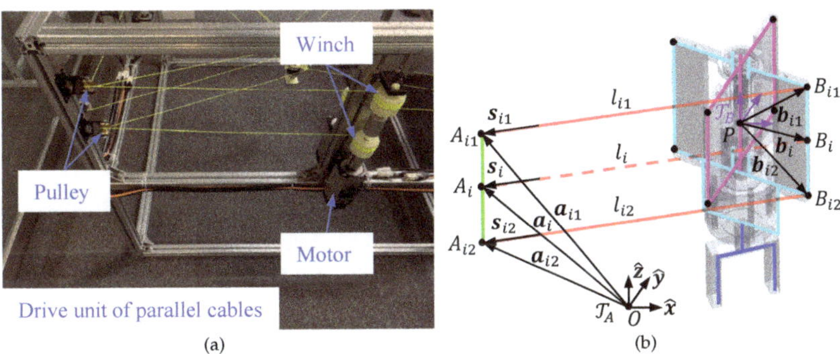

Figure 3. Working principle of parallel cables. (**a**) Drive unit. (**b**) Kinematic diagram.

2.2. Articulated Moving Platform

The moving platform of the robot is specially designed to extend the rotational capability of the robot about the vertical axis. Figure 4 shows the detailed architecture of the moving platform. The moving platform of the robot consists of two sub-platforms and one end-effector. The two sub-platforms are articulated together and coupled to the end-effector through a gearbox. The gearbox amplifies the relative motions between the two sub-platforms, thus extending the rotational capability of the end-effector. The two sub-platforms are named sub-platform 1 and sub-platform 2, respectively. Sub-platform 1 is shown in cyan color, and the cable attachment points on sub-platform 1 are indicated by $i \in \{1,3\}$. Sub-platform 2 is shown in the magenta color, and the cable attachment points on sub-platform 2 are indicated by $i \in \{2,4\}$. To describe the orientation of the moving platform, we attached a moving frame on each part of the moving platform at point P. Sub-platform 1 is attached with frame $\mathcal{T}_{p1}$, and the orientation of frame $\mathcal{T}_{p1}$ with respect to frame $\mathcal{T}_A$ is denoted as θ_1. Sub-platform 2 is attached with frame $\mathcal{T}_{p2}$, and the orientation of frame $\mathcal{T}_{p2}$ with respect to frame $\mathcal{T}_A$ is denoted as θ_2. Let θ represent the angle between the two sub-platforms, and assume the x axis of frame $\mathcal{T}_B$ locates on the bisectrix of θ. The orientation of frame $\mathcal{T}_B$ with respect to frame $\mathcal{T}_A$ is denoted as ψ. Frame $\mathcal{T}_{ee}$ is attached on the end-effector of the robot, and the orientation of frame $\mathcal{T}_{ee}$ with respect to frame $\mathcal{T}_A$ is denoted as ϕ. The articulated structure of the moving platform introduces one internal DOF into the robot. We define ψ and θ as the configuration of the moving platform. Then, the relations between θ_1, θ_2, ϕ and ψ, θ are given as

$$\theta_1 = \psi - \frac{1}{2}\theta, \tag{4}$$

$$\theta_2 = \psi + \frac{1}{2}\theta, \tag{5}$$

$$\phi = \psi + (\eta - \frac{1}{2})\theta + (\phi_0 - \theta_{10}) - \eta(\theta_{20} - \theta_{10}), \tag{6}$$

where θ_{10}, θ_{20}, ϕ_0 are the initial values of θ_1, θ_2, ϕ, and η represents the amplification ratio of the gearbox.

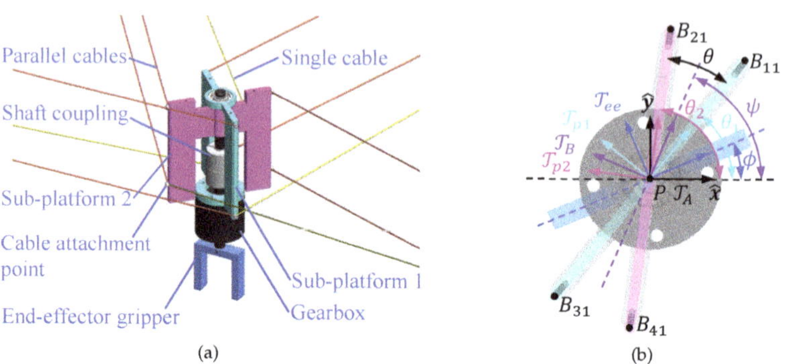

(a)

(b)

Figure 4. Architecture of the articulated moving platform. (**a**) Front view. (**b**) Top view.

3. Kinematics

For CDPRs, the mass and elasticity of the cables introduce a coupling relation between the kinematics and the cable tensions. Some researchers have presented approaches such as the kinetostatic modeling [23] and kinetics modeling [24] to consider these effects. However, these approaches are usually complicated and require numerical methods to solve the models. In this paper, in order to derive closed-form solutions for the kinematics of the novel CDPR, we assume the decoupling of the kinematics and the cable tensions by considering the cables as massless and inelastic straight lines. In this section, we present the kinematic modeling of the novel CDPR. Firstly, the inverse and forward kinematic problems of the robot are solved. Then, the Jacobian matrix of the robot is derived.

3.1. Inverse and Forward Kinematics

The inverse kinematic problem is to calculate the lengths of the driving cables according to the prescribed pose of the moving platform. The forward kinematic problem is to compute the pose of the moving platform according to the prescribed cable lengths. Based on the notations defined in Section 2, the pose of the moving platform in task space is defined by the generalized coordinate $\mathcal{X}$ as

$$\mathcal{X} = \left(\boldsymbol{p}^T \quad \psi \hat{\boldsymbol{z}}^T \quad \theta\right)^T = \left(x \quad y \quad z \quad 0 \quad 0 \quad \psi \quad \theta\right)^T, \tag{7}$$

where $\hat{z} = (0\ 0\ 1)^T$ represents the unit length vector along the z axis. The vector of the cable lengths is given as

$$\boldsymbol{l} = \left(l_1 \quad l_2 \quad l_3 \quad l_4 \quad l_{13} \quad l_{23} \quad l_{33} \quad l_{43}\right)^T. \tag{8}$$

For CDPRs, the inverse kinematic problem is straightforward and generally has closed-form solutions. According to the vector loops in Figure 2, the inverse kinematic equation for the robot is derived as

$$l_{ij} = ||\boldsymbol{a}_{ij} - \boldsymbol{b}_{ij} - \boldsymbol{p}||, \ i \in \{1,2,3,4\}, \ j \in \{1,2,3\}. \tag{9}$$

The forward kinematic problem of CDPRs is generally more involved than the inverse kinematics as there may exist multiple solutions. Numerical methods are usually adopted to solve the forward kinematics of CDPRs. By exploiting the special architecture of the robot, here we propose a geometrical approach to derive a closed-form solution for the forward kinematics of the robot. Figure 5 shows the kinematic diagram of sub-platform 1 used for the forward kinematics. Defining L_{A4} as the distance between point A_{12} and point A_{32}, we have

$$L_{A4} = \sqrt{L_{A1}^2 + L_{A2}^2}. \tag{10}$$

For triangle $A_{12}A_{32}B_{12}$, we define h_1 as the altitude from vertex B_{12} to side $A_{12}A_{32}$, and define w_1 as the distance between vertex A_{12} and the foot of altitude h_1 on side $A_{12}A_{32}$. Since the lengths of the cables are all known, we have

$$w_1 = \frac{l_{12}^2 + L_{A4}^2 - l_{13}^2}{2L_{A4}}, \tag{11}$$

$$h_1 = \sqrt{l_{12}^2 - w_1^2}. \tag{12}$$

Similarly, for triangle $A_{12}A_{32}B_{32}$, we define h_3 as the altitude from vertex B_{32} to side $A_{12}A_{32}$, and define w_3 as the distance between vertex A_{12} and the foot of altitude h_3 on side $A_{12}A_{32}$. Then, w_3 and h_3 are expressed as

$$w_3 = \frac{l_{33}^2 + L_{A4}^2 - l_{32}^2}{2L_{A4}}, \tag{13}$$

$$h_3 = \sqrt{l_{33}^2 - w_3^2}. \tag{14}$$

Since the robot performs Schönflies motions, line $B_{12}B_{32}$ is always parallel with the plane formed by vectors n_1 and n_2, where n_1 and n_2 are defined as

$$n_1 = \frac{a_{32} - a_{12}}{||a_{32} - a_{12}||}, \tag{15}$$

$$n_2 = n_1 \times \hat{z}. \tag{16}$$

To solve the positions of points B_{12} and B_{32}, we project the kinematic diagram onto the plane formed by vectors n_2 and $\hat{z}$, as shown in Figure 5. The kinematic diagram becomes a triangle on the projection plane. Defining h_{13} as the length of line $B_{12}B_{32}$ on the projection plane, we have

$$h_{13} = \sqrt{L_{B1}^2 - (w_1 - w_3)^2}. \tag{17}$$

Then, the two internal angles ξ_1 and ξ_3 of the triangle $A_{32}B_{12}B_{32}$ on the projection plane are derived as

$$\xi_1 = \arccos \frac{h_1^2 + h_{13}^2 - h_3^2}{2h_1 h_{13}}, \tag{18}$$

$$\xi_3 = \arccos \frac{h_3^2 + h_{13}^2 - h_1^2}{2h_3 h_{13}}. \tag{19}$$

Now, we define $r_{12} = p + b_{12}$ and $r_{32} = p + b_{32}$ as the position vectors of points B_{12} and B_{32}, respectively. According to Figure 5, r_{12} and r_{32} can be formulated as

$$r_{12} = \pm h_1 \cos \xi_1 n_2 - h_1 \sin \xi_1 \hat{z} + a_{12} + w_1 n_1, \tag{20}$$

$$r_{32} = \mp h_3 \cos \xi_3 n_2 - h_3 \sin \xi_3 \hat{z} + a_{12} + w_3 n_1. \tag{21}$$

Equations (20) and (21) determine two sets of positions for points B_{12} and B_{32} which are symmetrical about the plane formed by vectors n_1 and $\hat{z}$. After obtaining the positions of point B_{12} and point B_{32}, the position and orientation of sub-platform 1 can be derived as

$$p = \frac{1}{2}(r_{12} + r_{32}) + \frac{1}{2}L_{B2}\hat{z}, \tag{22}$$

$$\theta_1 = \arctan \frac{(r_{12} - r_{32}) \cdot \hat{y}}{(r_{12} - r_{32}) \cdot \hat{x}}, \tag{23}$$

where $\hat{x} = (1\ 0\ 0)^T$ and $\hat{y} = (0\ 1\ 0)^T$. Equations (22) and (23) also determine two sets of poses of sub-platform 1. In order to determine the pose of the entire moving platform, we use the same procedure to solve the poses of sub-platform 2. Since sub-platform 1 and sub-platform 2 are articulated together, we can take the intersection of the position vectors determined by sub-platform 1 and sub-platform 2 as the true position vector of the moving platform. Then, the orientation of the moving platform can be obtained by computing Equations (4) and (5).

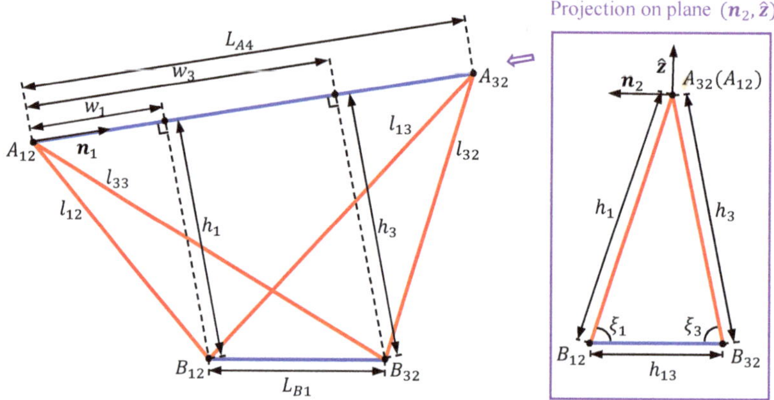

Figure 5. Kinematic diagram of sub-platform 1 for forward kinematics.

3.2. Jacobian Matrix

The Jacobian matrix plays an important role in the modeling and analysis of the robot as the Jacobian matrix maps the velocity of the moving platform to the velocity of the driving cables. Differentiating $\mathcal{X}$ and l with respect to time, we have the generalized velocity $\dot{\mathcal{X}}$ and the cable velocity $\dot{l}$ as

$$\dot{\mathcal{X}} = \begin{pmatrix} \dot{p}^T & \dot{\psi}\hat{z}^T & \dot{\theta} \end{pmatrix}^T = \begin{pmatrix} \dot{x} & \dot{y} & \dot{z} & 0 & 0 & \dot{\psi} & \dot{\theta} \end{pmatrix}^T, \tag{24}$$

$$\dot{l} = \begin{pmatrix} \dot{l}_1 & \dot{l}_2 & \dot{l}_3 & \dot{l}_4 & \dot{l}_{13} & \dot{l}_{23} & \dot{l}_{33} & \dot{l}_{43} \end{pmatrix}^T. \tag{25}$$

Define $\mathcal{V}_{p1}$ and $\mathcal{V}_{p2}$ as the twist vectors of sub-platform 1 and sub-platform 2, respectively, and $\mathcal{V}_{ee}$ as the twist vector of the end-effector. Based on Equations (4)–(6), we have

$$\mathcal{V}_{p1} = \begin{pmatrix} \dot{p} \\ \dot{\theta}_1\hat{z} \end{pmatrix} = \begin{pmatrix} \dot{p} \\ \dot{\psi}\hat{z} - \frac{1}{2}\dot{\theta}\hat{z} \end{pmatrix} = \underbrace{\begin{bmatrix} E_{6\times6} & \begin{matrix} \mathbf{0}_{3\times1} \\ -\frac{1}{2}\hat{z} \end{matrix} \end{bmatrix}}_{H_{p1}} \dot{\mathcal{X}}, \tag{26}$$

$$\mathcal{V}_{p2} = \begin{pmatrix} \dot{p} \\ \dot{\theta}_2\hat{z} \end{pmatrix} = \begin{pmatrix} \dot{p} \\ \dot{\psi}\hat{z} + \frac{1}{2}\dot{\theta}\hat{z} \end{pmatrix} = \underbrace{\begin{bmatrix} E_{6\times6} & \begin{matrix} \mathbf{0}_{3\times1} \\ \frac{1}{2}\hat{z} \end{matrix} \end{bmatrix}}_{H_{p2}} \dot{\mathcal{X}}, \tag{27}$$

$$\mathcal{V}_{ee} = \begin{pmatrix} \dot{p} \\ \dot{\phi}\hat{z} \end{pmatrix} = \begin{pmatrix} \dot{p} \\ \dot{\psi}\hat{z} + (\eta - \frac{1}{2})\dot{\theta}\hat{z} \end{pmatrix} = \underbrace{\begin{bmatrix} E_{6\times6} & \begin{matrix} \mathbf{0}_{3\times1} \\ (\eta - \frac{1}{2})\hat{z} \end{matrix} \end{bmatrix}}_{H_{ee}} \dot{\mathcal{X}}, \tag{28}$$

where $E_{n \times n}$ represents the $n \times n$ identity matrix and $0_{n \times m}$ represents the $n \times m$ matrix of zeros. Differentiating Equation (9) with respect to time and considering the properties of parallel cables, we have

$$
-\underbrace{\begin{bmatrix} s_1^T & (b_1 \times s_1)^T \\ s_3^T & (b_3 \times s_3)^T \\ s_{13}^T & (b_{13} \times s_{13})^T \\ s_{33}^T & (b_{33} \times s_{33})^T \end{bmatrix}}_{J_{p1}} v_{p1} = \underbrace{\begin{pmatrix} i_1 \\ i_3 \\ i_{13} \\ i_{33} \end{pmatrix}}_{i_{p1}},
\tag{29}
$$

$$
-\underbrace{\begin{bmatrix} s_2^T & (b_2 \times s_2)^T \\ s_4^T & (b_3 \times s_4)^T \\ s_{23}^T & (b_{23} \times s_{23})^T \\ s_{43}^T & (b_{43} \times s_{43})^T \end{bmatrix}}_{J_{p2}} v_{p2} = \underbrace{\begin{pmatrix} i_2 \\ i_4 \\ i_{23} \\ i_{43} \end{pmatrix}}_{i_{p2}},
\tag{30}
$$

where $b_i = (b_{i1} + b_{i2})/2$, $i \in \{1,2,3,4\}$. Substituting Equations (26) and (27) into Equations (29) and (30), the velocity equation of the entire moving platform is derived as

$$
\begin{bmatrix} J_{p1} H_{p1} \\ J_{p2} H_{p2} \end{bmatrix} \dot{x} = \begin{pmatrix} i_{p1} \\ i_{p2} \end{pmatrix}.
\tag{31}
$$

Reshaping Equation (31), we have

$$
-\underbrace{\begin{bmatrix} s_1^T & (b_1 \times s_1)^T & -\frac{1}{2}(b_1 \times s_1)^T \hat{z} \\ s_2^T & (b_2 \times s_2)^T & \frac{1}{2}(b_2 \times s_2)^T \hat{z} \\ s_3^T & (b_3 \times s_3)^T & -\frac{1}{2}(b_3 \times s_3)^T \hat{z} \\ s_4^T & (b_4 \times s_4)^T & \frac{1}{2}(b_4 \times s_4)^T \hat{z} \\ s_{13}^T & (b_{13} \times s_{13})^T & -\frac{1}{2}(b_{13} \times s_{13})^T \hat{z} \\ s_{23}^T & (b_{23} \times s_{23})^T & \frac{1}{2}(b_{23} \times s_{23})^T \hat{z} \\ s_{33}^T & (b_{33} \times s_{33})^T & -\frac{1}{2}(b_{33} \times s_{33})^T \hat{z} \\ s_{43}^T & (b_{43} \times s_{43})^T & \frac{1}{2}(b_{43} \times s_{43})^T \hat{z} \end{bmatrix}}_{J} \dot{x} = i,
\tag{32}
$$

where J is denoted as the Jacobian matrix of the robot.

4. Dynamics

The dynamic equation of the robot describes the relation between the actuation torques and the motion of the moving platform. The dynamic model of the robot can be formulated using finite element methods such as the SPACAR software [25], which is a general tool for the dynamic analysis of flexible multi-body systems. However, the analytical dynamic model plays an important role in the analysis and control of the robot, which can facilitate the workspace determination and lay the foundation for model-based control approaches. In this paper, we assume that the cables are massless and inelastic straight lines to derive the analytical dynamic equation of the robot. In this section, we present the dynamic modeling of the robot based on the virtual power principle. Firstly, the dynamics of the moving platform is formulated. Then, the effect of the drive trains is derived.

4.1. Dynamics of Moving Platform

The dynamic equation of the moving platform can be derived based on the virtual power principle. The friction inside the articulated joints of the moving platform is neglected, because ball bearings are installed inside the joints to reduce the friction. We define

f_{ij} as the tension of cable l_{ij}, and $f_i = f_{i1} + f_{i2}$ as the total tension of parallel cables l_{i1} and l_{i2}. Then, the vector of the cable tensions is dented as

$$f_c = \begin{pmatrix} f_1 & f_2 & f_3 & f_4 & f_{13} & f_{23} & f_{33} & f_{43} \end{pmatrix}^T. \tag{33}$$

For sub-platform 1, let m_{p1} be the total mass, r_{p1} be the position vector of the center of mass, I_{p1} be the rotational inertia matrix, and ω_{p1} be the angular velocity vector. For sub-platform 2, let m_{p2} be the total mass, r_{p2} be the position vector of the center of mass, I_{p2} be the rotational inertia matrix, and ω_{p2} be the angular velocity vector. Similarly, for the end-effector, let m_{ee} be the total mass, r_{ee} be the position vector of the center of mass, I_{ee} be the rotational inertia matrix, and ω_{ee} be the angular velocity vector. According to the virtual power principle [26], we have the dynamic equation of the moving platform as

$$\begin{aligned}
&(m_{p1}\ddot{r}_{p1} - m_{p1}g) \cdot \delta r_{p1} + (I_{p1}\dot{\omega}_{p1} + \omega_{p1} \times I_{p1}\omega_{p1}) \cdot \delta\omega_{p1} + \\
&(m_{p2}\ddot{r}_{p2} - m_{p2}g) \cdot \delta r_{p2} + (I_{p2}\dot{\omega}_{p2} + \omega_{p2} \times I_{p2}\omega_{p2}) \cdot \delta\omega_{p2} + \\
&(m_{ee}\ddot{r}_{ee} - m_{ee}g) \cdot \delta r_{ee} + (I_{ee}\dot{\omega}_{ee} + \omega_{ee} \times I_{ee}\omega_{ee}) \cdot \delta\omega_{ee} = f_c \cdot \delta l,
\end{aligned} \tag{34}$$

where $g = (0 \; 0 \; -9.81)^T$ is the vector of gravitational acceleration. To formulate the dynamic equation as a function of the generalized coordinate $\mathcal{X}$, we need to derive the detailed expression of each component in Equation (34). For sub-platform 1, we define c_{p1} as the position vector of the center of mass in frame $\mathcal{T}_{p1}$, and I_{c1} as the rotational inertia matrix about the center of mass in frame $\mathcal{T}_{p1}$; then, we have

$$r_{p1} = p + R_z(\theta_1)c_{p1}, \tag{35}$$

$$I_{p1} = R_z(\theta_1)I_{c1}R_z(\theta_1)^T. \tag{36}$$

Differentiating Equation (35) with respect to time, the velocity and acceleration of sub-platform 1 are expressed as

$$\begin{aligned}
\dot{r}_{p1} &= \dot{p} + \dot{\theta}_1 \hat{z} \times R_z(\theta_1)c_{p1} \\
&= \underbrace{\begin{bmatrix} E_{3\times3} & -[R_z(\theta_1)c_{p1}]_\times & \tfrac{1}{2}[R_z(\theta_1)c_{p1}]_\times \hat{z} \end{bmatrix}}_{H_{c1}} \dot{\mathcal{X}},
\end{aligned} \tag{37}$$

$$\begin{aligned}
\ddot{r}_{p1} &= \ddot{p} + \ddot{\theta}_1 \hat{z} \times R_z(\theta_1)c_{p1} + \dot{\theta}_1^2 \hat{z} \times \hat{z} \times R_z(\theta_1)c_{p1} \\
&= H_{c1}\ddot{\mathcal{X}} + \dot{\mathcal{X}}^T H_{\theta1}^T \underbrace{[\hat{z}]_\times^2 R_z(\theta_1)c_{p1} \, H_{\theta1}}_{C_{c1}} \dot{\mathcal{X}},
\end{aligned} \tag{38}$$

where $[\cdot]_\times$ denotes the skew-symmetric matrix representation of a vector, and $H_{\theta1} = (0 \; 0 \; 0 \; 0 \; 0 \; 1 \; -\tfrac{1}{2})$. Since the robot performs Schönflies motions, the angular velocity and angular acceleration of sub-platform 1 are expressed as

$$\omega_{p1} = \dot{\theta}_1 \hat{z} = \underbrace{\begin{bmatrix} 0_{3\times3} & E_{3\times3} & -\tfrac{1}{2}\hat{z} \end{bmatrix}}_{H_{\omega1}} \dot{\mathcal{X}}, \tag{39}$$

$$\dot{\omega}_{p1} = \ddot{\theta}_1 \hat{z} = H_{\omega1}\ddot{\mathcal{X}}. \tag{40}$$

Then, for sub-platform 2, we define c_{p2} as the position vector of the center of mass in frame $\mathcal{T}_{p2}$, and I_{c2} as the rotational inertia matrix about the center of mass in frame $\mathcal{T}_{p2}$. Similar results can be obtained as follows:

$$r_{p2} = p + R_z(\theta_2)c_{p2}, \tag{41}$$

$$\dot{r}_{p2} = \underbrace{\begin{bmatrix} E_{3\times3} & -[R_z(\theta_2)c_{p2}]_\times & -\frac{1}{2}[R_z(\theta_2)c_{p2}]_\times \hat{z} \end{bmatrix}}_{H_{c2}} \dot{X}, \tag{42}$$

$$\ddot{r}_{p2} = H_{c2}\ddot{X} + \dot{X}^T H_{\theta2}^T \underbrace{[\hat{z}]_\times^2 R_z(\theta_2)c_{p2} H_{\theta2}}_{C_{c2}} \dot{X}, \tag{43}$$

$$H_{\theta2} = \begin{pmatrix} 0 & 0 & 0 & 0 & 0 & 1 & \frac{1}{2} \end{pmatrix}, \tag{44}$$

$$I_{p2} = R_z(\theta_2) I_{c2} R_z(\theta_2)^T, \tag{45}$$

$$\omega_{p2} = \underbrace{\begin{bmatrix} 0_{3\times3} & E_{3\times3} & \frac{1}{2}\hat{z} \end{bmatrix}}_{H_{\omega2}} \dot{X}, \tag{46}$$

$$\dot{\omega}_{p2} = H_{\omega2}\ddot{X}. \tag{47}$$

For the end-effector, we define c_{ee} as the position vector of the center of mass in frame $\mathcal{T}_{ee}$, and I_{ce} as the rotational inertia matrix about the center of mass in frame $\mathcal{T}_{ee}$. Then, we can derive the following results similar to sub-platform 1 and sub-platform 2:

$$r_{ee} = p + R_z(\phi)c_{ee}, \tag{48}$$

$$\dot{r}_{ee} = \underbrace{\begin{bmatrix} E_{3\times3} & -[R_z(\phi)c_{ee}]_\times & -(\eta - \frac{1}{2})[R_z(\phi)c_{ee}]_\times \hat{z} \end{bmatrix}}_{H_{ce}} \dot{X}, \tag{49}$$

$$\ddot{r}_{ee} = H_{ce}\ddot{X} + \dot{X}^T H_\phi^T \underbrace{[\hat{z}]_\times^2 R_z(\phi)c_{ee} H_\phi}_{C_{ce}} \dot{X}, \tag{50}$$

$$H_\phi = \begin{pmatrix} 0 & 0 & 0 & 0 & 0 & 1 & \eta - \frac{1}{2} \end{pmatrix}, \tag{51}$$

$$I_{ee} = R_z(\phi) I_{ce} R_z(\phi)^T, \tag{52}$$

$$\omega_{ee} = \underbrace{\begin{bmatrix} 0_{3\times3} & E_{3\times3} & (\eta - \frac{1}{2})\hat{z} \end{bmatrix}}_{H_{\omega e}} \dot{X}, \tag{53}$$

$$\dot{\omega}_{ee} = H_{\omega e}\ddot{X}. \tag{54}$$

Substituting the above equations into Equation (34), the dynamic equation of the moving platform can be formulated as a function of the generalized coordinate X. Then, we have the final form of the dynamic equation of the moving platform as

$$M(X)\ddot{X} + \dot{X}^T C(X)\dot{X} + G(X) = J^T f_c, \tag{55}$$

where

$$\begin{aligned}
M(X) = \; & m_{p1} H_{c1}^T H_{c1} + H_{\omega1}^T R_z(\theta_1) I_{c1} R_z(\theta_1)^T H_{\omega1} + \\
& m_{p2} H_{c2}^T H_{c2} + H_{\omega2}^T R_z(\theta_2) I_{c2} R_z(\theta_2)^T H_{\omega2} + \\
& m_{ee} H_{ce}^T H_{ce} + H_{\omega e}^T R_z(\phi) I_{ce} R_z(\phi)^T H_{\omega e},
\end{aligned} \tag{56}$$

$$\begin{aligned}
C(X) = \; & m_{p1} H_{\theta1}^T H_{c1}^T C_{c1} H_{\theta1} + H_{\theta1}^T H_{\omega1}^T [\hat{z}]_\times R_z(\theta_1) I_{c1} R_z(\theta_1)^T H_{\omega1} + \\
& m_{p2} H_{\theta2}^T H_{c2}^T C_{c2} H_{\theta2} + H_{\theta2}^T H_{\omega2}^T [\hat{z}]_\times R_z(\theta_2) I_{c2} R_z(\theta_2)^T H_{\omega2} + \\
& m_{ee} H_\phi^T H_{ce}^T C_{ce} H_\phi + H_\phi^T H_{\omega e}^T [\hat{z}]_\times R_z(\phi) I_{ce} R_z(\phi)^T H_{\omega e},
\end{aligned} \tag{57}$$

$$G(X) = -m_{p1} H_{c1}^T g - m_{p2} H_{c2}^T g - m_{ee} H_{ce}^T g. \tag{58}$$

4.2. Dynamics of Drive Train

The drive unit of the robot contains a servo motor as the actuator and a winch which is directly coupled to the servo motor to drive the cable. For all the drive units of the robot, we define τ_m as the vector of the motor torques, θ_m as the vector of the motor rotation angles, τ_f as the vector of the frictional torques, I_m as the diagonal matrix of the moments of inertia of the winches, and r_w as the radius of the winches. Based on Newton's second law, we have

$$\tau_m + \tau_f + r_w f_c = I_m \ddot{\theta}_m. \tag{59}$$

We use the Coulomb and viscous model [27] to formulate the frictional torques, and note that $r_w \dot{\theta}_m = \dot{l}$. Then, the dynamic equation of the drive units of the robot is derived as

$$\tau_m = \frac{I_m}{r_w} \ddot{l} + K_c \mathrm{sign}(\frac{\dot{l}}{r_w}) + \frac{K_v}{r_w} \dot{l} - r_w f_c, \tag{60}$$

where K_c represents the diagonal matrix of the Coulomb friction coefficients, and K_v represents the diagonal matrix of the viscous friction coefficients.

5. Workspace Analysis

In this section, we evaluate the dynamic feasible workspace of the robot. The dynamic workspace is a set of poses where a prescribed set of moving platform accelerations can be achieved by applying feasible cable tensions. The cable tensions required for generating the prescribed accelerations can be obtained from Equation (55). For simplicity, we assume that the vectors c_{p1}, c_{p2}, and c_{ee} are all parallel with the z axis of frame $\mathcal{T}_A$; we then have $C(\mathcal{X}) = 0$. Defining $\ddot{\mathcal{X}} = (\ddot{p}^T\ 0\ 0\ \ddot{\alpha}^T)^T$, $\ddot{p} = (\ddot{x}\ \ddot{y}\ \ddot{z})^T$, and $\ddot{\alpha} = (\ddot{\psi}\ \ddot{\theta})^T$, the set of the required moving platform accelerations is formulated as

$$[\ddot{\mathcal{X}}]_r = \left\{ \ddot{\mathcal{X}} \mid \ddot{p}_{\min} \preceq \ddot{p} \preceq \ddot{p}_{\max}, \ddot{\alpha}_{\min} \preceq \ddot{\alpha} \preceq \ddot{\alpha}_{\max} \right\}, \tag{61}$$

where $\preceq$ means the component-wise inequality. Similarly, the set of the admissible cable tensions is defined as

$$[f_c]_a = \left\{ f_c \mid f_{\min} \preceq f_c \preceq f_{\max} \right\}. \tag{62}$$

Thus, the dynamic feasible workspace of the robot is defined as

$$[\mathcal{X}]_{dyn} = \left\{ \mathcal{X} \mid \forall \ddot{\mathcal{X}} \in [\ddot{\mathcal{X}}]_r, \exists f_c \in [f_c]_a, M(\mathcal{X})\ddot{\mathcal{X}} + G(\mathcal{X}) = J^T f_c, n_c = 0 \right\}, \tag{63}$$

where n_c represents the number of collisions inside the robot, which is determined by calculating the distances between two cables and between the cable and the moving platform [28]. In order to investigate the effects of different acceleration values on the dynamic feasible workspace, we solve the dynamic feasible workspace of the robot in the following four scenarios:

$$\text{Scenario 1} \quad \ddot{p}_{\max} = -\ddot{p}_{\min} = \begin{pmatrix} 0 \\ 0 \\ 0 \end{pmatrix} \mathrm{mm/s^2}, \quad \ddot{\alpha}_{\max} = -\ddot{\alpha}_{\min} = \begin{pmatrix} 0 \\ 0 \end{pmatrix} \mathrm{rad/s^2},$$

$$\text{Scenario 2} \quad \ddot{p}_{\max} = -\ddot{p}_{\min} = \begin{pmatrix} 1500 \\ 1500 \\ 1500 \end{pmatrix} \mathrm{mm/s^2}, \quad \ddot{\alpha}_{\max} = -\ddot{\alpha}_{\min} = \begin{pmatrix} 0 \\ 0 \end{pmatrix} \mathrm{rad/s^2},$$

$$\text{Scenario 3} \quad \ddot{p}_{\max} = -\ddot{p}_{\min} = \begin{pmatrix} 0 \\ 0 \\ 0 \end{pmatrix} \mathrm{mm/s^2}, \quad \ddot{\alpha}_{\max} = -\ddot{\alpha}_{\min} = \begin{pmatrix} 150 \\ 150 \end{pmatrix} \mathrm{rad/s^2},$$

$$\text{Scenario 4} \quad \ddot{p}_{\max} = -\ddot{p}_{\min} = \begin{pmatrix} 1500 \\ 1500 \\ 1500 \end{pmatrix} \mathrm{mm/s^2}, \quad \ddot{\alpha}_{\max} = -\ddot{\alpha}_{\min} = \begin{pmatrix} 150 \\ 150 \end{pmatrix} \mathrm{rad/s^2}. \tag{64}$$

The kinematic and dynamic parameters of the robot are summarized in Table 1. To visualize the dynamic workspace of the robot in Cartesian space, the orientation of the moving platform is set as $\psi = 0$ and $\theta = \pi/2$. The dynamic feasible workspace of the robot is calculated using the differential workspace hull approach [20], which approximates the hull of the workspace using a triangulation representation. Firstly, a unity sphere located at the estimated workspace center is subdivided into a set of triangular faces. Then, the boundary of the workspace is calculated along the vertices of each triangular face using a line search method. At each position, the workspace criterion is investigated for a discrete set of moving platform configurations and a Boolean result is yielded to amend the searching direction. Finally, the algorithm ends when the hull of the workspace is approximated within a certain accuracy. After determining the workspace hull, the volume of the workspace can be obtained by computing the volumes of the tetrahedrons formed by all the triangular faces.

Table 1. Kinematic and dynamic parameters of the robot.

Parameter	Notation	Value	Unit
Length of the base	L_{A1}	1200	mm
Width of the base	L_{A2}	1200	mm
Height of the base	L_{A3}	800	mm
Length of the moving platform	L_{B1}	120	mm
Height of the moving platform	L_{B2}	120	mm
Offset of the moving platform height	ΔL_B	17.5	mm
Amplification ratio of the gearbox	η	4.25	-
Initial value of θ_1	θ_{10}	$-\pi/4$	rad
Initial value of θ_2	θ_{20}	$\pi/4$	rad
Initial value of ϕ	ϕ_0	0	rad
Mass of sub-platform 1	m_{p1}	0.6	kg
Center of mass position of sub-platform 1	c_{p1}	$(0\ 0\ -75)^T$	mm
Rotational inertia matrix of sub-platform 1	I_{c1}	$\mathrm{diag}(1.2\ 2.5\ 4.4)$	$\mathrm{kg\cdot cm^2}$
Mass of sub-platform 2	m_{p2}	0.35	kg
Center of mass position of sub-platform 2	c_{p2}	$(0\ 0\ -75)^T$	mm
Rotational inertia matrix of sub-platform 2	I_{c2}	$\mathrm{diag}(0.5\ 0.9\ 1.5)$	$\mathrm{kg\cdot cm^2}$
Mass of the end-effector	m_{ee}	0.18	kg
Center of mass position of the end-effector	c_{ee}	$(0\ 0\ -100)^T$	mm
Rotational inertia matrix of the end-effector	I_{ce}	$\mathrm{diag}(0.1\ 0.1\ 0.2)$	$\mathrm{kg\cdot cm^2}$
Coulomb friction coefficient of the drive unit	K_c	0.042	-
Viscous friction coefficient of the drive unit	K_v	5.3×10^{-4}	-
Upper bound of the cable tension	$f_{\max}$	50	N
Lower bound of the cable tension	$f_{\min}$	0	N

Figure 6 shows the obtained dynamic feasible workspace in each scenario. The volume of the dynamic feasible workspace in each scenario is summarized in Table 2. In Scenario 1, where the required acceleration set is empty, the dynamic feasible workspace is actually equivalent to the static feasible workspace. The volume of the dynamic feasible workspace in Scenario 1 is the largest among the four scenarios, and the shape of the workspace is like a cuboid similar to the base. In Scenario 2, the moving platform of the robot only needs to perform linear accelerations. The shape of the dynamic workspace in Scenario 2 is like a frustum whose cross-section increases along the z direction. The workspace volume in Scenario 2 is reduced by 46.69% compared with Scenario 1. In Scenario 3 where only angular accelerations are required, the dynamic feasible workspace is reduced by 20.46% compared with Scenario 1. Comparing Scenario 3 with Scenario 2, it is observed that the angular accelerations have a weaker effect on the volume of the dynamic feasible workspace. In Scenario 4, the moving platform is required to perform both linear and angular accelerations. The dynamic feasible workspace in Scenario 4 is the smallest among the four scenarios, and its volume is reduced by 62.82% compared with Scenario 1. The shape of the dynamic workspace in Scenario 4 is similar to that of the workspace in Scenario 2, which indicates that the values of linear accelerations play a decisive role in determining the dynamic feasible workspace of the robot.

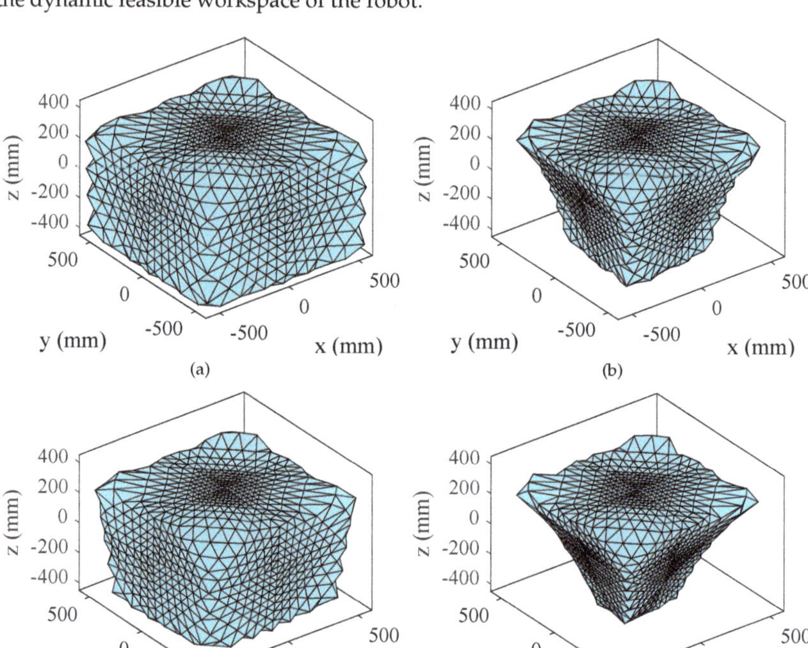

Figure 6. Dynamic feasible workspace of the robot in different scenarios. (**a**) Scenario 1. (**b**) Scenario 2. (**c**) Scenario 3. (**d**) Scenario 4.

Table 2. Comparison of the dynamic feasible workspace in different scenarios.

Scenario	Scenario 1	Scenario 2	Scenario 3	Scenario 4
Volume of the robot covered by the dynamic feasible workspace	71.08%	37.89%	56.54%	26.43%

6. Experiment

To validate the correctness of the previously derived models and workspace, in this section, we performed experiments on the robot prototype shown in Figure 1. The main parameters of the prototype are summarized in Table 1. The control system diagram of the prototype which demonstrates the working principle and main components of the prototype system is shown in Figure 7. The control algorithms were implemented in the real-time kernel of Simulink Desktop Real-Time on an industrial computer. A proportional-integral-derivative (PID) controller was deigned in joint space with the sample rate of 1 kHz to control the prototype. At the current stage, the robot prototype was not equipped with tension sensors; the cable tensions were therefore not available in the experimental results.

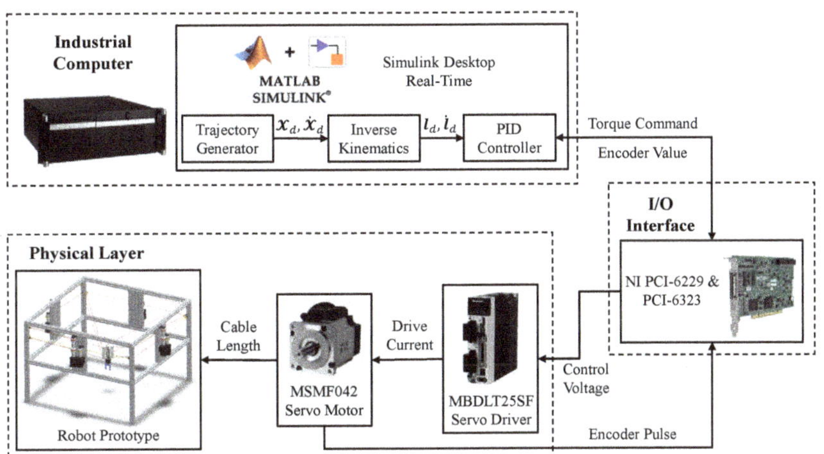

Figure 7. Control system diagram of the robot prototype.

In the experiments, the robot prototype was controlled to track a predefined trajectory inside the workspace. A circular trajectory located near the boundary of the dynamic feasible workspace determined in Section 5 was designed as

$$
\begin{aligned}
x_d &= 300\cos{(2\pi s)}\,\text{mm}, & \psi_d &= 10\sin{(2\pi s)}\,^\circ, \\
y_d &= 300\sin{(2\pi s)}\,\text{mm}, & \theta_d &= 30\sin{(2\pi s)} + 90\,^\circ, \\
z_d &= 100\,\text{mm}, & s &= 10(\frac{t}{T})^3 - 15(\frac{t}{T})^4 + 6(\frac{t}{T})^5,
\end{aligned}
\tag{65}
$$

where $T = 5\,\text{s}$ is the period of the trajectory. The trajectory was successfully tracked by the robot prototype and no cable slackness was observed, which demonstrates that the robot can perform dynamic trajectories with feasible cable tensions inside the dynamic feasible workspace. Figure 8 shows the trajectory tracking performance of the robot prototype in joint space. The desired cable lengths were obtained by calculating the inverse kinematics based on the robot trajectory in Equation (65). These values were fed into the robot controller as reference signals. The cable length errors were measured with the encoders on the servo motors. The maximum error among all the cables was 1.26 mm, which demonstrates that the robot has a high tracking accuracy in joint space. Based on the measured cable lengths, the task space trajectory performed by the robot prototype can be obtained by calculating the forward kinematics of the robot. Figure 9 presents the comparison of the desired trajectory and the calculated trajectory from the forward kinematics in Cartesian space. The average tracking error and the maximum tracking error along the Cartesian space trajectory were 0.35 mm and 0.99 mm, respectively. These results verify the consistency of the inverse and forward kinematic models of the robot derived in Section 3. However, the real trajectory performed by the robot may deviate from the calculated trajectory because the unmodeled uncertainties and the manufacturing errors were neglected. To evaluate the rotational accuracy of the robot, an LPMS-IG1 inertial measurement unit (IMU) was installed on the end-effector of the robot prototype to measure the rotation angle. Figure 10 shows the rotation angle of the end-effector along the testing trajectory. The desired value of the rotation angle was calculated using Equation (6) based on the robot trajectory, and the measured value was obtained from the IMU sensor. The rotation error of the end-effector varied from $-2.05\,^\circ$ to $2.81\,^\circ$, with an average value of $0.54\,^\circ$. Taking into account the manufacturing and assembly errors of the prototype, these values can be considered acceptable. The experimental results in this section validate the correctness of the previously derived models and workspace, and demonstrate the feasibility of the proposed robot in generating Schönflies motions.

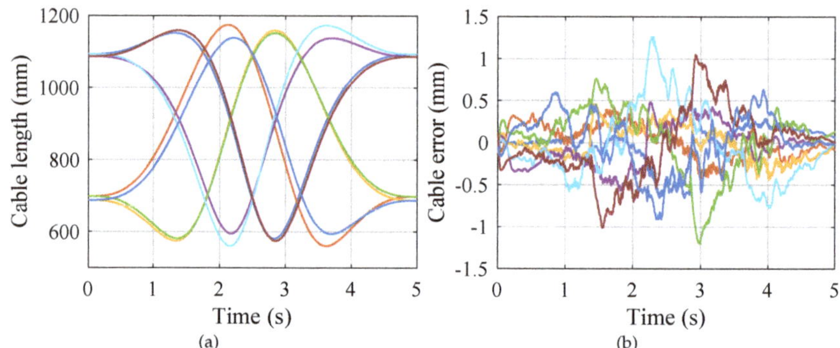

Figure 8. Trajectory tracking performance in joint space. (**a**) Desired cable lengths. (**b**) Cable length errors.

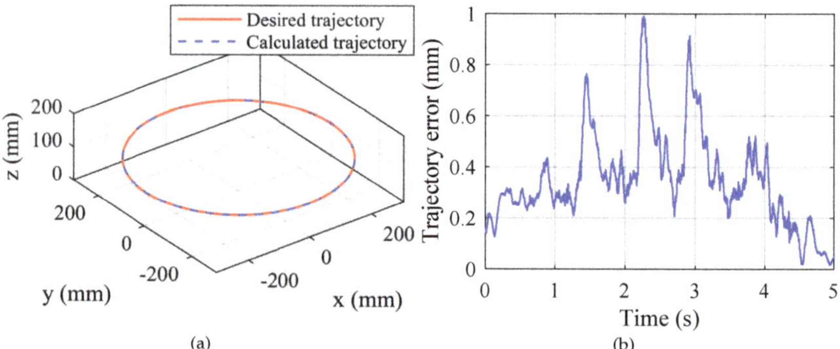

Figure 9. Trajectory tracking performance in Cartesian space. (**a**) Desired trajectory and calculated trajectory. (**b**) Trajectory error.

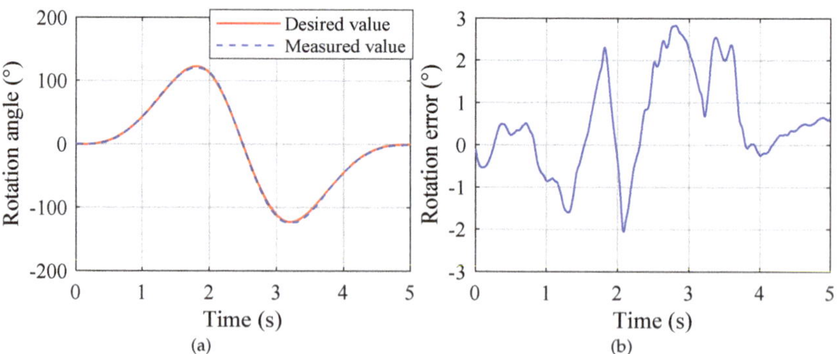

Figure 10. Rotation angle of the end-effector. (**a**) Desired value and measured value. (**b**) Rotation error.

7. Conclusions

In this paper, kinematic and dynamic modeling and a workspace analysis of a novel suspended CDPR for Schönflies motions were presented. The inverse and forward kinematic problems of the robot were solved through a geometrical approach. The dynamic equation of the robot was derived by separately considering the moving platform and the drive trains. Based on the dynamic equation, the dynamic feasible workspace of the robot was determined under different values of accelerations. The correctness of the derived models and the workspace was verified through experiments on a prototype of the robot. In future works, we will develop a cable tension measurement system and use a motion capture system to measure the poses of the robot for further validation of the models derived in this paper.

Author Contributions: Conceptualization and methodology, R.W.; validation, Y.X. and X.C.; formal analysis, Y.L.; writing—original draft preparation, R.W.; writing—review and editing, Y.X. and X.C.; supervision, project administration, and funding acquisition, Y.L. All authors have read and agreed to the published version of the manuscript.

Funding: This research was funded by the Research Grants Council of Hong Kong under Grant No. PolyU 15213719.

Institutional Review Board Statement: Not applicable.

Informed Consent Statement: Not applicable.

Data Availability Statement: Not applicable.

Conflicts of Interest: The authors declare no conflict of interest.

References

1. Gosselin, C. Cable-driven parallel mechanisms: State of the art and perspectives. *Mech. Eng. Rev.* **2014**, *1*, DSM0004. [CrossRef]
2. Tang, X. An overview of the development for cable-driven parallel manipulator. *Adv. Mech. Eng.* **2014**, *6*, 823028. [CrossRef]
3. Tempel, P.; Schnelle, F.; Pott, A.; Eberhard, P. Design and programming for cable-driven parallel robots in the German Pavilion at the EXPO 2015. *Machines* **2015**, *3*, 223–241. [CrossRef]
4. Jiang, X.; Barnett, E.; Gosselin, C. Dynamic point-to-point trajectory planning beyond the static workspace for six-DOF cable-suspended parallel robots. *IEEE Trans. Robot.* **2018**, *34*, 781–793. [CrossRef]
5. Jamshidifar, H.; Rushton, M.; Khajepour, A. A reaction-based stabilizer for nonmodel-based vibration control of cable-driven parallel robots. *IEEE Trans. Robot.* **2020**, *37*, 667–674. [CrossRef]
6. Zarebidoki, M.; Dhupia, J.S.; Xu, W. A review of cable-driven parallel robots: Typical configurations, analysis techniques, and control methods. *IEEE Robot. Autom. Mag.* **2022**, 2–19. [CrossRef]
7. Paty, T.; Binaud, N.; Caro, S.; Segonds, S. Cable-driven parallel robot modelling considering pulley kinematics and cable elasticity. *Mech. Mach. Theory* **2021**, *159*, 104263. [CrossRef]
8. Zhang, Z.; Xie, G.; Shao, Z.; Gosselin, C. Kinematic calibration of cable-driven parallel robots considering the pulley kinematics. *Mech. Mach. Theory* **2022**, *169*, 104648. [CrossRef]
9. Mishra, U.A.; Caro, S. Forward kinematics for suspended under-actuated cable-driven parallel robots with elastic cables: A neural network approach. *ASME J. Mech. Robot.* **2022**, *14*, 041008 . [CrossRef]
10. Zi, B.; Zhang, L.; Zhang, D.; Qian, S. Modeling, analysis, and co-simulation of cable parallel manipulators for multiple cranes. *Proc. Inst. Mech. Eng. Part C J. Mech. Eng. Sci.* **2015**, *229*, 1693–1707. [CrossRef]
11. Zi, B.; Sun, H.; Zhang, D. Design, analysis and control of a winding hybrid-driven cable parallel manipulator. *Robot. Comput.-Integr. Manuf.* **2017**, *48*, 196–208. [CrossRef]
12. Ferravante, V.; Riva, E.; Taghavi, M.; Braghin, F.; Bock, T. Dynamic analysis of high precision construction cable-driven parallel robots. *Mech. Mach. Theory* **2019**, *135*, 54–64. [CrossRef]
13. Wang, R.; Li, Y. Analysis and multi-objective optimal design of a planar differentially driven cable parallel robot. *Robotica* **2021**, *39*, 2193–2209. [CrossRef]
14. Abbasnejad, G.; Eden, J.; Lau, D. Generalized ray-based lattice generation and graph representation of wrench-closure workspace for arbitrary cable-driven robots. *IEEE Trans. Robot.* **2018**, *35*, 147–161. [CrossRef]
15. Gouttefarde, M.; Daney, D.; Merlet, J.P. Interval-analysis-based determination of the wrench-feasible workspace of parallel cable-driven robots. *IEEE Trans. Robot.* **2010**, *27*, 1–13. [CrossRef]
16. Gagliardini, L.; Gouttefarde, M.; Caro, S. Determination of a dynamic feasible workspace for cable-driven parallel robots. In *Advances in Robot Kinematics*; Springer: Cham, Switzerland, 2018; pp. 361–370.
17. Zhang, Z.; Shao, Z.; Wang, L. Optimization and implementation of a high-speed 3-DOFs translational cable-driven parallel robot. *Mech. Mach. Theory* **2020**, *145*, 103693. [CrossRef]
18. Zhang, Z.; Shao, Z.; Peng, F.; Li, H.; Wang, L. Workspace analysis and optimal design of a translational cable-driven parallel robot with passive springs. *ASME J. Mech. Robot.* **2020**, *12*, 051005. [CrossRef]
19. Tang, X.; Wang, W.; Tang, L. A geometrical workspace calculation method for cable-driven parallel manipulators on minimum tension condition. *Adv. Robot.* **2016**, *30*, 1061–1071. [CrossRef]
20. Pott, A. Efficient computation of the workspace boundary, its properties and derivatives for cable-driven parallel robots. In *Computational Kinematics*; Springer: Cham, Switzerland, 2018; pp. 190–197.

21. Wang, Y.; Belzile, B.; Angeles, J.; Li, Q. Kinematic analysis and optimum design of a novel 2PUR-2RPU parallel robot. *Mech. Mach. Theory* **2019**, *139*, 407–423. [CrossRef]
22. Meng, Q.; Liu, X.J.; Xie, F. Design and development of a Schönflies-motion parallel robot with articulated platforms and closed-loop passive limbs. *Robot. Comput.-Integr. Manuf.* **2022**, *77*, 102352. [CrossRef]
23. Gouttefarde, M.; Nguyen, D.Q.; Baradat, C. Kinetostatic analysis of cable-driven parallel robots with consideration of sagging and pulleys. In *Advances in Robot Kinematics*; Springer: Cham, Swizterland, 2014; pp. 213–221.
24. Hu, X.; Cao, L.; Luo, Y.; Chen, A.; Zhang, E.; Zhang, W. A novel methodology for comprehensive modeling of the kinetic behavior of steerable catheters. *IEEE/ASME Trans. Mechatron.* **2019**, *24*, 1785–1797. [CrossRef]
25. Zhang, W.; Werf, K. Automatic communication from a neutral object model of mechanism to mechanism analysis programs based on a finite element approach in a software environment for CADCAM of mechanisms. *Finite Elem. Anal. Des.* **1998**, *28*, 209–239. [CrossRef]
26. Yang, X.; Wu, H.; Li, Y.; Kang, S.; Chen, B. Computationally efficient inverse dynamics of a class of six-DOF parallel robots: Dual quaternion approach. *J. Intell. Robot. Syst.* **2019**, *94*, 101–113. [CrossRef]
27. Liu, Y.; Li, J.; Zhang, Z.; Hu, X.; Zhang, W. Experimental comparison of five friction models on the same test-bed of the micro stick-slip motion system. *Mech. Sci.* **2015**, *6*, 15–28. [CrossRef]
28. Nguyen, D.Q.; Gouttefarde, M. On the improvement of cable collision detection algorithms. In *Cable-Driven Parallel Robots*; Springer: Cham, Swizterland, 2015; pp. 29–40.

Kinematic Calibration of Parallel Robots Based on L-Infinity Parameter Estimation

Dayong Yu

School of Mechanical Engineering, University of Shanghai for Science and Technology, Shanghai 200093, China; yudayong@usst.edu.cn

Abstract: Pose accuracy is one of the most important problems in the application of parallel robots. In order to adhere to strict pose error bounds, a new kinematic calibration method is proposed, which includes a new pose error model with 60 error parameters and a different kinematic parameter error identification algorithm based on L-infinity parameter estimation. Parameter errors are identified by using linear programming to minimize the maximum difference between predictions and workspace measurements. Simulation results show that the proposed kinematic calibration has better kinematic parameter error estimation and fewer pose errors when measurement noise is less than kinematic parameter errors. Experimental results show that maximum position and orientation errors, respectively, based on the proposed method are decreased by 86.48% and 87.85% of the original values and by 14.32% and 18.23% of those based on the conventional least squares method. The feasibility and validity of the proposed kinematic calibration are verified by improved pose accuracy of the parallel robot.

Keywords: kinematic calibration; parallel robot; parameter estimation; error model; pose accuracy

Citation: Yu, D. Kinematic Calibration of Parallel Robots Based on L-Infinity Parameter Estimation. *Machines* **2022**, *10*, 436. https://doi.org/10.3390/machines10060436

Academic Editors: Zhufeng Shao, Dan Zhang and Stéphane Caro

Received: 24 April 2022
Accepted: 27 May 2022
Published: 1 June 2022

1. Introduction

Parallel robots have higher carrying capacity, greater structural rigidity and better dynamic response than traditional serial robots. Parallel robots have been widely applied in motion simulators, machine tools and medical devices. The pose accuracy of parallel robots is required to be higher and higher in the fields of motion simulation [1,2], mechanical manufacturing [3,4] and surgery [5,6]. The pose accuracy of parallel robots is one of the most important performance measures in the above fields.

The pose accuracy of parallel robots is affected by geometric errors [7–11] and nongeometric errors [12–16]. Geometric errors are mainly caused by manufacturing tolerances and assembly errors. Nongeometric errors might result from clearance, friction, deformation, and so on. Previous studies have shown that geometric errors are the dominant factor leading to pose inaccuracy of parallel robots. It is important for parallel robots to promote pose accuracy in practical application. Pose accuracy improvement in parallel robots is divided into accuracy analysis, synthesis and kinematic calibration.

Accuracy analysis evaluates whether the pose performance of parallel robots meets the design specifications and identifies sensitive factors affecting pose accuracy based on geometric error. An analytical method for the forward and inverse error bound analyses of a Stewart platform was developed by Kim et al. [17]. The relationship between the Stewart platform pose errors and the joint space errors is characterized by the kinematic error model. The forward and inverse error bound are obtained by solving two eigenvalue problems. Comprehensive accuracy modeling and analysis of a new type of lock-or-release mechanism was proposed by Ding et al. [18]. Two accuracy models were established and verified by Monte Carlo simulation and an experiment designed to influence factor sensitivities, and results show that the manufacturing tolerances of a lead screw are the most significant influence factor. Accuracy analysis of a parallel positioning mechanism with actuation redundancy was investigated by Ding et al. [19]. The effects of input uncertainty, components stiffness and redundant limbs were addressed; mean value and standard deviation of the pose errors were computed by optimal Latin hypercube sampling algorithm.

Accuracy synthesis optimally allocates component tolerances of parallel robots under different assembly indices according to the design specification. Accuracy synthesis of a multi-level hybrid positioning mechanism was studied by Tang et al. [20]. Three types of error influence factors are considered in the error model, and the error boundary of the multi-level hybrid positioning mechanism is obtained by using the vector set theory and a linear algebra method. Accuracy synthesis was performed based on a nonlinear optimization algorithm. A comprehensive methodology for implementing the required pose accuracy of a 4-DOF parallel robot was presented by Huang et al. [21]. In this work, all possible geometric errors were separated as either identifiable or unidentifiable geometric errors. The unidentifiable geometric errors were restrained by tolerance design and assembly. Pose accuracy in the whole workspace was achieved by a linear and real-time error compensator. A systematic tolerance design method of parallel link robots was proposed by Takematsu et al. [22]. The standard deviations of the kinematic motions of the end effector were represented by the tolerance values of all joints and links. A suitable set the tolerance values for all joints and links was determined using an optimization algorithm.

Kinematic calibration achieves an inverse kinematic model that more closely matches the actual system in all possible configurations. In general, kinematic calibration can be divided into four steps: error modeling, pose measurement, parameter identification and error compensation. Kinematic calibration can be classified into two categories: external calibration and self-calibration. Kinematic calibration of a Stewart platform was presented by Zhuang et al. [23]. Kinematic error parameters of the Stewart platform were identified using the Gauss–Newton algorithm, and the kinematic error parameters of each leg were solved independently. However, precise pose measurement needs be performed in this approach. Daney [24] established a complete kinematic model of the Gough platform and a unified kinematic parameter identification scheme, and presented an original kinematic calibration method based on the above principle. The accuracy of the Hexapode 300 was experimentally improved by 99% using the original kinematic calibration. A novel identifiable parameter separation method for kinematic calibration of a 6-DOF parallel manipulator was proposed by Hu et al. [25]. The method can reduce the number of kinematic error parameters in the identification model and improve the convergence of the parameter identification algorithm by simple and direct measuring. A systematic kinematic calibration method of a 6-DOF hybrid polishing robot was presented by Huang et al. [26]. Ill-conditioning of the identification Jacobian was dealt with by establishing a linear regression model and implementing kinematic error parameter estimation and pose error compensation using a linear least squares algorithm and Liu estimator. A new error model based on a dimensionless error mapping matrix for kinematic calibration of a 5-axis parallel machining robot was proposed by Luo et al. [27]. Kinematic error parameters are unified into the same unit in the error model and are identified by an iterative least squares procedure based on full pose measurement with a laser tracker. A comprehensive error model for kinematic calibration of a non-fully symmetric parallel Delta robot was presented by Shen et al. [28]. Variations of the parallel Delta robot components and geometric parameters were considered in the error identification model, and the variations were identified by a least squares algorithm.

Kinematic error parameter identification in kinematic calibration can be treated as the best approximation of measurement data. Large amounts of research have been reported on kinematic error parameter identification based on various identification algorithms [29–32]. A novel geometric calibration of industrial robots was presented by Wu et al. [33]. The design of experiments was proposed and added to the conventional kinematic calibration procedure. The additional step is performed before pose measurement in order to obtain a set of optimal measurement poses that ensure the best robot positioning accuracy after kinematic calibration. A dedicated geometric parameter identification algorithm was described, and the identification procedure was divided into two steps. These two steps were repeated iteratively to achieve the desired geometric parameter identification accuracy. A robust kinematic calibration of serial robots based on separable nonlinear least squares was proposed by Mao et al. [34]. The optimal geometric parameter identification problem was converted into a separable nonlinear least squares problem by using the distinctive characteristic of the MDH model. Kinematic calibration of industrial robots based on distance measurement information was presented by Gao et al. [35]. A novel extended Kalman filter and regularized particle filter hybrid identification algorithm was adopted to identify the kinematic parameters of the linearized error model. The algorithm solved the problem with traditional optimization algorithms of being easily affected by measurement noise in high-dimension identification. However, there is little in the literature on reduction of the impact of measurement noise by selecting optimal measurement poses in kinematic calibration. In order to compare the different pose measurement schemes, several observability measures were presented in [36–40] and were used to choose an optimal pose measurement scheme. These measures are not directly related to the pose accuracy of kinematic calibration. A new industry-oriented performance measure is presented in [33] with the intent of ensuring the best robot positioning accuracy after geometric error compensation.

The least squares algorithm has been universally used to identifying kinematic error parameters of parallel robots from pose measurements. Kinematic error can be identified, analyzed and corrected to minimize the sum of squares of the difference between measured errors and computed errors. Although this algorithm is mathematically convenient and can achieve better average pose accuracy in a parallel robot workspace, it may result in pose accuracy not being evenly distributed in the workspace and may even lead to large pose errors outside of the subset.

In order to improve the uneven distribution of pose accuracy and to reduce large pose error, a new kinematic calibration method for parallel robots is presented based on L-infinity parameter estimation and applied to the spacecraft docking motion simulation system. The paper is organized as follows: An inverse kinematic model of the parallel robot is described, and a forward kinematic solution is presented in Section 2. A pose error model for kinematic calibration is established in Section 3. A new kinematic parameter error identification algorithm based on L-infinity parameter estimation is proposed in Section 4. Simulations and experiments are performed, and the results are shown in Section 5. Finally, some conclusions are given in Section 6.

2. Kinematic Model

A parallel robot model is composed of a moving platform, a base, and six identical hydraulic cylinders with variable lengths, as shown in Figure 1. The moving platform's position relative to the base can be controlled by varying the length of the six hydraulic cylinders. The parallel robot has six DOF. The base coordinate system O_B-xyz is located in the center of the base. The mobile coordinate system O_P-xyz is attached to the center of the moving platform. All vectors and matrices will be denoted in bold letters. The two coordinate systems O_B-xyz and O_P-xyz can be related to each other through a vector $\mathbf{q} = \begin{bmatrix} x & y & z & \phi & \theta & \psi \end{bmatrix}^T$ that describes the pose of the moving platform by its position (longitudinal (x), lateral (y) and vertical (z) displacements) and its orientation (Roll (ϕ), Pitch (θ) and Yaw (Ψ) angles). Thus, the position of the moving platform can be expressed by a position vector $\mathbf{t}$ as

$$\mathbf{t} = \begin{bmatrix} x & y & z \end{bmatrix}^T \tag{1}$$

and the orientation of the moving platform can be expressed by a rotation matrix $\mathbf{R}$ as

$$\mathbf{R} = \begin{bmatrix} \cos\psi\cos\theta & \cos\psi\sin\theta\sin\phi - \sin\psi\cos\phi & \cos\psi\sin\theta\cos\phi + \sin\psi\sin\phi \\ \sin\psi\cos\theta & \sin\psi\sin\theta\sin\phi + \cos\psi\cos\phi & \sin\psi\sin\theta\cos\phi - \cos\psi\sin\phi \\ -\sin\theta & \cos\theta\sin\phi & \cos\theta\cos\phi \end{bmatrix} \tag{2}$$

where ϕ, θ, and ψ are three Roll, Pitch and Yaw (RPY) angles chosen with respect to the x-axes, the y-axes and the z-axes, respectively, of the base coordinate system O_B-xyz.

Figure 1. The parallel robot.

2.1. Inverse Kinematics

Referring to Figures 2 and 3, $\mathbf{u}_i$ is the unit vector along the ith hydraulic cylinder direction, and l_i is the length of the ith hydraulic cylinder; $\mathbf{a}_i$ is the position vector from O_P to A_i and is represented in the mobile coordinate system O_P-xyz, and $\mathbf{b}_i$ is the position vector from O_B to B_i and is represented in the base coordinate system O_B-xyz. A vector chain equation can be expressed as

$$l_i\mathbf{u}_i = \mathbf{R}\mathbf{a}_i + \mathbf{t} - \mathbf{b}_i \; i = 1, 2, \ldots, 6 \tag{3}$$

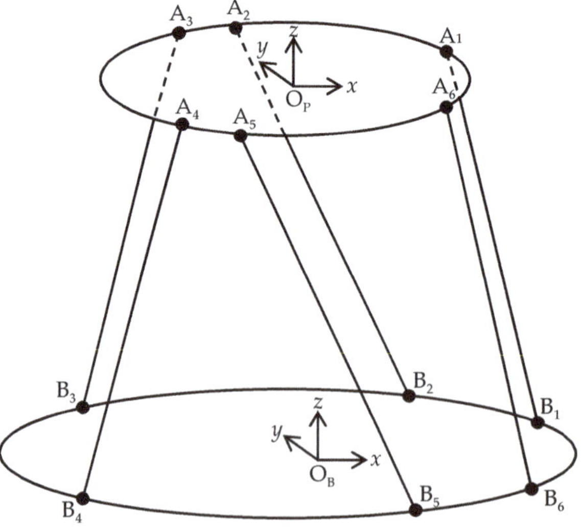

Figure 2. Coordinate system of the parallel robot.

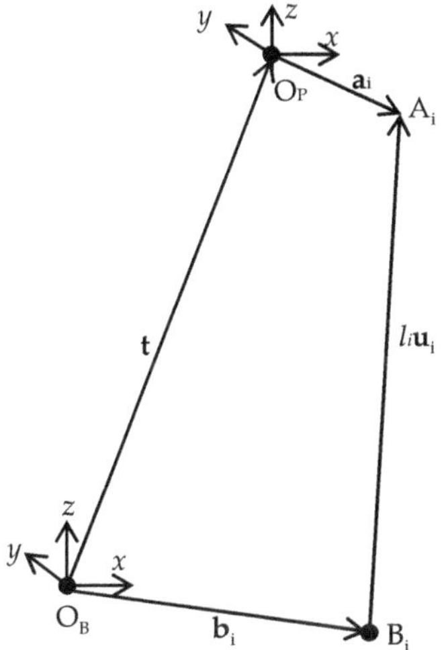

Figure 3. Vector chain for hydraulic cylinder *i*.

The vector chain equation is derived for the perfect (no errors) parallel robot. The length of the *i*th hydraulic cylinder can be computed from

$$l_i = f_i(\mathbf{R}, \mathbf{t}) = \sqrt{(\mathbf{Ra}_i + \mathbf{t} - \mathbf{b}_i)^T (\mathbf{Ra}_i + \mathbf{t} - \mathbf{b}_i)} \ i = 1, 2, \ldots, 6 \tag{4}$$

and the measured length of the *i*th hydraulic cylinder can be obtained by

$$s_i = l_i - l_{0,i} \ i = 1, 2, \ldots, 6 \tag{5}$$

where $l_{0,i}$ is the initial length of the *i*th hydraulic cylinder.

2.2. Forward Kinematics

The forward kinematics of the parallel robot compute the moving platform pose when the measured hydraulic cylinder lengths are given and the kinematic parameters are known. Although the inverse kinematics for the parallel robot can be expressed in a closed form, forward kinematics offer no analytical solution. Mapping the pose using the hydraulic cylinder lengths is complicated to solve (Equation (4)). Numerical methods are often employed to solve the forward kinematics for parallel robots. The following method for the forward kinematics of a parallel robot is based on the Newton–Raphson algorithm.

For solving the forward kinematics of a parallel robot, a vector function is defined to describe the difference between the estimated hydraulic cylinder length s_{ei} and the measured hydraulic cylinder length s_{ai}.

$$\mathbf{f} = \begin{bmatrix} f_1 \\ \vdots \\ f_6 \end{bmatrix} = \begin{bmatrix} s_{e1}^2 - s_{a1}^2 \\ \vdots \\ s_{e6}^2 - s_{a6}^2 \end{bmatrix} \tag{6}$$

The Newton–Raphson algorithm can be stated as:

(1) Measure s_{ai}, and select an initial guess for the pose, $\mathbf{q}$.
(2) Compute s_{ei} based on $\mathbf{q}$.
(3) Form $\mathbf{f}$.
(4) If $\mathbf{q}^T \mathbf{q} <$ tolerance$_1$, exit with $\mathbf{q}$ as the solution.
(5) Compute the partial derivative matrix $\mathbf{J} = \frac{\partial \mathbf{f}}{\partial \mathbf{q}}$ such that $\mathbf{J}_{i,j} = \frac{\partial f_i}{\partial q_j}$.

(6) Obtain update $\delta\mathbf{q}$ by solving $\mathbf{J}\delta\mathbf{q} = -\mathbf{f}$.

(7) If $\delta\mathbf{q}^T\delta\mathbf{q} < \text{tolerance}_2$, exit with $\mathbf{q}$ as the solution.

(8) Update $\mathbf{q}$ by $\mathbf{q} = \mathbf{q} + \delta\mathbf{q}$ and go to step (2).

The accuracy and rate convergence for the Newton–Raphson algorithm depend on several factors. The algorithm rapidly converges if the initial guess is in the neighborhood of the solution, and the algorithm is fairly robust with the choice of the initial guess. If the second order terms are large, this first order approach will not be accurate, and the algorithm will converge very slowly. The existence of the Jacbian inverse is required in step (6), and thus the moving platform may not be near a singularity configuration. If convergence problems arise, or if speed is of paramount importance, the forward kinematics may require a different algorithm, such as those presented in [41–44].

3. Error Model

The kinematic parameters of the parallel robot might be different from those in the design specification due to imprecision in manufacturing and assembly of the joints and the initial length of the hydraulic cylinders. The difference will lead to pose error of the moving platform. An error model relating the kinematic parameter errors to the pose errors is derived in this section.

A vector differential error model is obtained by performing the following differentiation for Equation (3) as

$$\delta l_i\mathbf{u}_i + l_i\delta\mathbf{u}_i = \delta\mathbf{R}\mathbf{a}_i + \mathbf{R}\delta\mathbf{a}_i + \delta\mathbf{t} - \delta\mathbf{b}_i \quad i = 1, 2, \ldots, 6 \tag{7}$$

where δl_i is the length error of l_i, $\delta\mathbf{u}_i$ is the deviation vector of $\mathbf{u}_i$, $\delta\mathbf{R}$ is the deviation matrix of $\mathbf{R}$, $\delta\mathbf{a}_i$ is the position error vector of $\mathbf{a}_i$, $\delta\mathbf{t}$ is a deviation vector of $\mathbf{t}$, and $\delta\mathbf{b}_i$ is the position error vector of $\mathbf{b}_i$.

The deviation vector $\delta\mathbf{u}_i$ can be expressed as

$$\delta\mathbf{u}_i = \Delta_{\mathbf{u}_i}\mathbf{u}_i = \begin{bmatrix} 0 & -\delta u_{iz} & \delta u_{iy} \\ \delta u_{iz} & 0 & -\delta u_{ix} \\ -\delta u_{iy} & \delta u_{ix} & 0 \end{bmatrix} \begin{bmatrix} u_{ix} \\ u_{iy} \\ u_{iz} \end{bmatrix} \tag{8}$$

where $\Delta_{\mathbf{u}_i}$ is a skew symmetric matrix of $\delta\mathbf{u}_i$.

Let $\delta\omega$ be the angular error vector of the nominal RPY angles ϕ, θ and ψ, and be represented in the base coordinate system. The angular error vector $\delta\omega$ can be expressed as

$$\delta\omega = \begin{bmatrix} \delta\omega_x \\ \delta\omega_y \\ \delta\omega_z \end{bmatrix} = \begin{bmatrix} -\sin\psi\delta\theta + \cos\psi\cos\theta\delta\phi \\ \cos\psi\delta\theta + \sin\psi\cos\theta\delta\phi \\ -\sin\theta\delta\phi + \delta\psi \end{bmatrix} \tag{9}$$

The skew symmetric matrix of $\delta\omega$ can be written as

$$\Delta\omega = \begin{bmatrix} 0 & -\delta\omega_z & \delta\omega_y \\ \delta\omega_z & 0 & -\delta\omega_x \\ -\delta\omega_y & \delta\omega_x & 0 \end{bmatrix} \tag{10}$$

The deviation matrix $\delta\mathbf{R}$ can be given by

$$\delta\mathbf{R} = \Delta_\omega\mathbf{R} = \begin{bmatrix} 0 & \sin\theta\delta\phi - \delta\psi & \cos\psi\delta\theta + \sin\psi\cos\theta\delta\phi \\ -\sin\theta\delta\phi + \delta\psi & 0 & \sin\psi\delta\theta - \cos\psi\cos\theta\delta\phi \\ -\cos\psi\delta\theta - \sin\psi\cos\theta\delta\phi & -\sin\psi\delta\theta + \cos\psi\cos\theta\delta\phi & 0 \end{bmatrix}\mathbf{R} \tag{11}$$

Substituting Equations (8) and (11) into Equation (7) yields

$$\delta l_i\mathbf{u}_i + l_i\Delta_{\mathbf{u}_i}\mathbf{u}_i = \Delta_\omega\mathbf{R}\mathbf{a}_i + \mathbf{R}\delta\mathbf{a}_i + \delta\mathbf{t} - \delta\mathbf{b}_i \quad i = 1, 2, \ldots, 6 \tag{12}$$

Let $\mathbf{a}'_i = \mathbf{R}\mathbf{a}_i$, Equation (12) can be rewritten as

$$\delta l_i \mathbf{u}_i + l_i \Delta_{\mathbf{u}_i} \mathbf{u}_i = \Delta_{\boldsymbol{\omega}} \mathbf{a}'_i + \mathbf{R}\delta \mathbf{a}_i + \delta \mathbf{t} - \delta \mathbf{b}_i \; i = 1, 2, \ldots, 6 \tag{13}$$

Equation (13) can be expressed in matrix form as

$$\begin{bmatrix} \mathbf{I} & \Delta_{\mathbf{a}'_i}^T \end{bmatrix} \begin{bmatrix} \delta \mathbf{t} \\ \delta \boldsymbol{\omega} \end{bmatrix} = \begin{bmatrix} \mathbf{u}_i & l_i \Delta_{\mathbf{u}'_i}^T & -\mathbf{R} & \mathbf{I} \end{bmatrix} \begin{bmatrix} \delta l_i \\ \delta \mathbf{u}_i \\ \delta \mathbf{a}_i \\ \delta \mathbf{b}_i \end{bmatrix} \; i = 1, 2, \ldots, 6 \tag{14}$$

where $\mathbf{I}$ is 3×3 unit matrix, $\Delta_{\mathbf{a}'_i}$ is a skew symmetric matrix of $\mathbf{a}'_i$, and $\Delta_{\mathbf{u}'_i}$ is a skew symmetric matrix of $\mathbf{u}_i$.

Equation (14) can be rewritten as

$$\mathbf{J}_{\Omega_i} \delta \Omega = \mathbf{J}_i \delta \mathbf{p}_i \; i = 1, 2, \ldots, 6 \tag{15}$$

where

$$\delta \Omega = \begin{bmatrix} \delta \mathbf{t}^T & \delta \boldsymbol{\omega}^T \end{bmatrix}^T = \begin{bmatrix} \delta x & \delta y & \delta z & \delta \omega_x & \delta \omega_y & \delta \omega_z \end{bmatrix}^T \tag{16}$$

represents the pose error of the parallel robot, and the following matrices, $\mathbf{J}_{\Omega_i}$ and $\mathbf{J}_i$ are the inverse and forward error mapping components defined as

$$\mathbf{J}_{\Omega_i} = \begin{bmatrix} \mathbf{I} & \Delta_{\mathbf{a}'_i}^T \end{bmatrix} \tag{17}$$

$$\mathbf{J}_i = \begin{bmatrix} \mathbf{u}_i & l_i \Delta_{\mathbf{u}'_i}^T & -\mathbf{R} & \mathbf{I} \end{bmatrix} \tag{18}$$

and

$$\delta \mathbf{p}_i = \begin{bmatrix} \delta l_i & \delta u_{ix} & \delta u_{iy} & \delta u_{iz} & \delta a_{ix} & \delta a_{iy} & \delta a_{iz} & \delta b_{ix} & \delta b_{iy} & \delta b_{iz} \end{bmatrix}^T \tag{19}$$

represents the kinematic parameter errors in the individual vector chain.

Considering all six vector chains, Equation (14) can be expressed in the following matrix form

$$\begin{bmatrix} \mathbf{I} & \Delta_{\mathbf{a}'_1}^T \\ \mathbf{I} & \Delta_{\mathbf{a}'_2}^T \\ \mathbf{I} & \Delta_{\mathbf{a}'_3}^T \\ \mathbf{I} & \Delta_{\mathbf{a}'_4}^T \\ \mathbf{I} & \Delta_{\mathbf{a}'_5}^T \\ \mathbf{I} & \Delta_{\mathbf{a}'_6}^T \end{bmatrix} \begin{bmatrix} \delta \mathbf{t} \\ \delta \boldsymbol{\omega} \end{bmatrix} = \begin{bmatrix} \mathbf{J}_1 & & & & & \\ & \mathbf{J}_2 & & & & \\ & & \mathbf{J}_3 & & & \\ & & & \mathbf{J}_4 & & \\ & & & & \mathbf{J}_5 & \\ & & & & & \mathbf{J}_6 \end{bmatrix} \begin{bmatrix} \delta \mathbf{p}_1 \\ \delta \mathbf{p}_2 \\ \delta \mathbf{p}_3 \\ \delta \mathbf{p}_4 \\ \delta \mathbf{p}_5 \\ \delta \mathbf{p}_6 \end{bmatrix} \tag{20}$$

Equation (20) above can be rewritten as

$$\mathbf{J}_\Omega \delta \Omega = \mathbf{J}_\mathbf{p} \delta \mathbf{p} \tag{21}$$

where $\mathbf{J}_\Omega$ represents the inverse error mapping matrix of the parallel robot, $\mathbf{J}_\mathbf{p}$ represents the forward error mapping matrix, and $\delta \mathbf{p}$ represents the kinematic parameter errors for all the vector chains. The vector $\delta \mathbf{p}$ contains 60 linearly independent error parameters, and the jth element of the vector can be denoted as δp_j.

The pose error of the parallel robot can be computed by

$$\delta \Omega = \mathbf{J} \delta \mathbf{p} \tag{22}$$

where

$$\mathbf{J} = \left(\mathbf{J}_\Omega^T \mathbf{J}_\Omega \right)^{-1} \mathbf{J}_\Omega^T \mathbf{J}_\mathbf{p} \tag{23}$$

is defined as the error Jacobian matrix for the parallel robot, and its condition number will be used to choose the optimal pose measurement configurations.

The relationship between the pose errors of the parallel robot and the kinematic parameter errors is described by Equation (22). It is a linear equation in terms of the unknown kinematic parameter errors, which can be identified based on L-infinity parameter estimation once the pose errors of the parallel robot are measured.

4. Calibration Method

The least squares fit is universally used to identify kinematic parameter errors from measurement data in kinematic calibration. Kinematic error can be identified, analyzed and corrected to minimize the sum of squares of the difference between measured errors and computed errors. Thus, for a parallel robot using a control model compensated with kinematic parameter errors and measuring a number of poses in its workspace, nothing can be said of its accuracy at any one pose. If the sample of poses measured represents an unbiased sample of the workspace, the mean squares errors of the parallel robot at these poses is minimized. That is, the least squares fit does not minimize or bound the pose error between the measured pose errors and the computed pose errors based on the error model at a single pose.

The parallel robot is used with the spacecraft docking motion simulation system, so its pose accuracy will be evaluated not on the basis of average error of all poses on a simulated trajectory, but based on the error of each pose of a simulated trajectory meeting a given accuracy specification. In order to achieve the given accuracy requirement at any one pose in the whole workspace, a different kinematic parameter error identification algorithm based on L-infinity parameter estimation is selected. It identifies kinematic parameter errors of the parallel robot by minimizing the maximum difference between measured pose errors and computed pose errors based on an error model and can bound large pose errors and equalize pose errors across the workspace. Unknown kinematic parameter errors of the error model (Equation (22)) can be identified by the following formulation based on L-infinity parameter estimation:

$$\min \max |\delta \Omega_i| \tag{24}$$

where $\delta \Omega_i$ is computed by

$$\delta \Omega_i = \delta \Omega_i^m - \delta \Omega_i^c \quad i = 1, 2, \ldots, n \tag{25}$$

$\delta \Omega_i^m$ is the measured pose error, $\delta \Omega_i^c$ is the computed pose error based on Equation (22), and n represents the number of measurement poses in the workspace of the parallel robot.

Equation (22) is rewritten in terms of the kinematic parameter errors at the ith measurement pose as

$$\delta \Omega_{ki}^c = \sum_{j=1}^{60} J_{ki}^j \delta p_j \quad k = 1, 2, \ldots, 6 \tag{26}$$

The total number of identification equations will be six times the total number of measurement poses. The index assigned to the identification equations will be w, and it can take values between 1 and $6n$. Substituting Equations (25) and (26) into Equation (24), the following is obtained:

$$\min \max \left| \delta \Omega_w^m - \sum_{j=1}^{60} J_w^j \delta p_j \right| \quad w = 1, 2, \ldots, 6n \tag{27}$$

Equation (27) is subject to no restrictions. The L-infinity parameter identification problem can be converted to a linear programming problem with the introduction of the variable z, thus the following is obtained:

$$\min z = \sum_{j=1}^{60} O_j \delta p_j + y \tag{28}$$

subject to

$$y + \sum_{j=1}^{60} J_w^j \delta p_j \geq \delta \Omega_w^m \quad w = 1, 2, \ldots, 6n \tag{29}$$

$$y - \sum_{j=1}^{60} J_w^j \delta p_j \geq -\delta \Omega_w^m \quad w = 1, 2, \ldots, 6n \tag{30}$$

$$y = \max \left| \delta \Omega_w^m - \sum_{j=1}^{60} J_w^j \delta p_j \right| \quad w = 1, 2, \ldots, 6n \tag{31}$$

The variable y represents the absolute value of the maximum discrepancy between the measured pose errors and the computed pose errors based on the error model in the above linear program, and $\mathbf{O}$ represents a 1×60 zero vector. The unknown kinematic parameter errors can be identified by minimizing the variable z. The above linear programming problem can be solved by using the simplex method [45]. For identification of 60 kinematic parameter errors, it is expected to measure the poses that are located near or at the boundaries of the workspace, which can provide sufficient pose error vectors to expand the parameter space of the error model (Equation (22)). At least 30 measurement poses are required in kinematic calibration of the parallel robot.

5. Simulations and Experiments

5.1. Model Verification

In order to verify the pose error model derived in Section 3, a numerical simulation scheme is designed and performed by computer programs. The nominal kinematic parameters and the assumed kinematic parameter errors are listed in Tables 1 and 2, respectively. The procedure can be described as follows:

1. Select a set of desired poses evenly distributed in the workspace.
2. Compute the measured lengths of the six hydraulic cylinders by using inverse kinematics with the nominal kinematic parameters in Table 1.
3. Actuate the parallel robot to the selected poses in sequence with the measured lengths of the hydraulic cylinders, and compute the actual poses by using forward kinematics with the actual kinematic parameters (the nominal kinematic parameters plus the assumed kinematic parameter errors in Table 2).
4. Compute the actual pose errors, namely, subtract the selected poses from the actual poses.
5. Compute the pose errors by using the pose error model with the nominal kinematic parameters, the lengths and the unit vectors of the hydraulic cylinders, and the kinematic parameter errors.
6. Draw the contrasting curves of the position error and the orientation error for the above numerical simulation results in Figures 4 and 5.

Table 1. The nominal kinematic parameters.

	a_{ix} (mm)	a_{iy} (mm)	a_{iz} (mm)	b_{ix} (mm)	b_{iy} (mm)	b_{iz} (mm)	l_{ix} (mm)	l_{iy} (mm)	l_{iz} (mm)
1	1394.7	122.0	0	2049.3	3038.5	0	−654.6	−2916.5	3091.2
2	−591.7	1268.8	0	1606.8	3294.0	0	−2198.5	−2025.2	3091.2
3	−803.0	1146.8	0	−3656.1	255.5	0	2853.1	891.3	3091.2
4	−803.0	−1146.8	0	−3656.1	−255.5	0	2853.1	−891.3	3091.2
5	−591.7	−1268.8	0	1606.8	−3294.0	0	−2198.5	2025.2	3091.2
6	1394.7	−122.0	0	2049.3	−3038.5	0	−654.6	2916.5	3091.2

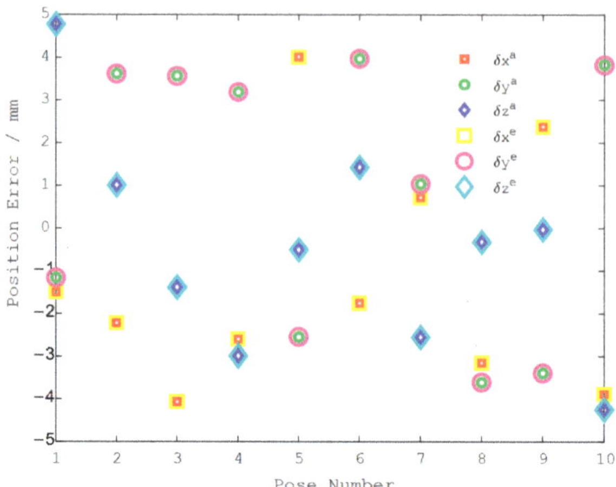

Figure 4. Comparison of the position error computation results.

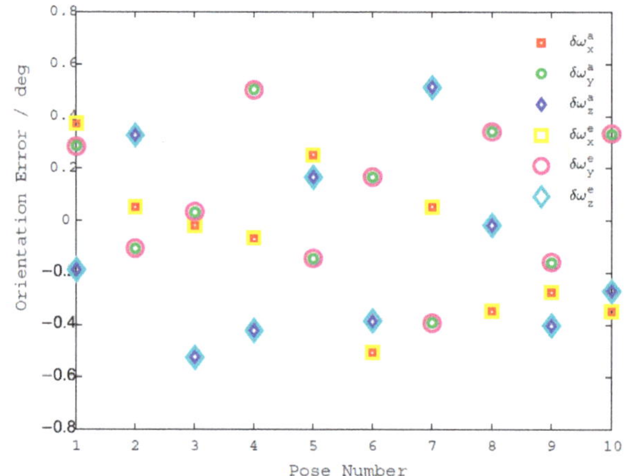

Figure 5. Comparison of the orientation error computation results.

Table 2. The assumed kinematic parameter errors.

	δa_{ix} (mm)	δa_{iy} (mm)	δa_{iz} (mm)	δb_{ix} (mm)	δb_{iy} (mm)	δb_{iz} (mm)	δl_{ix} (mm)	δl_{iy} (mm)	δl_{iz} (mm)
1	0.90	-0.09	0.84	-0.18	-0.72	-0.97	0.69	0.36	-0.39
2	-0.54	-0.96	0.48	0.79	-0.59	0.49	0.05	-0.24	-0.62
3	0.21	0.64	-0.65	-0.88	-0.60	-0.11	-0.59	0.66	-0.61
4	-0.03	-0.11	-0.19	-0.29	0.21	0.86	0.34	0.05	0.36
5	0.78	0.23	0.87	0.63	-0.46	-0.07	0.68	0.42	-0.39
6	0.52	0.58	0.83	-0.98	-0.61	-0.16	-0.96	-0.14	0.08

The following conclusions can be summarized from Figures 4 and 5.

1. Pose errors vary at different locations in the workspace. The pose errors are affected not only by kinematic parameter errors, but also by the pose of the parallel robot.
2. Pose errors computed by using the pose error model are basically consistent with the actual pose errors.

Therefore, the proposed pose error model is verified to be correct and to represent the kinematic parameter errors of the parallel robot.

5.2. Identification Simulations

Kinematic parameter error identifications were simulated with various kinematic parameter errors, measurement noise levels and pose configuration sets. The actual coordinates of the feature points of the moving platform were measured by a coordinate measuring machine, and the actual poses of the parallel robot were computed. Three kinematic parameter error sets were given, with the assumed kinematic parameter errors obtained from normal distributions with variances of 0.01 mm (set I), 0.1 mm (set II) and 1 mm (set III). These kinematic parameter error sets are shown in Tables 3–5, respectively. Gaussian noise with variances of 0.0001 mm, 0.001 mm, 0.01 mm and 0.1 mm was added to the coordinate measurements of the feature points of the moving platform to simulate measurement noise. Four different pose sets were used in the identification simulations. Pose set 1 contains 32 random poses. Pose set 2 contains 24 poses based on a full factorial exploration of the six pose variable limits. Pose set 3 contains 32 poses selected from the workspace using a coordinate exchange algorithm for optimal experimental design. Pose set 4 contains 64 poses selected using a coordinate exchange algorithm.

Table 3. The assumed kinematic parameter errors with variances of 0.01 mm.

	δa_{ix} (mm)	δa_{iy} (mm)	δa_{iz} (mm)	δb_{ix} (mm)	δb_{iy} (mm)	δb_{iz} (mm)	δl_{ix} (mm)	δl_{iy} (mm)	δl_{iz} (mm)
1	0.0538	0.1834	−0.2259	0.0862	0.0319	−0.1308	−0.0434	0.0343	0.3578
2	0.2769	−0.1350	0.3035	0.0725	−0.0063	0.0715	−0.0205	−0.0124	0.1490
3	0.1409	0.1417	0.0671	−0.1207	0.0717	0.1630	0.0489	0.1035	0.0727
4	−0.0303	0.0294	−0.0787	0.0888	−0.1147	−0.1069	−0.0809	−0.2944	0.1438
5	0.0325	−0.0755	0.1370	−0.1712	−0.0102	−0.0241	0.0319	0.0313	−0.0865
6	−0.0030	−0.0165	0.0628	0.1093	0.1109	−0.0864	0.0077	−0.1214	−0.1114

Table 4. The assumed kinematic parameter errors with variances of 0.1 mm.

	δa_{ix} (mm)	δa_{iy} (mm)	δa_{iz} (mm)	δb_{ix} (mm)	δb_{iy} (mm)	δb_{iz} (mm)	δl_{ix} (mm)	δl_{iy} (mm)	δl_{iz} (mm)
1	−0.0022	0.4847	−0.2434	0.1174	−0.0713	0.3533	−0.3444	0.0103	0.1747
2	0.3480	0.4883	0.0272	−0.4717	−0.2347	−0.3357	0.7433	−0.1947	0.2366
3	−0.0608	0.2810	−0.2419	−0.4434	−0.4498	0.1544	−0.0561	−0.0620	0.4488
4	0.0922	0.0626	0.5021	−0.2544	0.2203	0.2641	−0.0771	0.0682	−0.3687
5	−0.3630	0.0332	0.2284	0.8176	−0.2109	0.0592	−0.0261	−0.6113	−0.1388
6	−0.5675	0.2658	−0.2808	0.0317	−0.1722	0.0960	−0.1898	0.1549	0.2338

Table 5. The assumed kinematic parameter errors with variances of 1 mm.

	δa_{ix} (mm)	δa_{iy} (mm)	δa_{iz} (mm)	δb_{ix} (mm)	δb_{iy} (mm)	δb_{iz} (mm)	δl_{ix} (mm)	δl_{iy} (mm)	δl_{iz} (mm)
1	1.7119	−0.1941	−2.1384	−0.8396	1.3546	−1.0722	0.9610	0.1240	1.4367
2	−1.9609	−0.1977	−1.2078	2.9080	0.8252	1.3790	−1.0582	−0.4686	−0.2725
3	1.0984	−0.2779	0.7015	−2.0518	−0.3538	−0.8236	−1.5771	0.5080	0.2820
4	0.0335	−1.3337	1.1275	0.3502	−0.2991	0.0229	−0.2620	−1.7502	−0.2857
5	−0.8314	−0.9792	−1.1564	−0.5336	−2.0026	0.9642	0.5201	−0.0200	−0.0348
6	−0.7982	1.0187	−0.1332	−0.7145	1.3514	−0.2248	−0.5890	−0.2938	−0.8479

For each simulation, kinematic parameter errors were identified using L-infinity parameter estimation based on the LINPROG function of the MATLAB Optimization Toolbox. Kinematic parameter identification error was computed as the root mean square value of the difference between the actual kinematic parameter errors and the identified kinematic parameter errors. In order to evaluate the resulting pose accuracy improvement, pose errors were computed before and after kinematic calibration by using the pose error model given in the previous section in this paper. Position error was computed as the maximum absolute value of the error along the x, y and z axes at 100 evaluation poses randomly distributed in the workspace. Orientation error was computed as the maximum absolute value of the error around the x, y and z axes at the same poses as above. All identification simulations where the measurement noise level was less than the kinematic parameter errors resulted in better kinematic parameter error identification and higher pose accuracy.

The effects of pose selection on kinematic calibration are shown in Figures 6–8. Notice that identification of kinematic parameter error obtained by using pose set 1 are consistently worse than those obtained by using pose set 2, even though pose set 1 has more poses than pose set 2. This shows that choosing the poses is more important than the number of poses contained in the pose set. However, very little improvement can be obtained once the number of poses exceeds a certain limit. For the rest of the discussions, kinematic calibration using pose set 3 will be compared, since this pose set yielded good identification results with only 32 poses.

The effect of measurement noise on kinematic calibration is shown in Figures 9 and 10. Notice that pose accuracy improves as measurement noise is reduced, and kinematic parameter errors are perfectly identified when measurement noise is close to zero. This shows that measurement noise should be at least an order of magnitude lower than the desired pose accuracy.

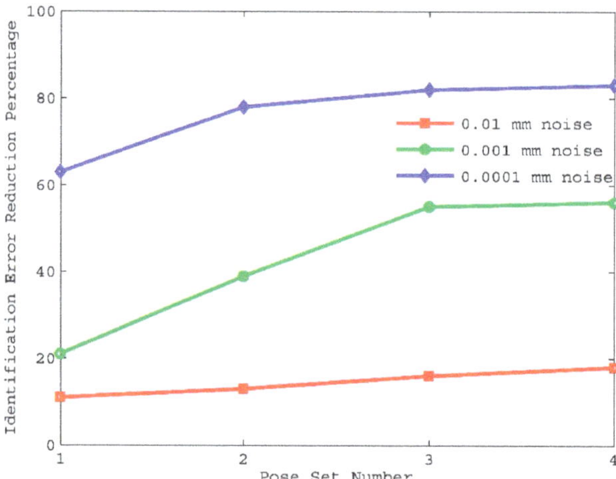

Figure 6. Identification error reduction percentage versus pose selection for set I.

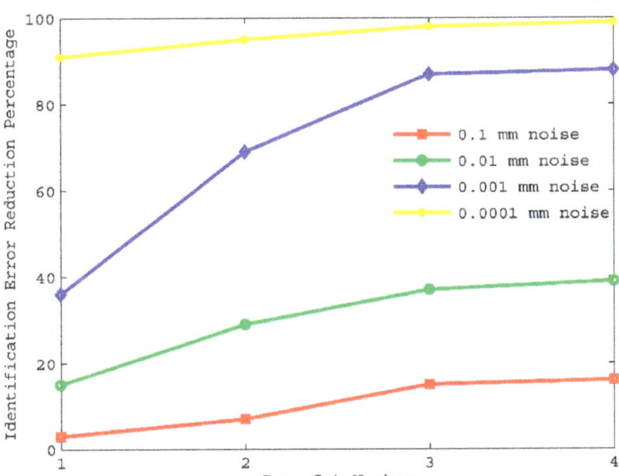

Figure 7. Identification error reduction percentage versus pose selection for set II.

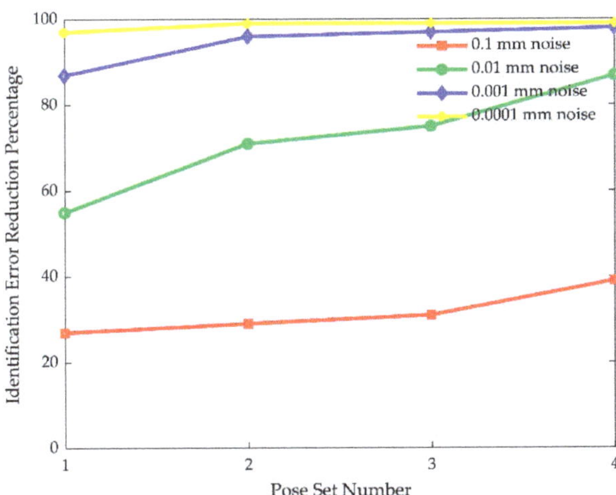

Figure 8. Identification error reduction percentage versus pose selection for set III.

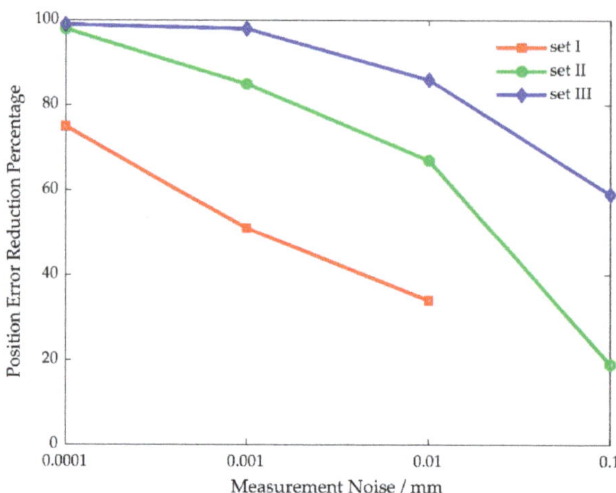

Figure 9. Position error reduction percentage versus measurement noise.

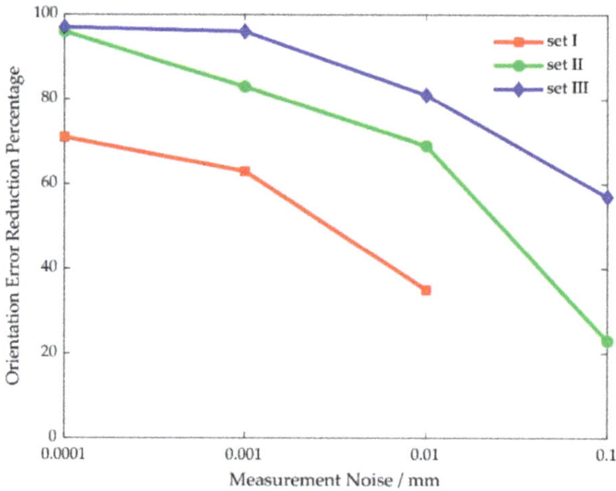

Figure 10. Orientation error reduction percentage versus measurement noise.

5.3. Comparison Experiments

The experimental system mainly consisted of a parallel robot, a three-dimensional coordinate-measuring machine and six standard spheres. The experimental system for kinematic calibration based on L-infinity parameter estimation is shown in Figure 11. Pose measurement of the parallel robot was done with a precise three-dimensional coordinate measuring machine, model 3000i manufactured by STAR Tech. The measuring machine has a point repeatability of 0.010 mm and a length accuracy of 0.016 mm in the 1.2 m × 1.2 m × 1.2 m measuring range. Three of these spheres were fixed at three specific locations of the moving platform, and the other three were fixed at three specific base locations. On this basis, pose measurement of the parallel robot was developed, which mainly measured the distances from three standard spheres on the moving platform to three standard spheres on the base by the coordinate-measuring machine. Then, the poses were computed using these distances, as shown in Figure 12. It is worth mentioning that the kinematic calibration experiments were performed in a limited area due to the length measurement limitation of the coordinate-measuring machine.

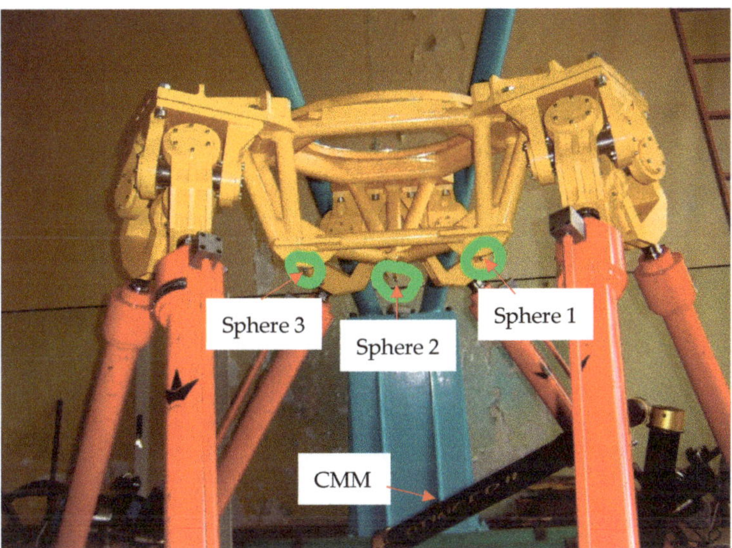

Figure 11. The experimental system.

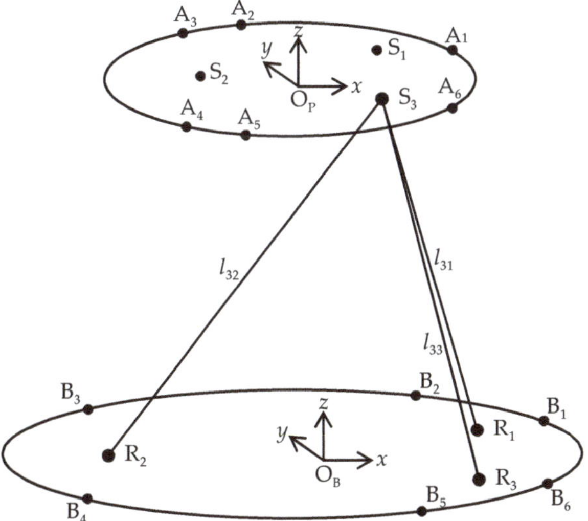

Figure 12. The schematic diagram of pose measurement.

 The comparison experiments of kinematic calibration were performed using the conventional least squares algorithm and the proposed L-infinity parameter identification algorithm. The two results were compared according to the following four indicators: (1) maximum error, (2) range of error, (3) average error and (4) root mean square error. According to the pose error model, full pose measurement is needed to solve kinematic parameter errors in the two kinematic calibrations. The full pose could be obtained by using a mobile, flexible triad coordinate measuring machine. On the foundation of the abovementioned identification simulations, a measured pose selection rule was determined to make the comparison more effective and to better carryout the experiment: 32 measurement poses based on pose set 3 were chosen to cover the whole workspace by using the union of a full factorial exploration and a coordinate exchange algorithm in the comparison experiments. Meanwhile, 24 verification poses evenly distributed in the workspace of the parallel robot were also collected to validate pose accuracy improvement in the experiment.

Pose error measurement and kinematic parameter error identification were accomplished by using the original 32 pose errors at the measurement poses listed in Table 6. Then a new round of pose measurement was performed to evaluate pose accuracy after the two kinematic calibrations. The pose errors at verification poses based on the two kinematic calibrations were obtained and are shown in Figures 13 and 14, and the four pose error indicators before and after kinematic calibration are listed in Table 7. The maximum position error after kinematic calibration by using conventional least squares algorithm was 1.0980 mm, and the maximum position error after kinematic calibration by using the proposed L-infinity parameter identification algorithm was 0.9408 mm. The maximum orientation errors were 0.1037 and 0.0848 degree, respectively. The maximum position error based on least squares was reduced by 84.22%, and the maximum position error based on L-infinity was reduced by 86.48%. The maximum orientation errors were reduced by 85.14% and 87.85%, respectively. The range of position error based on least squares is reduced by 85.76%, and the range of position errors based on L-infinity was reduced by 87.01%. The range of orientation errors were reduced by 87.67% and 88.99%, respectively. The percentage reductions of average error and root-mean-square error for a conventional least squares algorithm were almost identical at 83.37% and 83.42%, respectively. The percentage reductions of average error and root-mean-square error for the L-infinity parameter identification algorithm was similar to the conventional least squares algorithm. The two errors were 85.29% and 85.34%, respectively. Clearly, both kinematic calibrations can improve pose accuracy, while kinematic calibration based on the L-infinity parameter identification algorithm is much better than the conventional least squares algorithm. This verifies that the proposed kinematic calibration based on the L-infinity parameter identification algorithm is effective and can achieve strict bounds on the pose errors produced by the parallel robot.

Table 6. The pose errors at measurement poses before kinematic calibration.

	δx (mm)	δy (mm)	δz (mm)	$\delta\omega_x$ (deg)	$\delta\omega_y$ (deg)	$\delta\omega_z$ (deg)
1	0.7286	−6.3683	3.9254	−0.6993	0.0787	−0.5509
2	3.2461	3.7276	−4.6233	0.5205	0.1881	−0.4997
3	0.4247	−3.5500	5.2487	0.1637	−0.6549	−0.4653
4	7.1205	−0.7135	7.0661	0.6958	0.1667	0.1756
5	−3.8926	2.7736	0.3100	0.0440	−0.1890	0.1089
6	−5.4966	−1.8954	5.5656	−0.0239	−0.6302	−0.6266
7	5.4412	3.4634	1.3559	0.4299	−0.0097	0.6130
8	−6.0964	−1.3912	−4.8010	−0.3787	−0.4286	0.3274
9	−1.2509	2.7113	−4.1599	0.0023	0.5265	0.3404
10	−0.6286	3.0045	−1.2172	0.5702	−0.4103	−0.6106
11	−1.8018	−0.7148	3.6391	0.1103	−0.4934	0.5132
12	3.8494	−6.7218	4.7315	0.4917	−0.4334	0.6175
13	1.9224	−2.2985	4.2254	0.3415	−0.6399	0.6880
14	3.9698	−0.9706	−2.4738	0.1262	0.1956	0.51111

Table 6. *Cont.*

	δx (mm)	δy (mm)	δz (mm)	$\delta \omega_x$ (deg)	$\delta \omega_y$ (deg)	$\delta \omega_z$ (deg)
15	6.2558	−3.1595	0.5891	−0.3521	−0.3026	0.4076
16	6.8226	−4.1999	−5.7218	0.2396	0.0594	0.0239
17	−4.2713	4.6767	−5.4127	−0.5823	0.2802	−0.4496
18	−5.0266	−0.8908	−5.0633	0.1826	0.0038	−0.1380
19	2.8939	5.6152	2.6436	0.2319	0.0555	−0.5112
20	−5.6668	−1.4413	0.0365	0.3290	−0.0723	−0.6564
21	0.4660	3.9291	−4.3042	0.5560	−0.5253	0.6242
22	0.5362	−1.3616	0.0340	0.6850	−0.0086	−0.2752
23	5.2368	4.4890	−4.9025	0.3843	0.5027	−0.2833
24	−0.1102	3.7296	−6.2188	0.1198	0.5322	−0.2306
25	−1.4090	−1.6372	5.0886	0.6089	−0.3189	−0.0414
26	2.5410	−3.9304	0.9656	0.1179	−0.4061	0.2140
27	3.5333	4.2317	6.2097	−0.6761	0.0966	−0.6644
28	0.3899	6.4896	2.8996	−0.5296	0.2028	0.4875
29	−2.0590	−2.3453	1.2815	0.5164	−0.1120	0.0882
30	−4.8685	2.5387	4.5868	−0.0171	−0.4096	0.5043
31	1.3284	−0.7669	5.4908	0.4912	0.6366	−0.2095
32	−3.2749	4.8440	7.0524	−0.4047	−0.5843	−0.0711

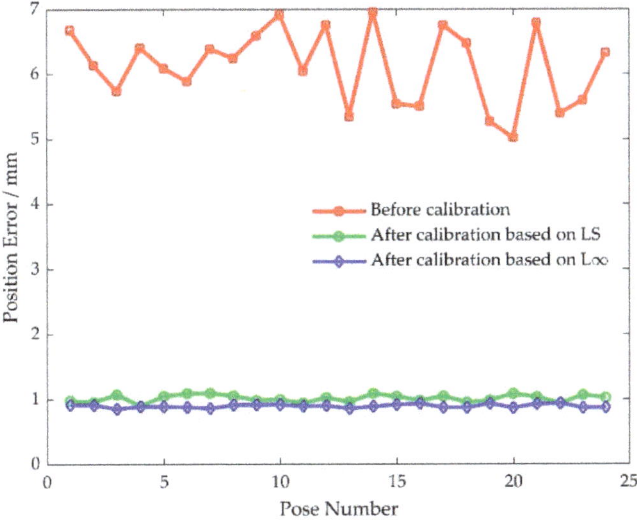

Figure 13. The absolute position errors at verification poses before and after calibration.

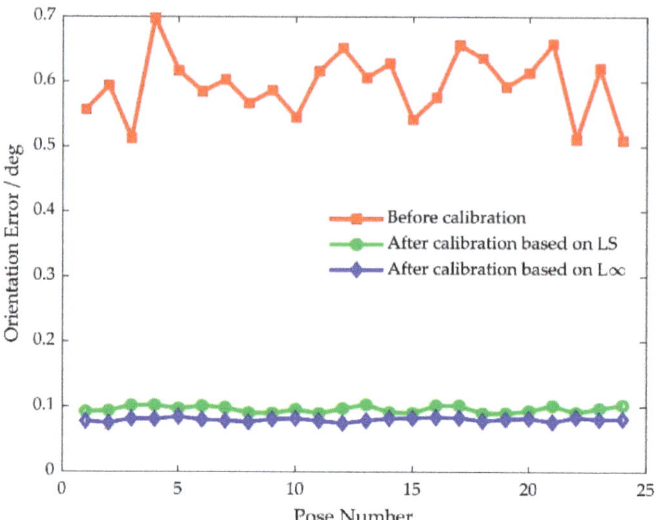

Figure 14. The absolute orientation errors at verification poses before and after calibration.

Table 7. Comparison of absolute pose errors at verification poses before and after calibration.

	Before Kinematic Calibration		After Kinematic Calibration			
	Position (mm)	Orientation (deg)	Based on L-Infinity		Based on Least Squares	
			Position (mm)	Orientation (deg)	Position (mm)	Orientation (deg)
Maximum error	6.9595	0.6977	0.9408	0.0848	1.0980	0.1037
Range of error	14.2670	1.4721	1.8534	0.1620	2.0313	0.1815
Average error	6.1194	0.5952	0.9001	0.0805	1.0174	0.0964
Root-mean-square error	6.1455	0.5971	0.9005	0.0806	1.0189	0.0965

6. Conclusions

A new kinematic parameter error identification algorithm for the kinematic calibration of a parallel robot using L-infinity parameter estimation is developed in this paper. The kinematic parameter error identification procedure is transformed into a linear programming problem that computes kinematic parameter errors for a pose error model of a parallel robot so that the maximum difference between the predictions and measurements across its workspace is minimized. A strict bound on the pose errors produced by the parallel robot is given in the kinematic calibration based on L-infinity parameter estimation. The experimental results show a 14.32% reduction in maximum position errors and a 18.23% reduction in maximum orientation errors by using L-infinity parameter estimation compared to least squares estimation. The comparison results show an 8.76% reduction in range of position errors and a 10.74% reduction in range of orientation errors by using L-infinity parameter estimation compared to least squares estimation. Therefore, this validates that the proposed kinematic calibration method can effectively improve pose accuracy of the parallel robot and determine the range of the pose error. It should be noted that this kinematic calibration method can be used when pose measurement errors are tightly restricted and measurement noise is low.

Funding: This research received no external funding.

Data Availability Statement: Not applicable.

Conflicts of Interest: The author declares no conflict of interest.

References

1. Yu, S.M.; Han, J.W.; Yang, Y.; Xu, D.M.; Qu, Z.Y. Force and Moment Compensation Method Based on Three Degree-of-Freedom Stiffness-Damping Identification for Manipulator Docking Hardware-In-The-Loop Simulation System. *IEEE Access* **2018**, *6*, 63452–63467. [CrossRef]
2. Hu, Y.; Gao, F.; Zhao, X.C.; Yang, T.H.; Shen, H.R.; Qi, C.K.; Cao, R. A parameter dimension reduction-based estimation approach to enhance the kinematic accuracy of a parallel hardware-in-the-loop docking simulator. *Robotica* **2021**, *39*, 959–974. [CrossRef]
3. Wang, L.P.; Zhang, Z.K.; Shao, Z.F.; Tang, X.Q. Analysis and optimization of a novel planar 5R parallel mechanism with variable actuation modes. *Robot. Comput. Integr.* **2019**, *56*, 178–190. [CrossRef]
4. Song, Y.M.; Zhang, J.T.; Lian, B.B.; Sun, T. Kinematic calibration of a 5-DoF parallel kinematic machine. *Precis. Eng. J. Int. Soc. Precis. Eng. Nanotechnol.* **2016**, *45*, 242–261. [CrossRef]
5. Bai, S.P.; Teo, M.Y. Kinematic calibration and pose measurement of a medical parallel manipulator by optical position sensors. *J. Robot. Syst.* **2003**, *20*, 201–209. [CrossRef]
6. Saafi, H.; Laribi, M.A.; Zeghloul, S. Optimal torque distribution for a redundant 3-RRR spherical parallel manipulator used as a haptic medical device. *Robot. Auton. Syst.* **2017**, *89*, 40–50. [CrossRef]
7. Lian, B. Geometric error modeling of parallel manipulators based on conformal geometric algebra. *Adv. Appl. Clifford Algebras* **2018**, *28*, 1–24. [CrossRef]
8. Simas, H.; Di Gregorio, R. Geometric error effects on manipulators' positioning precision: A general analysis and evaluation method. *J. Mech. Robot.* **2016**, *8*, 061016. [CrossRef]
9. Sun, T.; Wang, P.F.; Lian, B.B.; Liu, S.D.; Zhai, Y.P. Geometric accuracy design and error compensation of a one-translational and three-rotational parallel mechanism with articulated traveling plate. *Proc. Inst. Mech. Eng. Part B J. Eng. Manuf.* **2018**, *232*, 2083–2097. [CrossRef]
10. Zhang, Z.K.; Xie, G.Q.; Shao, Z.F.; Gosselin, C. Kinematic calibration of cable-driven parallel robots considering the pulley kinematics. *Mech. Mach. Theory* **2022**, *169*, 104648. [CrossRef]
11. Yang, S.H.; Lee, D.M.; Lee, H.H.; Lee, K.I. Sequential measurement of position-independent geometric errors in the rotary and spindle axes of a hybrid parallel kinematic machine. *Int. J. Precis. Eng. Manuf.* **2020**, *21*, 2391–2398. [CrossRef]
12. Shao, Z.F.; Tang, X.Q.; Chen, X.; Wang, L.P. Research on the inertia matching of the Stewart parallel manipulator. *Robot. Comput. Integr.* **2012**, *28*, 649–659. [CrossRef]
13. Chebbi, A.H.; Affi, Z.; Romdhane, L. Prediction of the pose errors produced by joints clearance for a 3-UPU parallel robot. *Mech. Mach. Theory* **2009**, *44*, 1768–1783. [CrossRef]
14. Cammarata, A. A novel method to determine position and orientation errors in clearance-affected over constrained mechanisms. *Mech. Mach. Theory* **2017**, *118*, 247–264. [CrossRef]
15. Yu, D.Y. A new pose accuracy compensation method for parallel manipulators based on hybrid artificial neural network. *Neural Comput. Appl.* **2021**, *33*, 909–923. [CrossRef]
16. Zhang, F.; Shang, W.W.; Li, G.J.; Cong, S. Calibration of geometric parameters and error compensation of non-geometric parameters for cable-driven parallel robots. *Mechatronics* **2021**, *77*, 102595. [CrossRef]
17. Kim, H.S.; Choi, Y.J. The kinematic error bound analysis of the Stewart platform. *J. Robot. Syst.* **2000**, *17*, 63–73. [CrossRef]
18. Ding, J.; Liu, J.G.; Zhang, R.P.; Zhang, L.; Hao, G.B. Accuracy modeling and analysis for a lock-or-release mechanism of the Chinese Space Station Microgravity Platform. *Mech. Mach. Theory* **2018**, *130*, 552–566. [CrossRef]
19. Ding, J.Z.; Wang, C.J.; Wu, H.Y. Accuracy analysis of a parallel positioning mechanism with actuation redundancy. *J. Mech. Sci. Technol.* **2019**, *33*, 403–412. [CrossRef]
20. Tang, X.Q.; Chai, X.M.; Tang, L.W.; Shao, Z.F. Accuracy synthesis of a multi-level hybrid positioning mechanism for the feed support system in FAST. *Robot. Comput. Integr. Manuf.* **2014**, *30*, 565–575. [CrossRef]
21. Huang, T.; Bai, P.J.; Mei, J.P.; Chetwynd, D.G. Tolerance design and kinematic calibration of a four-degrees-of-freedom pic-and-Place parallel robot. *J. Mech. Robot.* **2016**, *8*, 061018. [CrossRef]
22. Takematsu, R.; Satonaka, N.; Thasana, W.; Iwamura, K.; Sugimura, N. A study on tolerances design of parallel link robots based on mathematical models. *J. Adv. Mech. Des. Syst. Manuf.* **2018**, *12*, JAMDSM0015. [CrossRef]
23. Zhuang, H.Q.; Yan, J.H.; Masory, O. Calibration of Stewart platforms and other parallel manipulators by minimizing inverse kinematic residuals. *J. Robot. Syst.* **1998**, *15*, 395–405. [CrossRef]
24. Daney, D. Kinematic calibration of the Gough platform. *Robotica* **2003**, *21*, 677–690. [CrossRef]
25. Hu, Y.; Gao, F.; Zhao, X.C.; Wei, B.C.; Zhao, D.H.; Zhao, Y.N. Kinematic calibration of a 6-DOF parallel manipulator based on identifiable parameters separation (IPS). *Mech. Mach. Theory* **2018**, *126*, 61–78. [CrossRef]
26. Huang, T.; Zhao, D.; Yin, F.W.; Tian, W.J.; Chetwynd, D.G. Kinematic calibration of a 6-DOF hybrid robot by considering multicollinearity in the identification Jacobian. *Mech. Mach. Theory* **2019**, *131*, 371–384. [CrossRef]
27. Luo, X.; Xie, F.G.; Liu, X.J.; Xie, Z.H. Kinematic calibration of a 5-axis parallel machining robot based on dimensionless error mapping matrix. *Robot. Comput. Integr. Manuf.* **2021**, *70*, 102115. [CrossRef]
28. Shen, H.P.; Meng, Q.M.; Li, J.; Deng, J.M.; Wu, G.L. Kinematic sensitivity, parameter identification and calibration of a non-fully symmetric parallel Delta robot. *Mech. Mach. Theory* **2021**, *161*, 104311. [CrossRef]

29. Jiang, Z.X.; Huang, M.; Tang, X.Q.; Guo, Y.X. A new calibration method for joint-dependent geometric errors of industrial robot based on multiple identification spaces. *Robot. Comput. Integr. Manuf.* **2021**, *71*, 102175. [CrossRef]

30. Sun, T.; Liu, C.Y.; Lian, B.B.; Wang, P.F.; Song, Y.M. Calibration for precision kinematic control of an articulated serial robot. *IEEE Trans. Ind. Electron.* **2021**, *68*, 6000–6009. [CrossRef]

31. Zhang, D.; Gao, Z. Optimal kinematic calibration of parallel manipulators with pseudoerror theory and cooperative coevolutionary network. *IEEE Trans. Ind. Electron.* **2012**, *59*, 3221–3231. [CrossRef]

32. Klimchik, A.; Wu, Y.; Caro, S.; Furet, B.; Pashkevich, A. Geometric and elastostatic calibration of robotic manipulator using partial pose measurements. *Adv. Robot.* **2014**, *28*, 1419–1429. [CrossRef]

33. Wu, Y.; Klimchik, A.; Caro, S.; Furet, B.; Pashkevich, A. Geometric calibration of industrial robots using enhanced partial pose measurements and design of experiments. *Robot. Comput. Integr. Manuf.* **2015**, *35*, 151–168. [CrossRef]

34. Mao, C.T.; Chen, Z.W.; Li, S.; Zhang, X. Separable nonlinear least squares algorithm for robust kinematic calibration of serial robots. *J. Intell. Robot. Syst.* **2021**, *101*, 2. [CrossRef]

35. Gao, G.B.; Li, Y.; Liu, F.; Han, S.C. Kinematic calibration of industrial robots based on distance information using a hybrid identification method. *Complexity* **2021**, *2021*, 8874226. [CrossRef]

36. Borm, J.H.; Meng, C.H. Determination of optimal measurement configurations for robot calibration based on observability measure. *J. Robot. Syst.* **1991**, *10*, 51–63. [CrossRef]

37. Khalil, W.; Gautier, C.D.; Enguehard, C. Identifiable parameters and optimum configurations for robot calibration. *Robotica* **1991**, *9*, 63–70. [CrossRef]

38. Zhuang, H.Q.; Wang, K.; Roth, Z.S. Optimal selection of measurement configurations for robot calibration using simulated annealing. In Proceedings of the 1994 IEEE International Conference on Robotics and Automation, San Diego, CA, USA, 8–13 May 1994; 1994; pp. 393–398. [CrossRef]

39. Zhuang, H.Q.; Wu, J.; Huang, W. Optimal planning of robot calibration experiments by genetic algorithms. *J. Robot. Syst.* **1997**, *14*, 741–752. [CrossRef]

40. Daney, D.; Papegay, Y.; Madeline, B. Choosing measurement poses for robot calibration with the local convergence method and tabu search. *Int. J. Robot. Res.* **2005**, *24*, 501–518. [CrossRef]

41. Dasgupta, B.; Mruthyunjaya, T.S. A canonical formulation of the direct position kinematics for a general 6-6 Stewart platform. *Mech. Mach. Theory* **1994**, *29*, 819–827. [CrossRef]

42. Dasgupta, B.; Mruthyunjaya, T.S. A constructive predictor-corrector algorithm for the direct position kinematics for a general 6-6 Stewart platform. *Mech. Mach. Theory* **1996**, *31*, 799–811. [CrossRef]

43. Husty, M.L. An algorithm for solving the direct kinematics of general Stewart-Gough Platform. *Mech. Mach. Theory* **1996**, *31*, 365–380. [CrossRef]

44. Pratik, P.; Sarah, L. A hybrid strategy to solve the forward kinematics problem in parallel manipulators. *IEEE Trans. Robot.* **2005**, *21*, 18–25.

45. Chvatal, V. *Linear Programming*; W. H. Freeman: New York, NY, USA, 1983.

Article

Analysis of 3-DOF Cutting Stability of Titanium Alloy Helical Milling Based on PKM and Machining Quality Optimization

Xuda Qin, Mengrui Shi, Zhuojie Hou, Shipeng Li *, Hao Li and Haitao Liu

Key Laboratory of Mechanism Theory and Equipment Design of Ministry of Education, Tianjin University, Tianjin 300072, China; qxd@tju.edu.cn (X.Q.); shimengrui@tju.edu.cn (M.S.); houzhuojie@tju.edu.cn (Z.H.); haolitju@tju.edu.cn (H.L.); liuht@tju.edu.cn (H.L.)
* Correspondence: shipengli@tju.edu.cn

Abstract: Aiming at the requirements of titanium alloy holes in aircraft industry, the 3-DOF cutting stability and surface quality optimization of parallel kinematic manipulator (PKM) are studied. The variation of natural frequencies with the end-effector position of the PKM is analyzed. The cutting force model of titanium alloy helical milling based PKM is developed, and the cutting force coefficients are identified. The prediction model for 3-DOF the stability of helical milling based on the PKM is established through a Semi-Discrete method, and the stability lobes are obtained. The correctness of the stability lobes is verified by subjecting the cutting force signal to time-frequency transformation and roughness detection. The step-cutter is used for machining process improvement to enhance the stability domain. The method proposed in this paper can provide a reference for further optimization of the prediction and optimization of the milling process of difficult-to-process materials based on PKM in the future.

Keywords: PKM; helical milling; cutting stability; surface quality; process optimization

Citation: Qin, X.; Shi, M.; Hou, Z.; Li, S.; Li, H.; Liu, H. Analysis of 3-DOF Cutting Stability of Titanium Alloy Helical Milling Based on PKM and Machining Quality Optimization. *Machines* **2022**, *10*, 404. https://doi.org/10.3390/machines10050404

Academic Editors: Zhufeng Shao, Dan Zhang and Stéphane Caro

Received: 2 May 2022
Accepted: 19 May 2022
Published: 21 May 2022

Publisher's Note: MDPI stays neutral with regard to jurisdictional claims in published maps and institutional affiliations.

1. Introduction

The quality requirements of the aircraft components are stringent because of their harsh working environment [1]. Besides, the aircraft components have the characteristics of large size, high requirements of precision and surface quality, high mechanical properties and high difficulty of material processing, which have posed great challenges to the traditional processing equipment and technology. Industrial robots have great advantages in milling process on the large aircraft components due to its high reconfigurablity. The hybrid PKM combines the advantages of high flexibility of series manipulator and large working space, high stiffness precision of parallel manipulator. The processing of hole making is an important part of aircraft assembly process in the aircraft industry. Compared with the traditional drilling process, helical milling has the advantages of less cutting force, simple process and low tool loss [2], which makes it more advantageous in the machining process of titanium alloy and other difficult-to-process materials in the aircraft industry. Therefore, it is necessary for the research of helical milling based on the hybrid PKM to improve the precision of hole making and improve the assembly efficiency. However, it is easy to occur the machining instability when machining the difficult-to-process materials, since the stiffness of the hybrid PKM is less than five-axis machine tool. Besides, the parameters of the hybrid PKM may vary with the position. Therefore, it is of great significance to analyze the machining performance of the hybrid PKM under different positions, study its cutting stability and machining quality and optimize the machining parameters, which may improve the machining quality.

Many scholars have carried out some related studies in cutting stability and machining quality. Altintas et al. [3] established a virtual prototype with the structure of the hybrid machine tool, which was used to study the stiffness, dynamics, vibration characteristics and cutting stability of the mechanism. The precision and machining performance of the hybrid machine tool were systematically analyzed on the basis of the analysis results. Piras et al. [4] analyzed the dynamic characteristics of the planar parallel mechanism with the simplified limbs with the help of the Finite Element Method. The variation of the natural frequency and the mode under different positions and attitude were obtained. Luo et al. [5] established an elastodynamics model which could predict the frequency response function (FRF) of 3-RPS parallel manipulator quickly based on the substructure synthesis technology. The natural frequency, frequency response and vibration mode of the parallel manipulator were simulated, and the variation trend of these parameters among the workspace was given. Law et al. [6] studied FRFs at the end of the system under different configurations based on the PKM by adopting modal polycondensation and substructure synthesis, and solved the cutting stability under different configurations by using a reduced order model. The relationship of dynamic characteristics, cutting stability and configurations was studied. Mousavi et al. [7] established a parameterized model to predict the dynamic characteristics and stability limit along the machining trajectory of the manipulator to achieve the maximum stability domain of the machining process. Li et al. [8] studied the dynamic behavior of the machine milling system and the chatter stability domain according to the stiffness model and the dynamic milling force model. The stability lobes of the whole system were obtained, and the experiments were carried out to verify the results. Shi et al. [9] took TriMule hybrid PKM as an example, and established the overall cutting dynamics model considering the hybrid PKM by analytical method. The cutting stability of TriMule was analyzed in the working space according to the cutting dynamics model. The trend of the 2-dimensional stability lobe diagram in the plane of workpiece with the variation of the position of the end-effector of hybrid PKM was obtained. Cao et al. [10] investigated the interaction mechanism between the spindle-tool system. Based on the process stability and surface quality of workpiece, a cutting parameter selection method is proposed from the perspective of machining stability. Mejri et al. [11] quantify the dynamic behavior s variation of an ABB IRB 6660 robot equipped with a high-speed machining (HSM) spindle in its workspace, and analyze the consequences in terms of machining stability. The results show that an adjustment of the cutting conditions must accompany the change of robot posture during machining to ensure stability.

Titanium alloy is widely used in the aircraft industry due to its excellent material characteristics. However, its surface quality directly affects its mechanical properties and practical life. Therefore, in terms of the surface machining quality and process of titanium alloy, Polini et al. [12] carried out cutting experiments on Ti6-Al4-V using tools of different materials. The influence of cutting force and tool wear on the machining surface quality were analyzed. It is found that TiAlN single-coated carbide tool had the better surface quality and longer tool life. Zain et al. [13] had given the optimal solution of cutting conditions with genetic algorithm to obtain the minimum surface roughness on the basis of radial rake processing parameters. The regression model was established according to the practical machining example. An optimal regression model was established to determine the fitness function of the genetic algorithm. The minimum surface roughness of the samples was reduced by about 27%. Munoz-escalona et al. [14] proposed a geometric model for predicting the surface roughness based on geometric analysis of tool path. The accuracy of surface roughness prediction reached 98% without considering the side wear of the tool, which was suitable for arbitrary diameter, arbitrary number of teeth and arbitrary nose radius of the tool. Liu et al. [15] established a mathematical model to calculate the uncut thickness and to describe the dynamic behavior of micro-milling force, including the combined influences of tool runout, minimum uncut thickness and edge plowing. Zhang et al. [16] designed the two-step milling (roughing and then finishing) experiments of Ti-6Al-4V titanium alloy to analyze the effect of different roughing parameters on the cutting force of roughing and finishing, the residual stress of finishing surface and the microstructure. The microstructural characteristics in terms of residual stress, XRD patterns, phase composition and content and nano-scale crystallite size of machined surface layer were characterized to reveal the machined surface layer quality adjustment and controlling mechanism from the prospective of the microscopic scale.

The static and dynamic stiffness of hybrid PKM is weaker than that of a 5-axis machine tool due to its structural characteristics. In order to meet the requirements of aircraft industry, it is necessary to study the cutting stability considering the PKM, when the PKM is applied to titanium alloy helical milling. The surface quality is analyzed to optimize the machining parameters and machining processing according to the above analysis. Few of the above studies have concerned this aspect. Therefore, this paper will arrange it as follows: in Section 2, a modal analysis of the PKM is performed and cutting forces model of helical milling based on PKM are established. The cutting forces coefficients are identified and the 3-DOF cutting dynamic model of helical milling-based PKM is established accordingly. In Section 3, the cutting experimental platform and modal experimental platform based on the PKM named TriMule are built, and cutting experiments and hammering modal experiments are performed. In Section 4, the experimentally obtained data are analyzed and the stability lobes for titanium alloy helical milling based PKM are obtained and validated. The step-cutter is used to optimize the machining process to improve the stability domain, and the optimization effect is demonstrated experimentally. The conclusion is given in Section 5.

2. Cutting Stability Analysis Based on Hybrid PKM

2.1. Modal Analysis of PKM

The natural frequency of the PKM is the basis for the study of its vibration characteristics. The PKM should be avoided from being excited by its natural frequency as far as possible to ensure the stability in the machining process [17]. The FEA method and experimental method are used to analyze the modal. The accuracy of the results is verified by experiments. The vibration equation of PKM can be expressed as

$$\mathbf{M\ddot{X}} + \mathbf{C\dot{X}} + \mathbf{KX} = \mathbf{F} \tag{1}$$

where $\mathbf{M}$, $\mathbf{C}$ and $\mathbf{K}$ are the mass matrix, damping matrix and stiffness matrix of the PKM respectively; $\mathbf{X}$, $\mathbf{\dot{X}}$ and $\mathbf{\ddot{X}}$ are the displacement matrix, velocity matrix and acceleration matrix respectively; $\mathbf{F}$ is the external excitation load at the end-effector of the PKM.

The damping and external load of the PKM in Equation (1) are zero when free vibration occurs, so the differential equation of free vibration of the PKM can be expressed as:

$$\mathbf{M\ddot{X}} + \mathbf{KX} = 0 \tag{2}$$

Equation (2) can be solved to obtain the vibration characteristics of the PKM in free vibration state

$$\mathbf{X} = \mathbf{X}_0 \sin(\omega t + \psi) \tag{3}$$

where $\mathbf{X}_0$ is the vibration amplitude of the PKM response; ω, t and ψ are the response frequency, time and original phase angel, respectively.

Substitute Equation (3) into Equation (2):

$$\left(\mathbf{K} - \omega^2 \mathbf{M}\right)\mathbf{X}_0 = 0 \tag{4}$$

Solve the characteristic equation of Equation (4) and obtain n different characteristic values ω_i ($I = 1 \sim n$), which are arranged in ascending order as follows

$$\omega_1 < \omega_2 < \omega_3 < \cdots < \omega_i < \cdots < \omega_n$$

where ω_i is the ith-order natural frequency of the PKM, which are related to the mass matrix and stiffness matrix.

The external excitation of the PKM in the process of machining is simple harmonic force, which can be express as

$$\mathbf{F} = \mathbf{F}_0 \cos(\omega t) \tag{5}$$

where $\mathbf{F}$ is the external excitation; $\mathbf{F}_0$ is the amplitude of the external excitation.

The response of the PKM to harmonic force can be express as:

$$\mathbf{X} = \mathbf{X}_0 \sin(\omega t) \tag{6}$$

Substitute Equations (5) and (6) into Equation (1), and $\mathbf{H}(\omega)$ represents the Frequency Response Runction (FRF) of the end-effector of the PKM, which is also depending on the position of the PKM:

$$\mathbf{H}(\omega)^{-1} = -\omega^2 \mathbf{M} + j\omega \mathbf{C} + \mathbf{K} \tag{7}$$

The PKM used in this paper is a novel 5-DOF hybrid PKM named TriMule, which have been proposed by Tian et al. [18]. As shown in Figure 1, the PKM is composed of a 3-DOF R(2-RPS&RP)&UPS parallel mechanism plus a A/C wrist. Here, R, P, U and S represent revolute, prismatic, universal and spherical joints, respectively, and P denotes an actuated prismatic joint. RPS branch is denoted as limb i (i = 1, 2); UPS branch is denoted as limb 3; and RP branch is denoted as limb 4. It is necessary to simplify the original complex components and establish their own coordinate system to describe the kinematics model of hybrid PKM. $\mathbf{w}_i$ is defined to represent the direction vector of stretch of limb i. The fixed coordinate system κ is established with the midpoint A4 of the base link as the origin, and the coordinate system κ_i is established at the intersection of the joint axes of R pair (or U pair) and P pair in limb i. $\mathbf{R}_i$ is the attitude matrix of coordinate system κ_i with respect to fixed coordinate system κ, which can be expressed as:

$$\mathbf{R}_i = \begin{bmatrix} \mathbf{u}_i & \mathbf{v}_i & \mathbf{w}_i \end{bmatrix} = \begin{bmatrix} \cos\theta_{2i} & 0 & \sin\theta_{2i} \\ \sin\theta_{1i}\sin\theta_{2i} & \cos\theta_{1i} & -\sin\theta_{1i}\cos\theta_{2i} \\ -\cos\theta_{1i}\sin\theta_{2i} & \sin\theta_{1i} & \cos\theta_{1i}\cos\theta_{2i} \end{bmatrix} \tag{8}$$

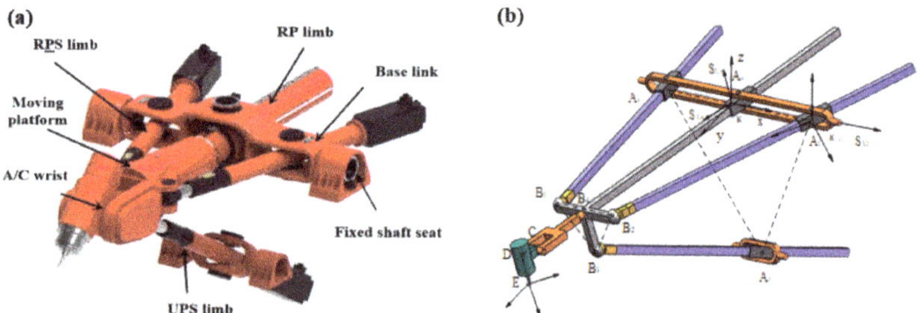

Figure 1. The PKM named TriMule: (**a**) CAD model (**b**) structure diagram with coordinates of TriMule.

Due to the structural characteristics of the PKM, its dynamic performance is varied with the end-effector position of the PKM. Therefore, seven representative positions are selected in the workspace to analyze the dynamic performance in the coordinate of O_m-$x_m y_m z_m$. The origin of the coordinate system O_m-$x_m y_m z_m$ is established at the intersection of the A/C wrist, and each axis is parallel to each axis of the coordinate system κ. Figure 2 shows the selected representative positions in the workspace. The coordinates of them in the fixed coordinate system κ are shown in Table 1.

The modal simulation of the PKM is carried out by using FEA software on the basis of the above analysis. The positions of the end-effector of the PKM as shown in Table 1 were also selected to analyze. Figures 3 and 4, respectively, gave the first 4 modal modes of the PKM at the initial position (Position 1 in Table 1) and the simulation values of natural frequencies at the 7 representative positions. As shown in Figure 4, the natural frequency of the PKM is strongly correlated with the end-effector position. The higher order modes are relatively more influenced by the end-effector position.

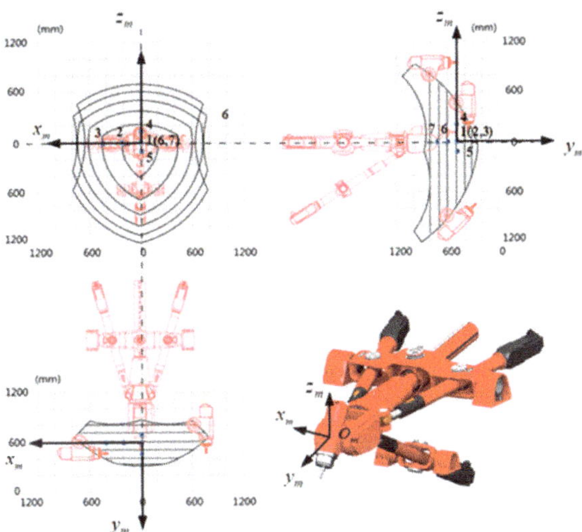

Figure 2. Seven representative positions in the workspace of TriMule.

Table 1. Coordinate of selected position points in fixed coordinate system κ.

Number of Positon	x (mm)	y (mm)	z (mm)	Angle of A/C Wrist
1	0	0	0	0
2	200	0	0	0
3	400	0	0	0
4	0	0	100	0
5	0	0	−50	0
6	0	−50	0	0
7	0	100	0	0

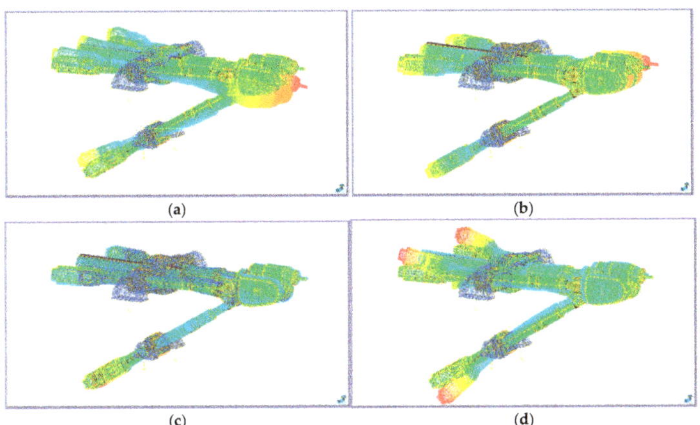

Figure 3. The mode shapes of the first 4 order simulation modes of the PKM at initial position. (**a**) 1st mode shape of vibration; (**b**) 2nd mode shape of vibration; (**c**) 3rd mode shape of vibration; and (**d**) 4th mode shape of vibration.

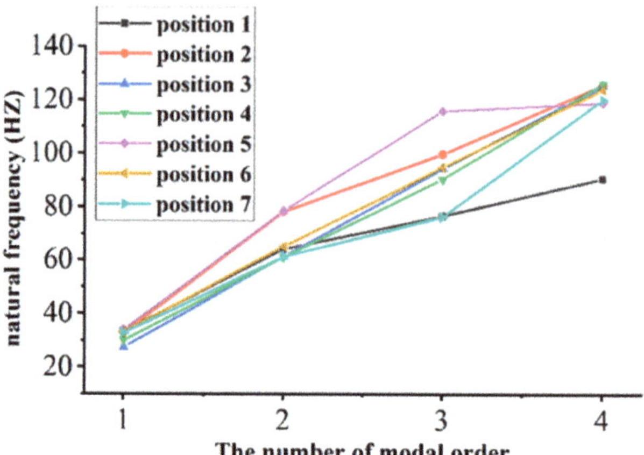

Figure 4. The natural frequency simulation values of the PKM at position 7.

2.2. Cutting Forces Modeling and Identification of Cutting Forces Coefficients of Helical Milling Based on PKM

In order to analyze the 3-DOF cutting stability of helical milling, the cutting forces of the helical milling based on PKM will be established first in this section. Both the side and the bottom edges of the tool are involved in the cutting process. The working state of the side edge can be approximated as an ordinary milling process, while the working state of bottom edge can be approximated as a drilling process. Therefore, the feed motion of helical milling can be divided into two parts, which are axial feed motion v_c and circumferential feed motion n_c. The feed trajectory is a spiral line, whose radius is the eccentric distance e of the tool and the hole, the pitch is the axial feed a_p per revolution, the spindle speed is n, and the number of tool edges is N. The schematic diagram of cutting thickness of helical milling is shown in Figure 5.

The cutting force model of helical milling based on PKM is established according to Literature [8]. According to the cosine theorem, Pc can be expressed as:

$$P_cO_w{}^2 = R_t{}^2 + e^2 - 2R_te\cos(\pi - \varphi) \tag{9}$$

$$P_cO_c = P_cO_w{}^2 + e^2 - 2P_cO_we\cos\left(\frac{\varphi_k}{2}\right) \tag{10}$$

φ_k can be express as:

$$\varphi_k = 2ar\cos\left(\frac{e - R_t\cos(\varphi)}{\sqrt{e^2 + R_t{}^2 - 2R_te\cos(\varphi)}}\right) \tag{11}$$

Therefore, the axial depth of side edge $b(\varphi)$ in the helical milling can be expressed as:

$$b(\varphi) = a_p - \frac{a_p}{\pi}ar\cos\left(\frac{e - R_t\cos(\varphi)}{\sqrt{e^2 + R_t{}^2 - 2R_te\cos(\varphi)}}\right) \tag{12}$$

k_r, k_t and k_a are the shear effect coefficients of the side edge of tool in radial direction, tangential direction and axial direction in the cutting force model of helical milling, which are the main sources of dynamic cutting force; k_{re}, k_{te}, k_{ae} are the friction effect coefficients of side edge of tool in radial direction, tangential direction and axial direction, which are mainly related to static cutting force and can be ignored to analyze the stability; k_{Fea}, k_{Fec} are the shear effect coefficient and friction effect coefficient of the bottom edge in the axial direction, respectively, which are not related to the dynamic cutting chips and stability of the process. Therefore, only k_r, k_t and k_a are considered in the cutting stability analysis.

As shown in Figure 5, radial, tangential and axial cutting forces in relation to the instantaneous cutting chips thickness during the instantaneous cutting process can be expressed as:

$$\begin{aligned}
F_{rj}(\varphi) &= k_r b(\varphi_j) h_j(\varphi_j) \\
F_{tj}(\varphi) &= k_t b(\varphi_j) h_j(\varphi_j) \\
F_{aj}(\varphi) &= k_a b(\varphi_j) h_j(\varphi_j)
\end{aligned} \tag{13}$$

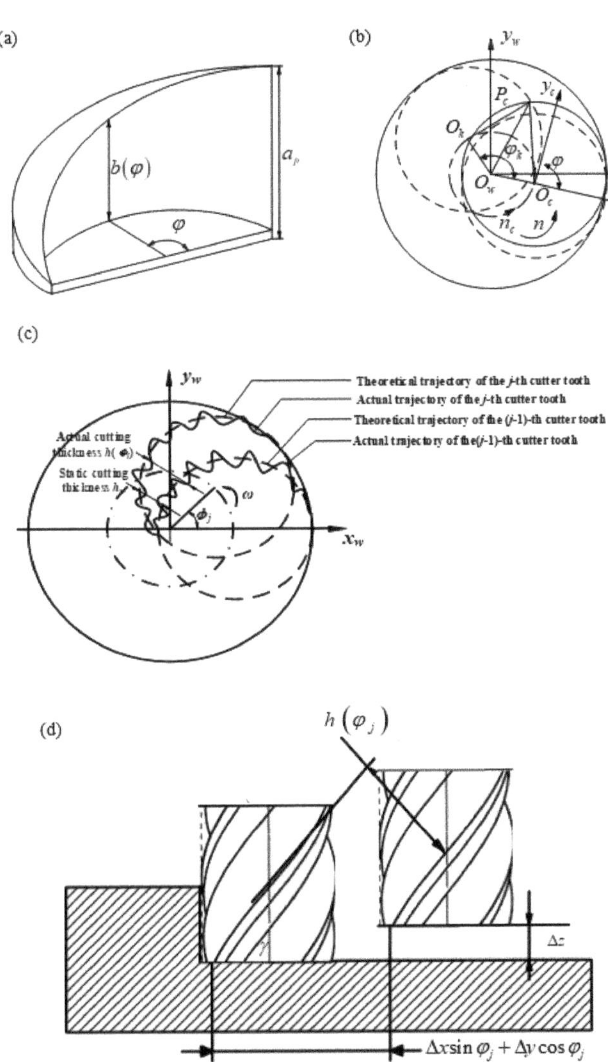

Figure 5. The schematic diagram of cutting thickness of helical milling: (**a**) the chip of helical milling; (**b**) the process of helical milling (top view); (**c**) dynamic cutting thickness of the j-th cutter tooth in the x_w direction and y_w direction of helical milling; and (**d**) dynamic cutting thickness of the j-th cutter tooth in the z_w direction.

Neglecting the effect of static cutting forces, which means neglecting the effect of static cutting thickness, the cutting thickness can be expressed as

$$h(\varphi_j) = \left[(\Delta x \sin \varphi_j + \Delta y \cos \varphi_j) \sin \gamma - \Delta z \cos \gamma \right] g(\varphi_j) \tag{14}$$

where Δx, Δy and Δz are the displacements generated by two adjacent cutter teeth cutting through the workpiece in each direction. γ is the cutter helix angle and $g(\varphi_j)$ is the function indicating whether the cutter teeth are cutting in, which can be expressed as:

$$g(\varphi_j) = \begin{cases} 1 & \varphi_{st} < \varphi_j < \varphi_{ex} \\ 0 & \varphi_j < \varphi_{st} \text{or } \varphi_j > \varphi_{ex} \end{cases} \tag{15}$$

Therefore, the differentiation of the radial, tangential and axial forces of helical milling can be expressed as:

$$\begin{bmatrix} dF_{rj} \\ dF_{tj} \\ dF_{aj} \end{bmatrix} = b(\theta) \begin{bmatrix} k_r \\ k_t \\ k_a \end{bmatrix} \begin{bmatrix} \sin \varphi_j \sin \gamma & \cos \varphi_j \sin \gamma & -\cos \gamma \end{bmatrix} \begin{bmatrix} \Delta x \\ \Delta y \\ \Delta z \end{bmatrix} \tag{16}$$

Converting the Equation (16) into workpiece coordinate system, it can be expressed as:

$$\begin{bmatrix} dF_{xj} \\ dF_{yj} \\ dF_{zj} \end{bmatrix} = \begin{bmatrix} -\sin \gamma \sin \varphi_j & -\cos \varphi_j & -\cos \gamma \sin \varphi_j \\ -\sin \gamma \cos \varphi_j & \sin \varphi_j & -\cos \gamma \cos \varphi_j \\ \cos \gamma & 0 & -\sin \gamma \end{bmatrix} \begin{bmatrix} dF_{rj} \\ dF_{tj} \\ dF_{aj} \end{bmatrix} \tag{17}$$

Therefore, the cutting forces can be expressed as:

$$\mathbf{F} = \sum_{j=0}^{N} \mathbf{F}_j \tag{18}$$

Translating the Equation (18) into matrix form, it can be expressed as

$$\mathbf{F} = \mathbf{A}b(\varphi)h(\varphi) \tag{19}$$

where $\mathbf{A}$ is the direction coefficient matrix, which can be expressed as:

$$\mathbf{A} = \begin{bmatrix} h_{xx}(t) & h_{xy}(t) & h_{xz}(t) \\ h_{yx}(t) & h_{yy}(t) & h_{yz}(t) \\ h_{zx}(t) & h_{zy}(t) & h_{zz}(t) \end{bmatrix} \tag{20}$$

In order to analyze the stability of helical milling based on PKM, the cutting force coefficients need to be identified. According to the method given in the Literature [9], the cutting force coefficients of the side edges are identified by means of slot milling experiments. The average cutting force in each direction in the workpiece coordinate system during slot milling (the entry angle is 0 and the exit angle is π) can be expressed as:

$$\begin{cases} \overline{F}_{wx} = \frac{1}{4} N k_r a_p s_t + \frac{1}{\pi} N k_{re} a_p \\ \overline{F}_{wy} = \frac{1}{4} N k_t a_p s_t + \frac{1}{\pi} N k_{te} a_p \\ \overline{F}_{wz} = -\frac{1}{4} N k_a a_p s_t - \frac{1}{2} N k_{ae} a_p \end{cases} \tag{21}$$

The cutting force coefficient can be identified by linear regression method according to Equation (21).

2.3. 3-DOF Cutting Stability Analysis of Helical Milling Based on the PKM

The improved Semi-Discrete Time Domain Method is used to analyze the cutting stability of titanium alloy helical milling at 7 positions of TriMule. The differential equations of 3-DOF cutting dynamics for helical milling of TriMule can be expressed as

$$\begin{pmatrix} \ddot{x} \\ \ddot{y} \\ \ddot{z} \end{pmatrix} + 2 \begin{pmatrix} \omega_{nx} & & \\ & \omega_{ny} & \\ & & \omega_{nz} \end{pmatrix} \begin{pmatrix} \xi_x \\ \xi_y \\ \xi_z \end{pmatrix} \begin{pmatrix} \dot{x} \\ \dot{y} \\ \dot{z} \end{pmatrix} + \left[\begin{pmatrix} \omega_{nx}^2 & & \\ & \omega_{ny}^2 & \\ & & \omega_{nz}^2 \end{pmatrix} + Q \right] \begin{pmatrix} x \\ y \\ z \end{pmatrix} = Q \begin{pmatrix} x(t-\tau) \\ y(t-\tau) \\ z(t-\tau) \end{pmatrix} \tag{22}$$

$$Q = \begin{pmatrix} \dfrac{a_p h_{xx}(t)}{m_{tx}} & \dfrac{a_p h_{xy}(t)}{m_{tx}} & \dfrac{a_p h_{xz}(t)}{m_{tx}} \\ \dfrac{a_p h_{yx}(t)}{m_{ty}} & \dfrac{a_p h_{yy}(t)}{m_{ty}} & \dfrac{a_p h_{yz}(t)}{m_{ty}} \\ \dfrac{a_p h_{zx}(t)}{m_{tz}} & \dfrac{a_p h_{zy}(t)}{m_{tz}} & \dfrac{a_p h_{zz}(t)}{m_{tz}} \end{pmatrix} \tag{23}$$

where ω_{nx}, ω_{ny} and ω_{nz} are the angular natural frequency of the PKM in the direction of x, y and z respectively; m_{tx}, m_{ty} and m_{tz} are the modal mass of the PKM in the direction of x, y and z respectively; and ξ_x, ξ_y and ξ_z are the damping ratio. $x(t-\tau)$ is the time delay due to the regeneration effect, which is equal to the time between the cutter teeth cutting into and cutting out the workpiece. T is the time delay of ordinary cutter with equal pitch. Since the revolution speed n_c is far less than the spindle speed n, T can be expressed as:

$$T = \frac{60}{Nn} \tag{24}$$

Transforming kinematics equation into state space equation

$$\dot{u}(t) = A_i u(t) + \omega_a B_i u_{i-m+1} + \omega_b B_i u_{i-m} \tag{25}$$

where A_i is the constant matrix; B_i is the state term; u_i is the space term. As the time delay is equal to the period, $\omega_a = \omega_b = 1/2$. m is the integer introduced:

$$m = \text{int}\left(\frac{\tau + \Delta t/2}{\Delta t} \right) \tag{26}$$

The dynamic equation of regenerative cutting can be expressed as:

$$u_{i+1} = P_i u_i + \omega_a R_i u_{i-m+1} + \omega_b R_i u_{i-m} \tag{27}$$

$$P_i = \exp(A_i \Delta t) \tag{28}$$

$$R_i = (\exp(A_i \Delta t) - I)A_i^{-1}B_i \tag{29}$$

I represents the identity matrix; $(3m + 6)$ dimensional state vector can be expressed as:

$$z_i = col\begin{pmatrix} x_i & y_i & z_i & \dot{x}_i & \dot{y}_i & \dot{z}_i & x_{i-1} & y_{i-1} & z_{i-1} & \cdots & x_{i-m} & y_{i-m} & z_{i-m} \end{pmatrix}$$

Thus, the following mapping can be constructed

$$z_{i+1} = D_i z_i \tag{30}$$

where D_i is the $(3m + 6)$ dimensional coefficient matrix, which can be expressed as:

$$\mathbf{D}_i = \begin{pmatrix}
P_{i11} & P_{i12} & P_{i13} & P_{i14} & P_{i15} & P_{i16} & 0 & \cdots & 0 & \frac{R_{i11}}{2} & \frac{R_{i12}}{2} & \frac{R_{i13}}{2} & \frac{R_{i11}}{2} & \frac{R_{i12}}{2} & \frac{R_{i13}}{2} \\
\vdots & \vdots & \vdots & \vdots & \vdots & \vdots & \vdots & \ddots & \vdots & \vdots & \vdots & \vdots & \vdots & \vdots & \vdots \\
P_{i61} & P_{i62} & P_{i63} & P_{i64} & P_{i65} & P_{i66} & 0 & \cdots & 0 & \frac{R_{i61}}{2} & \frac{R_{i62}}{2} & \frac{R_{i63}}{2} & \frac{R_{i61}}{2} & \frac{R_{i62}}{2} & \frac{R_{i63}}{2} \\
1 & 0 & 0 & 0 & 0 & 0 & 0 & \cdots & 0 & 0 & 0 & 0 & 0 & 0 & 0 \\
0 & 1 & 0 & 0 & 0 & 0 & 0 & \cdots & 0 & 0 & 0 & 0 & 0 & 0 & 0 \\
0 & 0 & 1 & 0 & 0 & 0 & 0 & \cdots & 0 & 0 & 0 & 0 & 0 & 0 & 0 \\
0 & 0 & 0 & 0 & 0 & 0 & 1 & \cdots & 0 & 0 & 0 & 0 & 0 & 0 & 0 \\
\vdots & \vdots & \vdots & \vdots & \vdots & \vdots & \vdots & \ddots & \vdots & \vdots & \vdots & \vdots & \vdots & \vdots & \vdots \\
0 & 0 & 0 & 0 & 0 & 0 & 0 & \cdots & 1 & 0 & 0 & 0 & 0 & 0 & 0 \\
0 & 0 & 0 & 0 & 0 & 0 & 0 & \cdots & 0 & 1 & 0 & 0 & 0 & 0 & 0 \\
0 & 0 & 0 & 0 & 0 & 0 & 0 & \cdots & 0 & 0 & 1 & 0 & 0 & 0 & 0 \\
0 & 0 & 0 & 0 & 0 & 0 & 0 & \cdots & 0 & 0 & 0 & 1 & 0 & 0 & 0
\end{pmatrix}. \tag{31}$$

The state-transition matrix $\mathbf{\Phi}$ can be expressed as:

$$\mathbf{\Phi} = \mathbf{D}_{k-1}\mathbf{D}_{k-2}\cdots\mathbf{D}_1\mathbf{D}_0 \tag{32}$$

According to Floquent Theory, indicating that the matrix converges, the variation of dynamic cutting thickness has an upper limit when the eigenvalue of the state transition matrix is less than 1. The process of the PKM helical milling is in a stable state. Otherwise, the cutting process is unstable.

3. Experimental Platform Set-Up

3.1. Cutting Force Coefficients Identification and Stability Verification

Before analyzing the cutting stability of helical milling based on PKM, it is necessary to model the cutting forces when it is applied to helical milling. The cutting force model of titanium alloy helical milling based on TriMule is established by taking titanium alloy commonly used in aircraft industry as an example in this paper, as shown in Figure 6. Specific modeling methods and experimental methods have been introduced in detail in Literature [8]. In this paper, the angle of A/C wrist is adjusted to 0, and the cutting force coefficients can be identified from the platform.

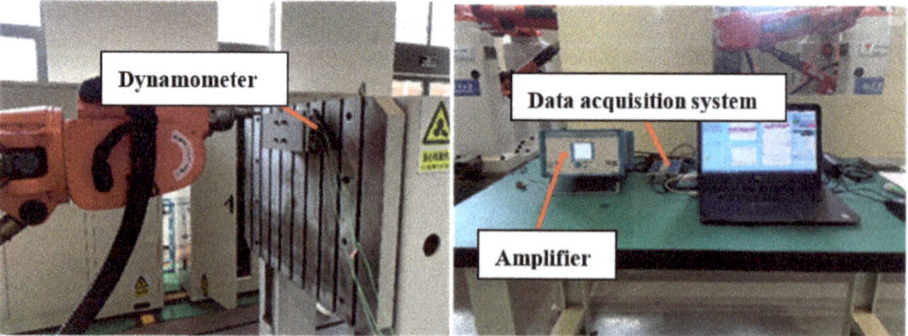

Figure 6. Experimental platform for cutting forces coefficients in TriMule based helical milling.

As shown in Figure 6, a Kistler three-direction dynamometer (9257A) with supporting charge amplifier (type 5070), and data acquisition system and Kistler software were utilized for the cutting force measurement. The workpiece material used in this paper is titanium alloy Ti6Al4V, the physical and mechanical properties of which are shown in Table 2 [19]. The diameter of the helical milling holes in this paper is 19.05 mm, and the carbide helical milling tool with a 12 mm diameter was selected for the helical milling tests. According to the selected tool diameter and the cutting conditions of titanium alloy, the spindle speed of 2000 rpm is selected to identify the cutting force parameters in order to control the tool wear during machining [20,21]. The helical milling tool is shown in Figure 7, and details about cutting tool are listed in Table 3. Cutting experiments were performed on the experimental platform according to the cutting experimental parameters shown in Table 4, and the cutting forces coefficients were identified.

Table 2. Physical and mechanical properties of titanium alloys used in the experiment.

Properties	Density (g/cm^3)	Tensile Strength (MPa)	Yield Strength (MPa)	Elastic Modulus (GPa)	Shear Modulus (MPa)	Poisson Ratio
Value	4.5	895	825	110	0.342	0.342

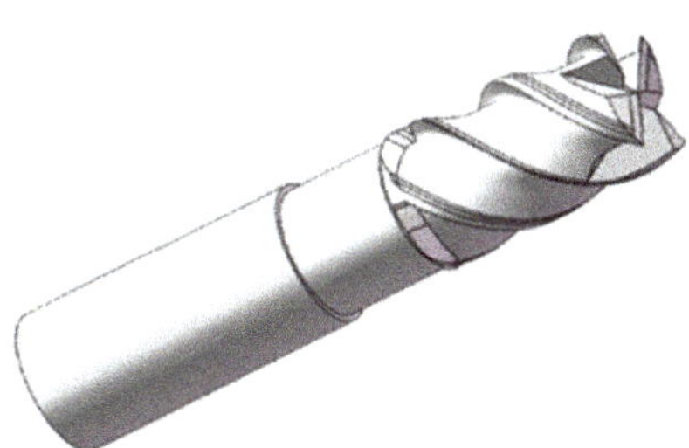

Figure 7. Helical milling tool and its 3D model.

Table 3. The geometric and property parameters of cutting tool used in the experiment.

Properties	Diameter	Rake Angle	Tool Clearance	Helix Angel	Number of Teeth	Materials
Values	12 mm	8°	15°	40°	4	K44 UF

Table 4. Cutting force identification experimental cutting parameters.

Spindle Speed (r/min)	a_p (mm)	Feed Engagement (mm)	Feed Speed (mm/min)
2000	0.2	0.02	160
2000	0.2	0.03	240
2000	0.2	0.04	320
2000	0.2	0.05	400
2000	0.4	0.02	160
2000	0.4	0.03	240
2000	0.4	0.04	320
2000	0.4	0.05	400
2000	0.6	0.02	160
2000	0.6	0.03	240
2000	0.6	0.04	320
2000	0.6	0.05	400

In order to verify the correctness of the stability prediction model of titanium alloy helical milling based on PKM, Position 1 and Position 2 shown in Table 1 are used as the validation positions considering the structural characteristics of the parallel mechanism. Since the change of spindle speed has little effect on the cutting force parameters and can be ignored, the cutting force parameters obtained at the spindle speed of 2000 rpm are used for stability verification experiments [22–24]. The cutting parameters shown in Table 5 are used for the machining experiments.

Table 5. Cutting parameters of stability verification experiment.

Number	Spindle Speed (r/min)	a_p (mm)	Feed Engagement (mm)	Feed Speed (mm/min)
1	1000	0.2	0.02	80
2	2000	0.2	0.02	160
3	3000	0.2	0.02	240
4	1000	0.4	0.02	80
5	2000	0.4	0.02	160
6	3000	0.4	0.02	240
7	1000	0.6	0.02	80
8	2000	0.6	0.02	160
9	3000	0.6	0.02	240
10	1000	0.8	0.02	80
11	2000	0.8	0.02	160
12	3000	0.8	0.02	240

As the analysis mentioned above, a step-cutter is proposed to reduce the radial cutting depth to improve the stability of helical milling and optimize the machining quality.

The step-cutter is improved the main structure of the helical milling cutter which includes two parts of cutting edge, neck part and shank. The height difference between the two parts of the cutting edge is 1mm. Right-angle transition mode is adopted to reduce impact. The necking part prevents cutting edge of the second part from cutting the surface again. The cutter material is cemented carbide. The cutter structure is shown in Figure 8. The parameters of the cutter are shown in Table 6.

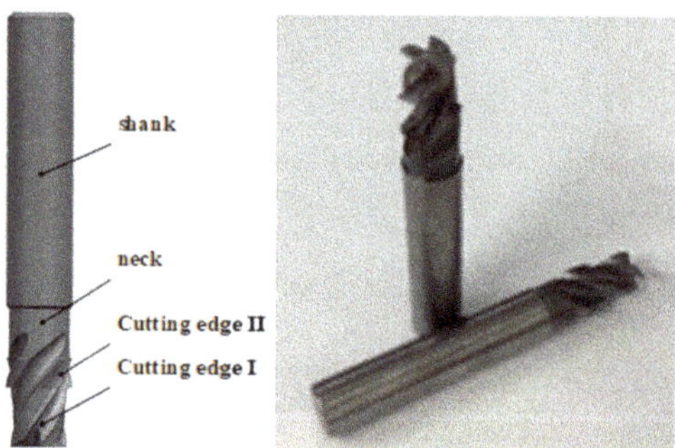

Figure 8. The structure of step-cutter.

Table 6. The parameters of step-cutter.

Number of Teeth	Longth of Cutting Edge I (mm)	Longth of Cutting Edge II (mm)	Rake Angel	Tool Clearance	Helix Angle	Transient Mode
4	11	5	8°	15°	40°	Right angle

3.2. Experimental Platform Set-Up to Indentify of Modal Parameters of TriMule

In order to verify the correctness of modal simulation, modal experiments are carried out on the 7 positions in Table 1 respectively by hammering experiment. The specific experimental methods and equipment have been described in detail in Literature [8]. The angle of A/C wrist of the PKM change to 0 this time, and the point-line model for hammering tests is re-established. Table 7 shows the modal parameters of the first 4 modes of the PKM. The experimental platform of TriMule helical milling modal tests is presented in Figure 9. The B&K impact hammer and PCB acceleration sensor were used to shock excitation and pick up the generated vibration signals. The input and output signals are processed through the data collection system of LMS, and the modal data processing is imported into the analysis software. In this paper, the modal tests are carried out by single point impact and single point vibration signal pickup method. As TriMule is a complex high-order system, multiple vibration pickup points were arranged, as shown in Figure 10. At the same time, the impact hammer strikes along different directions of each impact point. In order to reduce the random measurement error, the vibration measurement tests for every pickup point repeat five times. Taking the initial position as an example, the modal parameters of the PKM at this position are obtained as shown in Table 5. Figure 11 shows the variation of natural frequency of the PKM with different position in x, y and z directions to facility the analysis of variation of modal parameters with the configuration of the PKM.

Table 7. First 4 modal parameters of the initial position of the PKM.

Direction	Number of Mode	ω_n (Hz)	ξ	m (kg)
	1	23.62	9.15	2.31×10^{-8}
x	2	29.23	9.80	2.10×10^{-8}
	3	34.82	6.03	2.07×10^{-8}
	4	40.93	4.22	7.77×10^{-9}
	1	23.54	7.79	2.08×10^{-8}
y	2	33.56	6.59	1.058×10^{-8}
	3	56.52	4.19	3.86×10^{-8}
	4	68.07	2.76	3.63×10^{-10}
	1	23.76	7.52	2.79×10^{-8}
z	2	29.23	10.07	7.68×10^{-8}
	3	34.33	5.53	2.32×10^{-8}
	4	39.83	3.48	2.13×10^{-8}

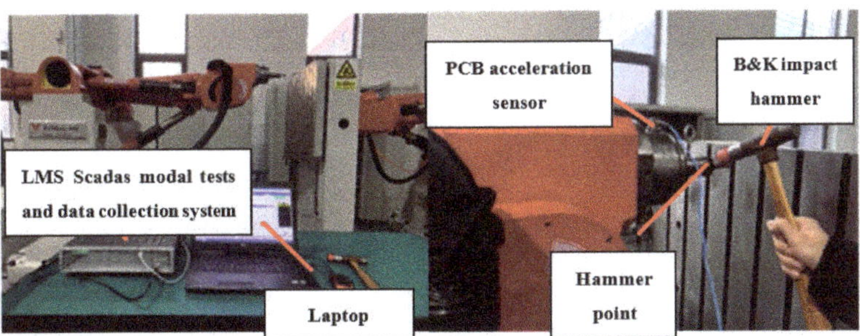

Figure 9. Experimental platform of TriMule based helical milling modal tests.

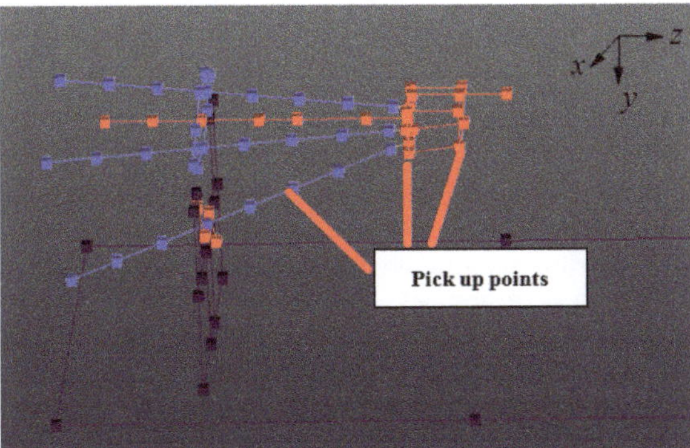

Figure 10. Arrangement of vibration pickup points on TriMule modal test.

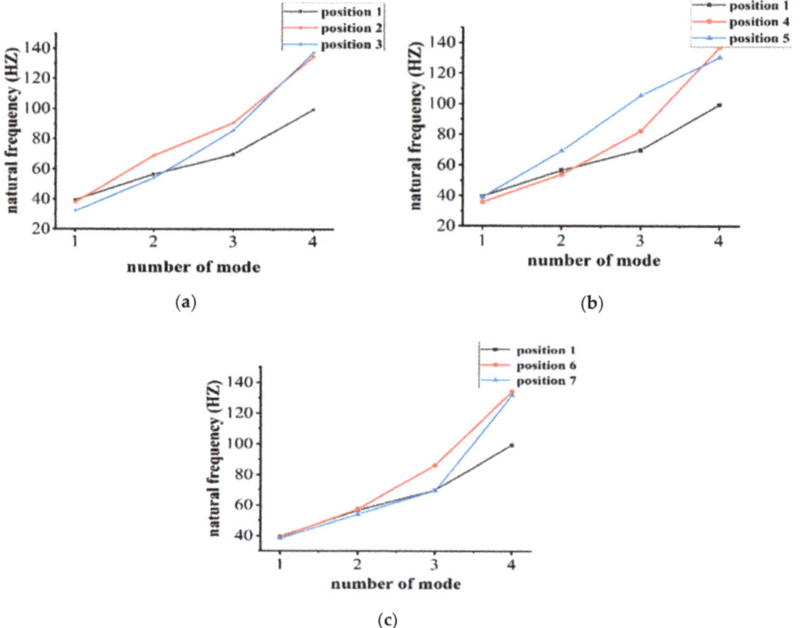

Figure 11. The first 4 order natural frequencies of the PKM in different positions. (**a**) Natural frequency in the x direction; (**b**) Natural frequency in the z direction; (**c**) Natural frequency in the y direction.

As shown in Figure 11, comparing Position 1 and Position 2 and Position 3, the 1st order natural frequency decreased gradually with the positive deviation of position along the x-axis. The bigger the offset is, the more decreased it is. Furthermore, the 3rd order and 4th order natural frequency from the Position 1 to Position 2 are risen sharply, and from the Position 2 to Position 3 are changed slightly. Comparing Position 1, Position 4 and Position 5, the 1st order natural frequency varies slightly along the z-axis. The 2nd, 3rd and 4th order natural frequencies all increase along the z-axis, and the fourth order natural frequencies increase most obviously. Comparing Position 1, Position 6 and Position 7, the natural frequencies of the first two orders are basically unchanged along the y-axis, while the natural frequencies of the 3rd and 4th orders increase significantly with the change of the y-axis. Therefore, it can be found that the 1st order natural frequency variation range is the lowest with the position variation. The value of 1st natural frequency decreases at most 18.4% from the initial position. While the 3rd and 4th orders natural frequency vary at most 51.2% from the initial position, which increases the most.

4. Results and Discussion

4.1. Cutting Stability Lobes of Titanium Alloy Helical Milling Based TriMule

According to the experimental platform established in Section 3, the cutting force coefficients of the PKM-based titanium alloy helical milling hole were identified and obtained, as shown in Table 8.

In addition, hammering modal tests are performed on the TirMule based on the established experimental platform. Due to the structural characteristics of the PKM, the modal parameters are strongly position-dependent. Therefore, modal tests are conducted for each of the 7 positions shown in Table 1, and the stability domain of the TriMule-based titanium alloy helical milling is analyzed by combining the identified modal parameters and cutting force coefficients.

Table 8. Cutting force coefficients of helical milling on titanium alloy based on TriMule.

Depth of Axial Cutting (mm)	k_r (N/mm^2)	k_t (N/mm^2)	k_a (N/mm^2)	k_{re} (N/mm)	k_{te} (N/mm)	k_{ae} (N/mm)
0.2	1044.5	477.4	287.9	299.1	54.3	32.5
0.4	1221.0	584.5	316.5	306.3	56.8	38.8
0.6	1383.1	648.8	357.2	311.5	58.5	48.2
The average	1216.2	570.2	320.5	305.6	56.5	39.8

According to the above analysis, the stability lobes of TriMule at 7 positions are drawn, as shown in Figure 12. The stability of other poses under the spindle speed of 1000 r/min is roughly determined by the 1st mode except Position 6 and 7. By comparing Position 1 and Position 2 and 3, the stability domain of the working condition above 2000 r/min increases significantly when the coordinate of end-effector deviation along x direction, and this variation is mainly affected by the 4th mode. When the deviation continues to 400 mm, the stability domain will decrease again. By comparing Position 1 and Position 4 and 5, the stability domain of the PKM decreases slightly at low spindle speed when the end-effector extends 100 mm forward along the z-axis, while the stability domain at high spindle speed has no significant change, which is mainly affected by the 4th mode. However, the end-effector retracted 50 mm along the z-axis will make the stability domain at high spindle speed significantly increase, which is mainly affected by the 3rd mode. By comparing Position 1 and Postion 6 and 7, the stability domain is mainly affected by the 1st and 3rd order modes after it is moved along the y-axis, and the rise of 100 mm along the y-axis will greatly improve the stability region at low spindle speed.

4.2. Validation of Cutting Stability Lobes of TriMule Based Titanium Alloy Helical Milling

It is necessary to verify the correctness of stability lobes of helical milling based TriMule. The acceleration signals in the machining process are collected to time-frequency transformation to determine whether the machining process was unstable. As mentioned in Section 3.1, the correctness of the obtained stability lobes is verified, with Position 1 and Position 2 as examples. The combination of cutting parameters is shown in Table 7. The spectrum of experimental acceleration signal is shown in Figure 13.

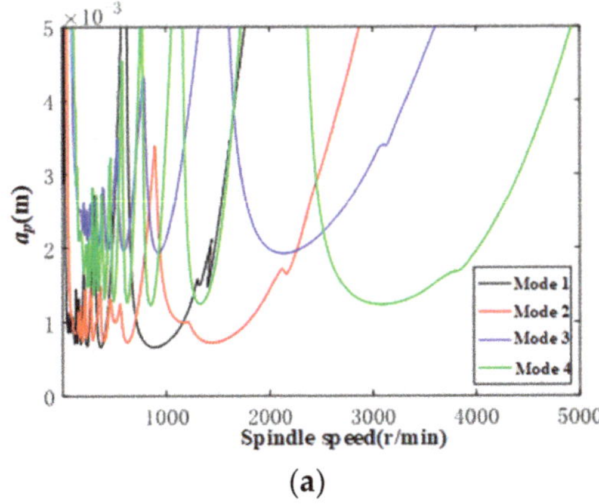

(a)

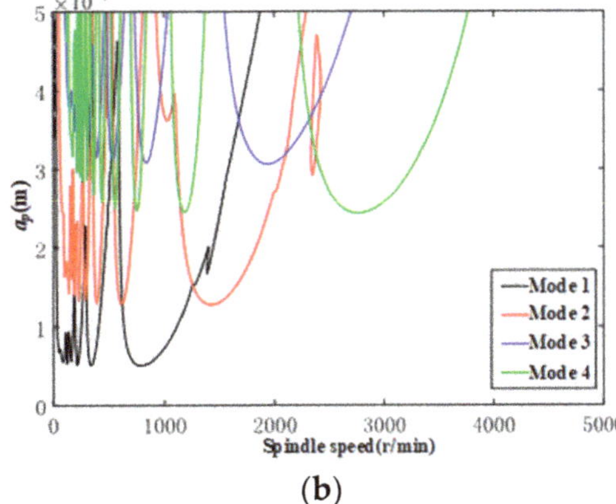

(b)

Figure 12. *Cont.*

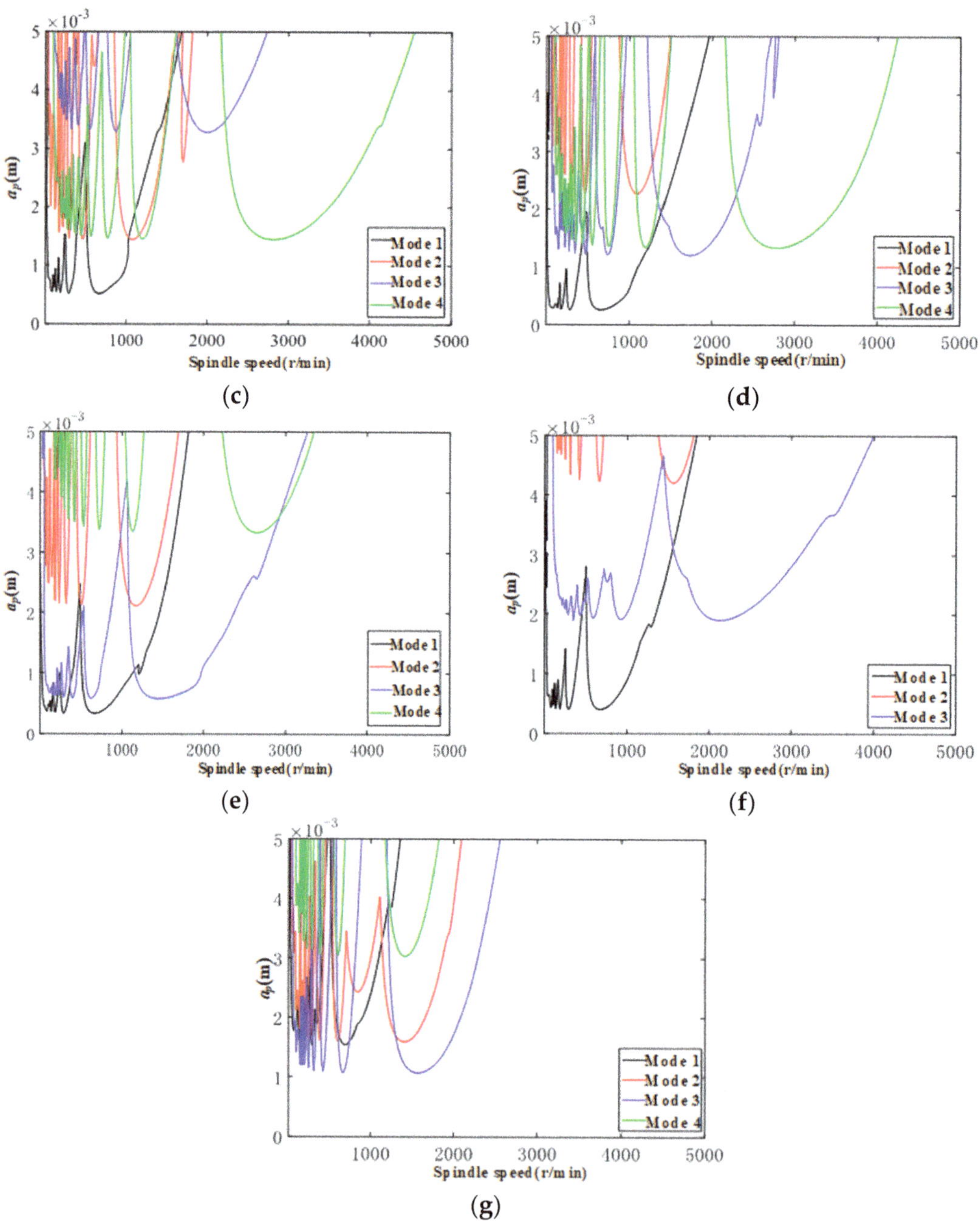

Figure 12. Stability lobes of helical milling based TriMule in 7 Positions. (**a**) Position 1; (**b**) Position 2; (**c**) Position 3; (**d**) Position 4; (**e**) Position 5; (**f**) Position 6; and (**g**) Position 7.

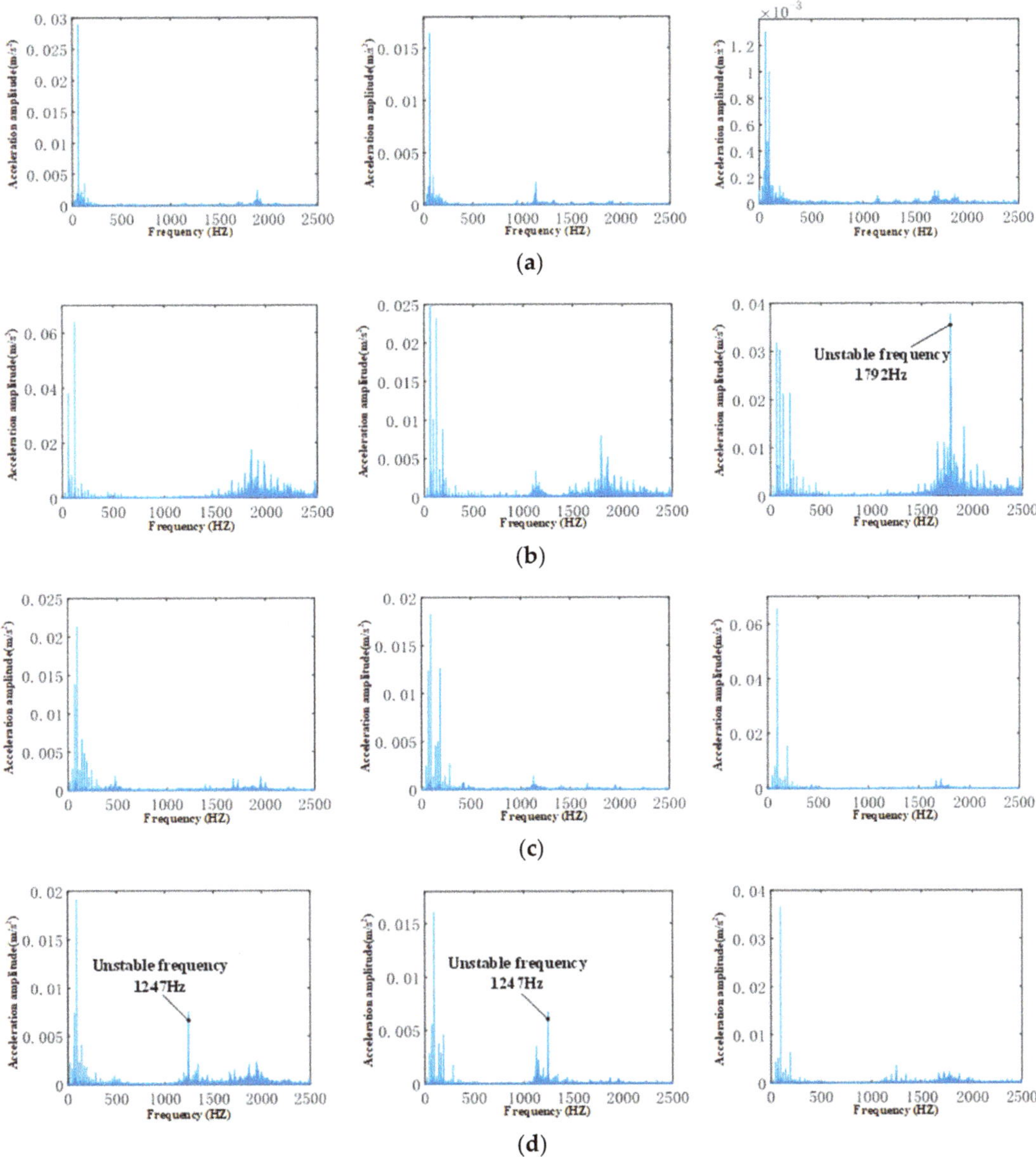

Figure 13. Spectrum of experimental acceleration signal. (**a**) Signal spectrum of acceleration along x, y and z under Position 1 (cutting parameter combine 1); (**b**) Signal spectrum of acceleration along x, y and z under Position 1 (cutting parameter combine 10); (**c**) Signal spectrum of acceleration along x, y and z under Position 2 (cutting parameter combine 1); and (**d**) Signal spectrum of acceleration along x, y and z under Position 2 (cutting parameter combine 10).

As shown in Figure 13, when the spindle speed is 1000 r/min and the axial feed is 0.8 mm, the helical milling of TriMule at Position 1 and Position 2 has a small vibration, and the machining stability is good under other processing parameters. It can be seen that the predicted results are in agreement with the experimental results. In addition, at Position 1, the main components of the acceleration signal spectrum are the passing frequency of the cutter tooth, and its frequency doubling when the spindle speed is 1000 r/min and the axial feed is 0.2 mm per revolution. It can be judged that no instability occurs in the milling process. When the axial feed is increased in the same position, it can be seen that vibration with a different amplitude occurs in the high frequency band of the acceleration signal spectrum in the three directions, and the vibration in the z direction is the most obvious. Therefore, it can be judged that instability occurs. At Position 2, when the rotational speed is 1000 r/min, with the increase of axial feed per revolution, the unstable frequency 1247 Hz of acceleration signal in frequency domain is significant, which will obviously affect the milling accuracy.

In addition, the cutting stability will also affect the roughness condition of the inner wall of the hole. Therefore, in verifying the correctness of the above stability leaflet diagram, the inner wall roughness of titanium alloy helical milling holes at different positions of TriMule is measured by the roughness measuring instrument. The results are shown in Figures 14 and 15.

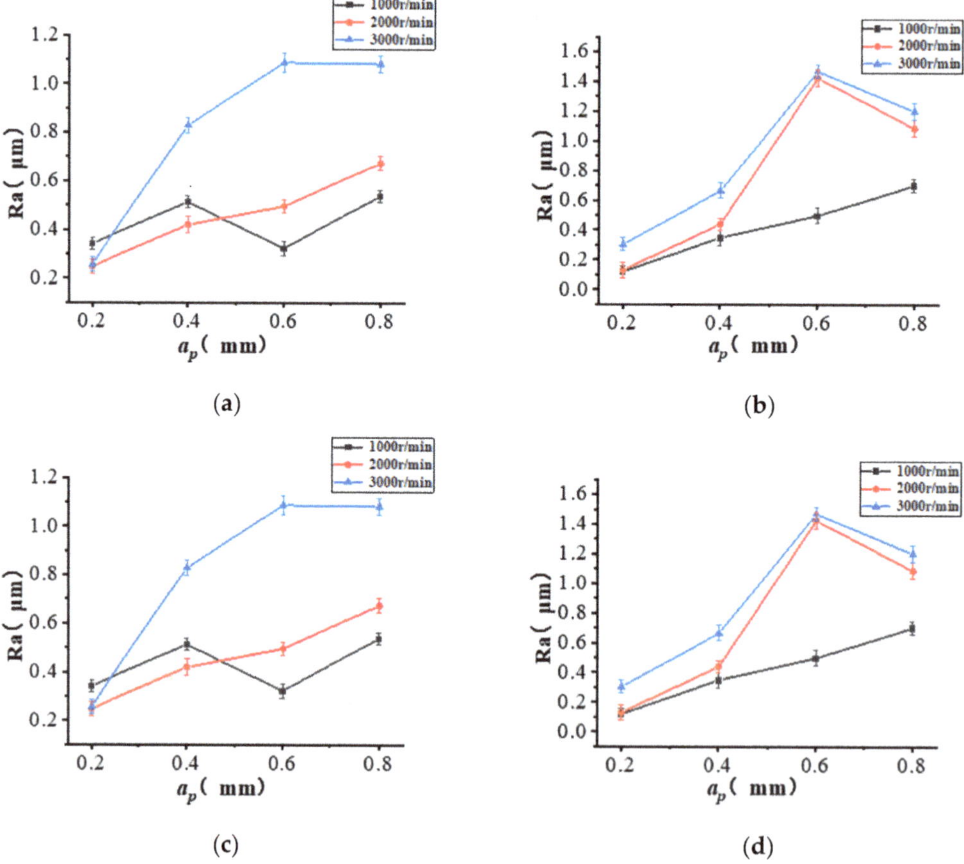

Figure 14. Surface roughness of helical milling hole with different processing parameters of TriMule. (**a**) Position 1; (**b**) Position 2; (**c**) Position 3; and (**d**) Position 4.

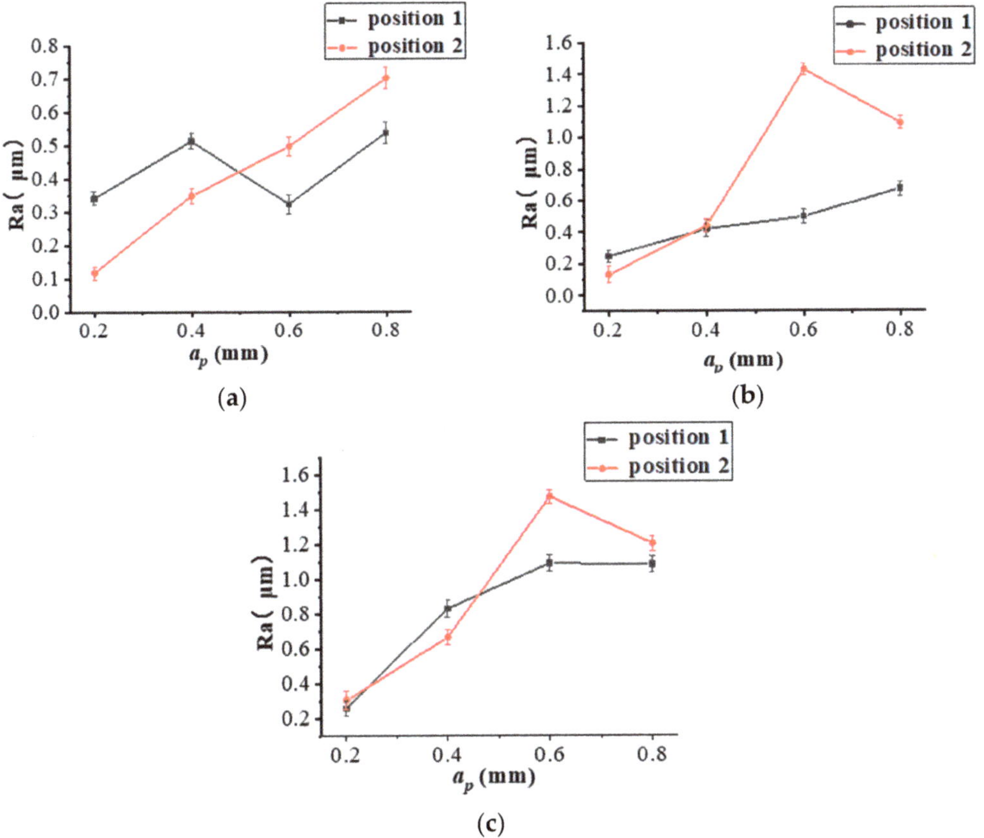

Figure 15. Surface roughness comparison of Position 1 and Position 2 at different spindle speeds. (**a**) n = 1000 r/min; (**b**) n = 2000 r/min; and (**c**) n = 3000 r/min.

As shown in Figure 14, the overall surface roughness increases with the axial feed per revolution increases. This is because the increase of axial feed leads to the increase of spiral feed track pitch, which makes the machining traces on the processed surface more serious, alongside the increase of roughness.

The comparison of the surface roughness of helical milling holes at Position 1 and 2 at each spindle speed shows Figure 15. It can be seen that when the axial feed per revolution is small (less than 0.4 mm), the surface roughness of holes at Position 2 is smaller and the machining quality is better. When the axial feed per revolution is larger (greater than 0.4 mm), the surface roughness of holes at position 1 of is smaller and the machining quality is better.

4.3. Optimization of Cutting Stability Lobes of Titanium Alloy Helical Milling Based TriMule

The influence of machining instability on machining quality can be reduced by optimizing the cutter. Since the dynamic cutting force that causes the milling instability is from the side edge of the tool, and the radial cutting depth of the side edge will directly affect the cutting force and continuous cutting time of the tool, the use of step-cutters can significantly reduce the radial cutting force to increase the cutting stability domain and reduce the cutting instability.

The ratio d of the radial cut depth to the tool diameter is used as a variable to optimize the cutting stability, which are set as 0.8, 0.6, 0.4 and 0.2, respectively. Taking TriMule at initial position as an example, the cutting stability of different radial cutting depths is shown in Figure 16. As shown in the Figure 16, when d decreases from 1 to 0.4, the limit cutting depth of the stability region increases slightly. When d decreases from 0.4 to 0.2, the ultimate cutting depth in the stability region has also increased, but the increase is small relative to the case where d decreases from 1 to 0.4.

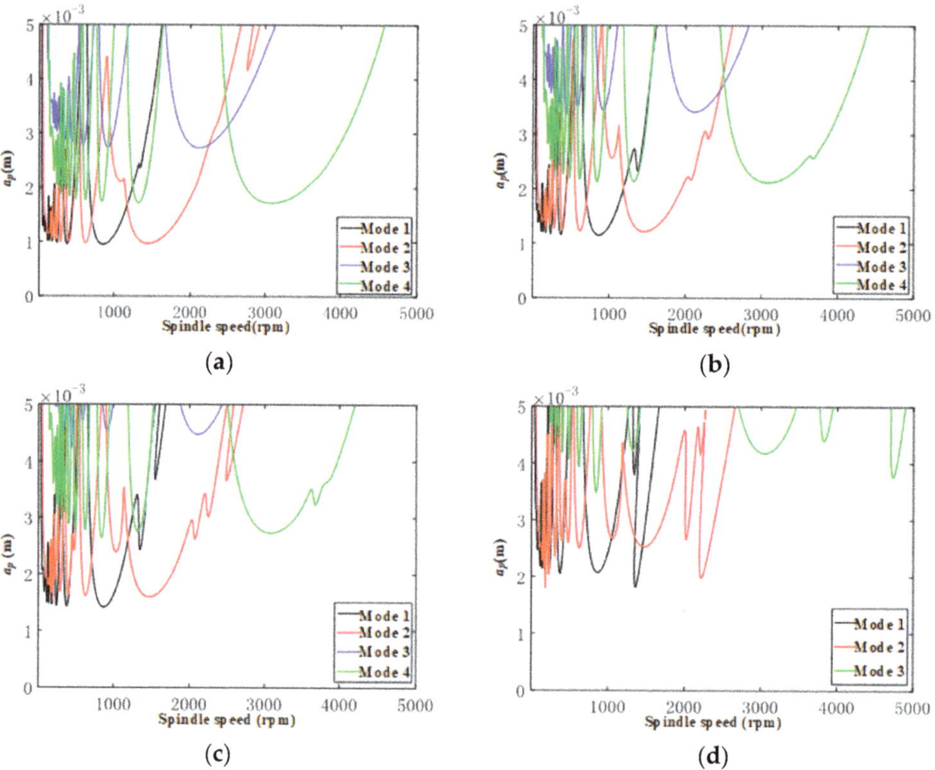

Figure 16. Stable lobe of TriMule with different radial cutting depths at Position 1. (**a**) d = 0.8. (**b**) d = 0.6. (**c**) d = 0.4. (**d**) d = 0.2.

The stability lobe diagrams of TriMule at the initial position with the ordinary helical milling cutter and the step-cutter were drawn respectively in Figure 17, and the cutting forces are collected in Figure 18. As shown in Figures 17 and 18, step-cutters significantly increase the cutting stability domain and reduce cutting forces. Furthermore, the decrease of the period over the same time is conducive to reducing the excitation of the cutter.

The aperture and roundness of the hole containing ordinary and step-cutters to characterize the precision of the milling hole. Figure 19 is shown the accuracy comparison of helical milling hole between ordinary cutter and step-cutter. By comparing the machining accuracy of helical milling with an ordinary cutter and a step-cutter, it can be concluded that the step-cutter can effectively reduce the aperture error under various parameter combinations. When the cutting depth is 0.4 mm and 0.6 mm with the spindle speed of 3000 r/min, the optimization effect is most obvious. The average aperture error decreased by 63.2%. The step-cutter can also effectively reduce the roundness error—particularly in the condition of large cutting depth and low spindle speed, the roundness error optimization is the most obvious. The average roundness error is reduced by 67.7%.

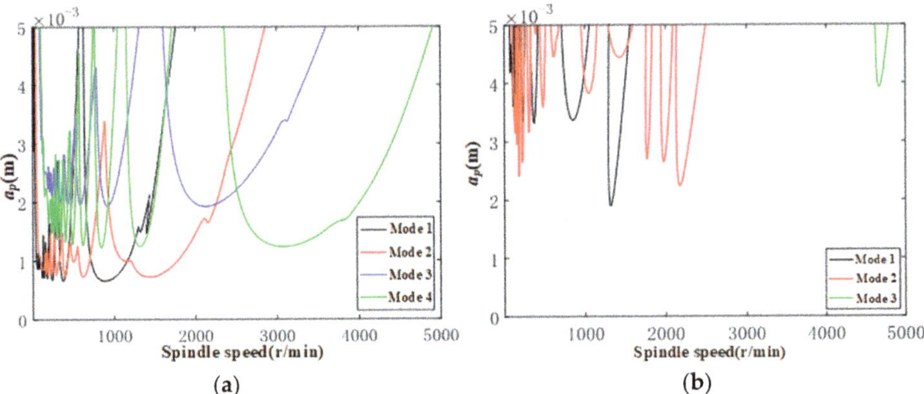

Figure 17. Stability lobes of initial position for ordinary helical milling cutter and step-cutter. (**a**) Ordinary helical milling cutter. (**b**) Step-cutter.

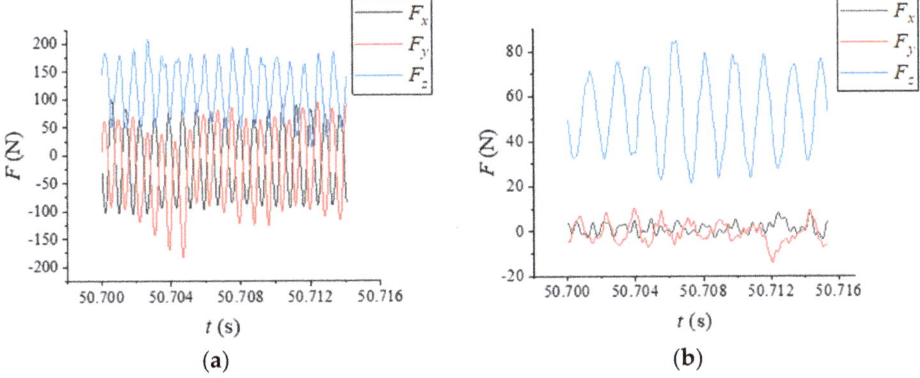

Figure 18. Cutting force of ordinary and step-cutter (Position 1, n = 1000 r/min, a_p = 0.2 mm). (**a**) Ordinary helical milling cutter. (**b**) Step-cutter.

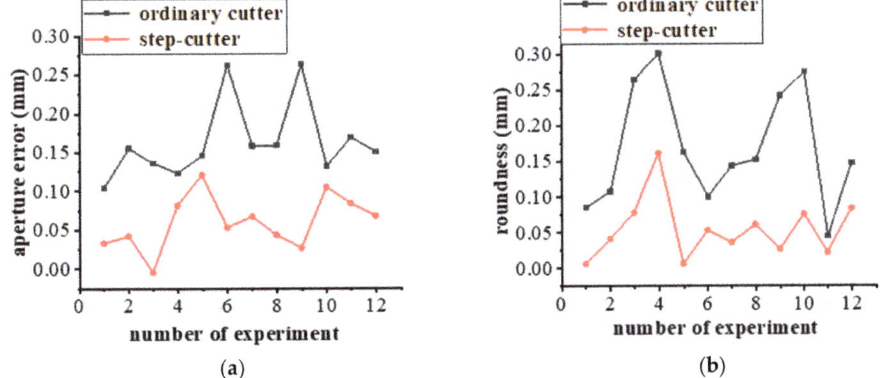

Figure 19. Accuracy comparison of helical milling hole between ordinary cutter and step-cutter. (**a**) Aperture error. (**b**) Roundness.

In order to verify that the step-cutter can effectively improve the quality of the machined surface, the roughness of the inner wall of the hole at different rotational speeds at the same location was measured using a roughness measuring instrument and compared with the roughness of the inner wall of the hole diameter machined with an ordinary cutter, as shown in Figure 20.

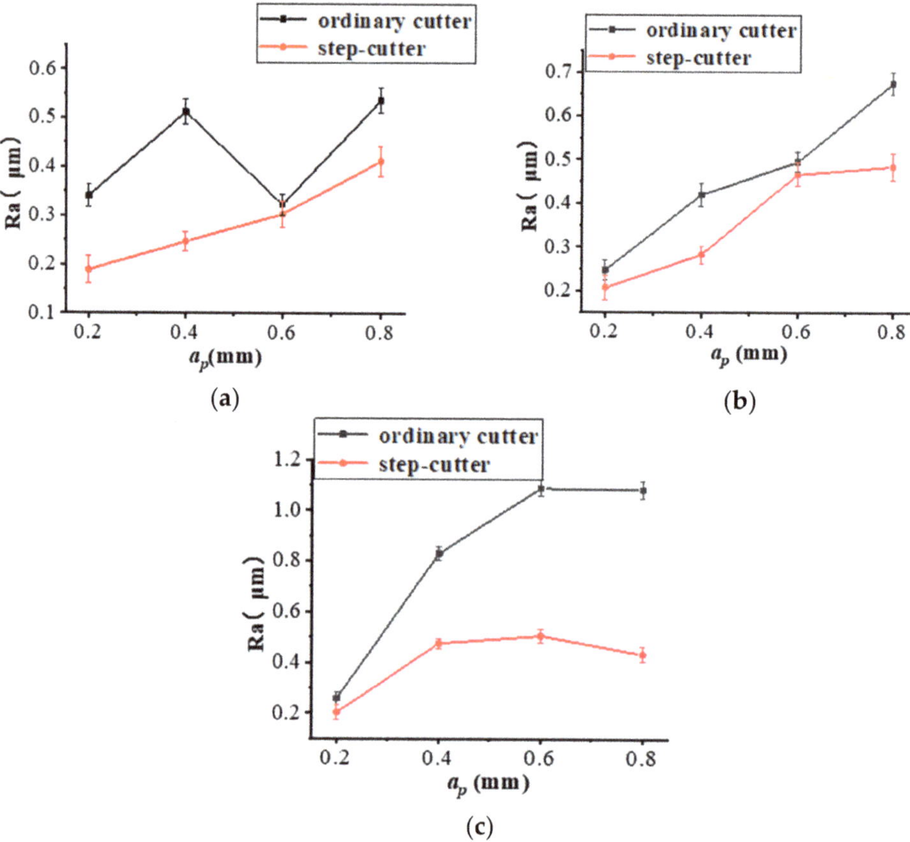

Figure 20. Roughness of helical milling holes surface with ordinary cutter and step-cutter. (**a**) n = 1000 r/min. (**b**) n = 2000 r/min. (**c**) n = 3000 r/min.

As shown in Figure 20, the surface roughness of helical milling holes is compared between an ordinary cutter and step-cutter at different spindle speeds. It can be found that with the increase of spindle speed, the optimization of step-cutter for surface roughness becomes more obvious. In addition, it can be seen from Figure 20c that the surface processed by step-cutter avoids the dramatic increase in roughness with the increase of axial feed per revolution, and still maintains good reliability. This is because the step-cutter can reduce the cutting force and increase the period of cutting force. The larger the machining parameters are, the greater the cutting force is, and the more obvious the optimization of step-cutter is. The surface roughness of titanium alloy helical milling hole is reduced by 30.7% on average.

5. Conclusions

The FEA software is used to analyze the variation of the static stiffness of the end-effector of the PKM at different positions. The cutting force model of titanium alloy helical milling based PKM is established, and the cutting force coefficients are identified. The static stiffness at the initial position is obviously better than that at other positions. The modal parameters and vibration modes of the PKM under different positions are obtained by simulation and hammer tests. The 1st order natural frequency is least affected by position, while the 3rd order natural frequency is most affected. A 3-DOF prediction model for the stability in helical milling based on the PKM is established, and the stability lobe diagram of seven positions and poses is obtained. The influence of position on stability region is analyzed. In order to reduce the influence of cutting instability on the machining quality, the machining process is optimized by using step-cutters. The step-cutter can significantly reduce the aperture error (63.2%), roundness (67.7%) and surface roughness (30.7%). The study of the stability of helical milling on titanium alloy provided a basis for further optimization of machining technology in the future.

Author Contributions: Conceptualization, M.S. and Z.H.; methodology, M.S.; software, Z.H.; validation, S.L., H.L. (Hao Li) and X.Q.; formal analysis, S.L.; investigation, M.S.; resources, X.Q.; data curation, S.L.; writing—original draft preparation, M.S.; writing—review and editing, Z.H.; visualization, H.L. (Hao Li); supervision, H.L. (Haitao Liu); project administration, X.Q.; and funding acquisition, X.Q., S.L. and H.L. (Hao Li). All authors have read and agreed to the published version of the manuscript.

Funding: This research was funded by National Natural Science Foundation of China (Grant numbers: 51775373 and 52075380), and Natural Science Foundation of Tianjin (Grant number: 19JCYBJC19000).

Institutional Review Board Statement: Not applicable.

Informed Consent Statement: Not applicable.

Data Availability Statement: Not applicable.

Conflicts of Interest: The authors declare no conflict of interest.

References

1. Mohd Yusuf, S.; Cutler, S.; Gao, N. Review: The Impact of Metal Additive Manufacturing on the Aerospace Industry. *Metals* **2019**, *9*, 1286. [CrossRef]
2. Pereira, R.B.D.; Brandão, L.C.; de Paiva, A.P.; Ferreira, J.R.; Davim, J.P. A review of helical milling process. *Int. J. Mach. Tools Manuf.* **2017**, *120*, 27–48. [CrossRef]
3. Altintas, Y.; Brecher, C.; Weck, M.; Witt, S. Virtual Machine Tool. *CIRP Ann.* **2005**, *54*, 115–138. [CrossRef]
4. Piras, G.; Cleghorn, W.L.; Mills, J.K. Dynamic finite-element analysis of a planar high-speed, high-precision parallel manipulator with flexible links. *Mech. Mach. Theory* **2005**, *40*, 849–862. [CrossRef]
5. Luo, H.-w.; Wang, H.; Zhang, J.; Li, Q. Rapid Evaluation for Position-Dependent Dynamics of a 3-DOF PKM Module. *Adv. Mech. Eng.* **2015**, *6*, 326–334. [CrossRef]
6. Law, M.; Ihlenfeldt, S.; Wabner, M.; Altintas, Y.; Neugebauer, R. Position-dependent dynamics and stability of serial-parallel kinematic machines. *CIRP Ann.* **2013**, *62*, 375–378. [CrossRef]
7. Mousavi, S.; Gagnol, V.; Bouzgarrou, B.C.; Ray, P. Dynamic modeling and stability prediction in robotic machining. *Int. J. Adv. Manuf. Technol.* **2016**, *88*, 3053–3065. [CrossRef]
8. Li, J.; Li, B.; Shen, N.; Qian, H.; Guo, Z. Effect of the cutter path and the workpiece clamping position on the stability of the robotic milling system. *Int. J. Adv. Manuf. Technol.* **2016**, *89*, 2919–2933. [CrossRef]
9. Shi, M.; Qin, X.; Li, H.; Shang, S.; Jin, Y.; Huang, T. Cutting force and chatter stability analysis for PKM-based helical milling operation. *Int. J. Adv. Manuf. Technol.* **2020**, *111*, 3207–3224. [CrossRef]
10. Cao, H.; Zhou, K.; Chen, X. Stability-based selection of cutting parameters to increase material removal rate in high-speed machining process. *Proc. Inst. Mech. Eng. Part B J. Eng. Manuf.* **2015**, *230*, 227–240. [CrossRef]
11. Mejri, S.; Gagnol, V.; Le, T.-P.; Sabourin, L.; Ray, P.; Paultre, P. Dynamic characterization of machining robot and stability analysis. *Int. J. Adv. Manuf. Technol.* **2015**, *82*, 351–359. [CrossRef]
12. Polini, W.; Turchetta, S. Cutting force, tool life and surface integrity in milling of titanium alloy Ti-6Al-4V with coated carbide tools. *Proc. Inst. Mech. Eng. Part B J. Eng. Manuf.* **2014**, *230*, 694–700. [CrossRef]
13. Zain, A.M.; Haron, H.; Sharif, S. Application of GA to optimize cutting conditions for minimizing surface roughness in end milling machining process. *Expert Syst. Appl.* **2010**, *37*, 4650–4659. [CrossRef]
14. Muñoz-Escalona, P.; Maropoulos, P.G. A geometrical model for surface roughness prediction when face milling Al 7075-T7351 with square insert tools. *J. Manuf. Syst.* **2015**, *36*, 216–223. [CrossRef]

15. Liu, H.; Sun, Y.; Geng, Y.; Shan, D. Experimental research of milling force and surface quality for TC4 titanium alloy of micro-milling. *Int. J. Adv. Manuf. Technol.* **2015**, *79*, 705–716. [CrossRef]
16. Zhang, R.; Li, A.; Song, X. Surface quality adjustment and controlling mechanism of machined surface layer in two-step milling of titanium alloy. *Int. J. Adv. Manuf. Technol.* **2022**, *119*, 2691–2707. [CrossRef]
17. Chen, C.; Peng, F.; Yan, R.; Tang, X.; Li, Y.; Fan, Z. Rapid prediction of posture-dependent FRF of the tool tip in robotic milling. *Robot. Comput.-Integr. Manuf.* **2020**, *64*. [CrossRef]
18. Huang, T.; Dong, C.; Liu, H. Five-Degree-of-Freedom Parallel Robot with Multi-Shaft Rotary Brackets. Patent No. WO/2017/005015, 1 December 2017.
19. Wang, H.; Qin, X.; Ren, C.; Wang, Q. Prediction of cutting forces in helical milling process. *Int. J. Adv. Manuf. Technol.* **2011**, *58*, 849–859. [CrossRef]
20. Guimu, Z.; Chao, Y.; Chen, S.R.; Libao, A. Experimental study on the milling of thin parts of titanium alloy (TC4). *J. Mater. Process. Technol.* **2003**, *138*, 489–493. [CrossRef]
21. Kao, Y.-C.; Nguyen, N.-T.; Chen, M.-S.; Su, S.-T. A prediction method of cutting force coefficients with helix angle of flat-end cutter and its application in a virtual three-axis milling simulation system. *Int. J. Adv. Manuf. Technol.* **2014**, *77*, 1793–1809. [CrossRef]
22. Peng, C.; Wang, L.; Liao, T.W. A new method for the prediction of chatter stability lobes based on dynamic cutting force simulation model and support vector machine. *J. Sound Vib.* **2015**, *354*, 118–131. [CrossRef]
23. Yang, Y.; Liu, Q.; Zhang, B. Three-dimensional chatter stability prediction of milling based on the linear and exponential cutting force model. *Int. J. Adv. Manuf. Technol.* **2014**, *72*, 1175–1185. [CrossRef]
24. Yi, J.; Chen, K.; Xu, Y. Microstructure, Properties, and Titanium Cutting Performance of AlTiN–Cu and AlTiN–Ni Coatings. *Coatings* **2019**, *9*, 818. [CrossRef]

Article

Elasto-Dynamic Modeling of an Over-Constrained Parallel Kinematic Machine Using a Beam Model

Hélène Chanal [1,*], Aurélie Guichard [2], Benoît Blaysat [1] and Stéphane Caro [3]

[1] Clermont Auvergne INP, Institut Pascal, CNRS, Université Clermont Auvergne, F-63000 Clermont-Ferrand, France; benoit.blaysat@uca.fr

[2] IUT Angers, Université Angers, 2 boulevard du Ronceray, F-49035 Angers, France; aurelie.guichard@univ-angers.fr

[3] CNRS, LS2N UMR CNRS 6004, 1 rue de la Noë, CEDEX 03, F-44321 Nantes, France; stephane.caro@ls2n.fr

[*] Correspondence: helene.chanal@sigma-clermont.fr

Abstract: This article deals with the development of a simple model to evaluate the first natural frequencies of over-constrained parallel kinematic machines (PKMs). The simplest elasto-dynamic models are based on multi-body approaches. However, these approaches require an expression of the Jacobian matrices that may be difficult to obtain for complex PKMs. Therefore, this paper focuses on the determination of the global mass and stiffness matrices of an over-constrained PKM in stationary configurations without the use of Jacobian matrices. The PKM legs are modeled by beams. Because the legs are connected to a moving platform and the mechanism is over-constrained, constraint equations between the parameters that model the deformation of each leg are determined according to screw theory. The first natural frequencies and associated modes can then be determined. It should be noted that the proposed method can be easily used at the conceptual design stage of PKMs. The Tripteor X7 machine is used as an illustrative example and is characterized experimentally.

Keywords: parallel kinematic machine tool; over-constrained system; elasto-dynamic model; screw theory

Citation: Chanal, H.; Guichard, A.; Blaysat, B.; Caro, S. Elasto-Dynamic Modeling of an Over-Constrained Parallel Kinematic Machine Using a Beam Model. *Machines* **2022**, *10*, 200. https://doi.org/10.3390/machines10030200

Academic Editor: Antonio J. Marques Cardoso

Received: 4 February 2022
Accepted: 8 March 2022
Published: 10 March 2022

Publisher's Note: MDPI stays neutral with regard to jurisdictional claims in published maps and institutional affiliations.

1. Introduction

Parallel Kinematic Robots are used in many fields such as medicinal, aerospace, rehabilitation, and astronomy [1]. In industry, parallel kinematic robots are mainly used for pick-and-place operations [2]. However, a few Parallel Kinematic Machine Tools (PKMs) are used in the automotive or aeronautical industries for High-Speed Machining (HSM) operations [3]. Their dynamic performances, in terms of acceleration, are better for an equivalent motorization than Serial Kinematic Machine Tools (SKMs) due their closed-loop architecture [4]. However, the design of these closed-loop architectures imposes to control leg size that reduces mobile masses and leads to a loss of stiffness compared to SKM [5,6].

One way to improve PKM stiffness behavior is to use an over-constrained architecture. To this aim, an over-constrained PKM, named Exechon, was designed by Neumann [7], and the PCI-SCEMM company chose to integrate an Exechon robot into the design of Tripteor X7 machine tool (Figure 1). An over-constrained system is considered to be a hyperstatic system which is defined by the IFToMM terminology as a system in which the distribution of internal forces depends on the material properties of the members of the system [8]. Thus, an over-constrained PKM is defined as a PKM with common or redundant constraints that can be removed without changing the kinematic properties of the mechanism [9]. Thus, the study of an over-constrained mechanism behavior can be complex due to the generation of static indeterminacy and geometric constraints [10].

Previous work highlights the impact on part quality of PKM vibrations during machining [11]. Depending on the machined part position, marks appear on the machined surfaces. To minimize these vibrations, it is of prime importance to express a parameterized elasto-dynamic model for over-constrained PKMs. Such a model enables the prediction of static and vibration behavior during the conceptual design stage. Moreover, as PKM behavior is anisotropic and changes regarding the tool pose, the elasto-dynamic model of the PKM should be fast to compute to quickly obtain its stiffness, its first natural frequencies and associated modes throughout the manipulator workspace.

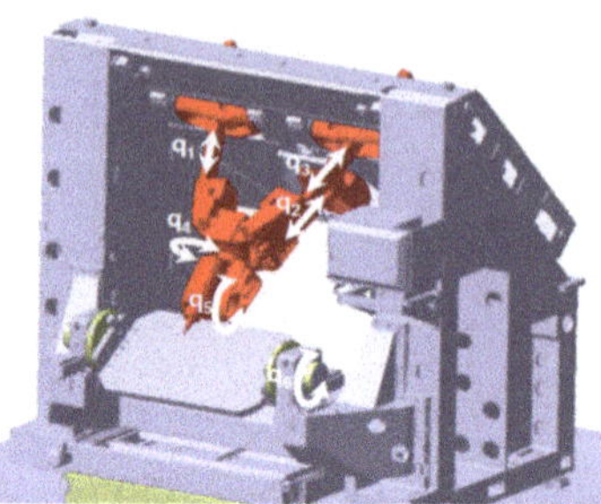

Figure 1. Tripteor X7 Parallel Kinematic Machine Tools.

Two main methods are classically used in the literature to express the elasto-dynamic model of an over-constrained PKM:

1. Finite Element Analysis (FEA): this method is the most accurate one, but it requires the exact definition of the mechanism components. It is usually used for validation purposes at the final design stage since it is time-consuming [12–15].
2. Multi-body approaches: PKM legs can be modeled, for example, using beam theory and their joint behavior relies on Virtual Joint Method (VJM) [16,17]. In the literature, the simplest elasto-dynamic models are based on multi-body approaches. This method can only be applied for robot architectures whose hypotheses are valid. Other formulations are introduced to consider large deformations of flexible manipulators [18] or to decrease simulation time for complex mechanisms such as Matrix Structural Analysis (MSA) [19]. The decreasing of the time simulation is generally based on the development of a methodology to merge stiffness and mass matrices of all elements of the mechanism [19,20].

In this article, multi-body approach, a robust but simple approach to obtain a parameterized elasto-dynamic model, is preferred to the use of complex and time-consuming FEA. The prediction of vibration phenomena requires the determination of the first natural frequencies and their associated modes. In [16,17], this is performed by computing the global mass matrix and the global stiffness matrix. However, the computation of mass and stiffness matrices is complex in the case of over-constrained mechanisms such as Exechon PKM. Indeed, the complex coupling between displacement vector components due to over-constrained properties must be considered during the sub-systems matrix assembly stage. Thus, the application of this approach to over-constrained PKMs requires the determination of Jacobian matrices to characterize kinematic dependencies due to closed-loop over-constrained mechanisms [16,21] or the definition of a complex methodology to merge stiffness matrices of each PKM element [19]. Those Jacobian matrices describe the kinematic behavior of PKMs by derivation of the geometric model as in Germain's [16] and Zhang's [17] works. Geometric model or closed-loop constraints of complex PKM, such as Exechon robot, can sometimes not be computed in a systematic and straightforward way [16,22]. For example, the expression of the geometric model of an Exechon robot is based on a system of nonlinear equations that are not easy to establish [23]. Moreover, its solving requires a numerical optimization with the Newton-Raphson method.

This paper aims to introduce a reliable and simple method to compute global mass and stiffness matrices for over-constrained PKMs, under the assumption of small displacements and using screw theory. This methodology is relevant for high-stiffness architectures such as machine tools and industrial robots where flexibility generates only small displacements.

These mathematical tools allow defining geometric dependencies of leg movements without the computation of Jacobian matrices [24]. Screw theory is used both:

1. To express the kinematic behavior of each PKM leg using the theory of Timoshenko beams [25] and
2. To determine the displacement constraints applied to each leg extremity due to over-constrained mechanism closed-loops [26].

There are different ways to parameterize the orientation of a rigid body such as quaternion [27]. The choice of screw theory ensures use of the same representation for the actuation and constraint wrenches applied to the moving platform by the actuators and the geometry of the mechanism.

This second point is the main contribution of our work and allows us to take into account stress due to over-constrained with geometric constraint equations extracted only from the screw theory application to the passive joints. With this method, the computation of global mass and stiffness matrices is easy to implement. It assumes only geometrical constraints between displacement vector components without the introduction of the Jacobian matrices. Global mass and stiffness matrices are then computed to estimate PKM Cartesian stiffness, the first natural frequencies and their associated modes. This work is relevant at the embodiment design stage of the robot when only a parameterized CAD modeling of the PKM under study is available. Indeed, at this stage, a fast computation time is required to assess the capability of the PKM in terms of vibration behavior and to determine the optimal architecture and dimensions of the robot under design. During the next design stages, when accurate simulation results are needed, FEA can be preferred although it is more computation intensive.

It should be noted that the revolute joints of the Tripteor legs are stiff; their stiffness is larger than 1.04×10^9 N·m^{-1} [28]. Moreover, the stiffness of the spherical wrist of the Tripteor is even more important. As a consequence, and for the sake of clarity, only the deflection of Tripteor X7 PKM legs is considered in what remains. The joint deflection can be taken into account by adding local stiffness such as with the Virtual Joint Method [29].

The paper is organized as follows: Section 2 presents the Tripteor X7 PKM and its parametrization. Section 3 describes the proposed method to express the elasto-dynamic model of PKMs and to compute their natural frequencies. As an illustrative example, the stiffness matrix, the natural frequencies and associated modes of the Tripteor X7 PKM are computed in Section 4. Finally, conclusions and future work are drawn in Section 5.

2. Tripteor X7 PKM

Tripteor X7 PKM is manufactured by the PCI-SCEMM company in France. It is a hybrid PKM. Its architecture is a combination of a parallel mechanism and a serial wrist mounted in series. Such a hybrid manipulator ensures the decoupling of translational and rotational motion of the end-effector. Moreover, hybrid manipulators usually have a better dexterity than fully parallel manipulators as explained in [30].

Tripteor X7 PKM is a six-axis PKM with the following actuated joint variables (q_1, q_2, q_3, q_4, q_5) shown in Figure 2 and one-DoF rotary table represented in Figure 1. The revolute joint variable of this rotary table is denoted as q_6. A parallel mechanism provides three Degrees of Freedom (DoFs) from the actuation of three ball screw systems (q_1, q_2 and q_3) and a serial wrist with two DoFs from the actuation of two revolute joints with direct and belt drives (q_4 and q_5). The elongations q_1, q_2 and q_3 of Leg 1, Leg 2 and Leg 3 provide translational motions of the moving platform along axis $\mathbf{x}_b$, $\mathbf{y}_b$ and axis $\mathbf{z}_b$ with induced rotational motions about axis $\mathbf{x}_b$ and axis $\mathbf{y}_m$. The serial wrist ensures rotational motions of the end-effector.

The paper aims to introduce a new methodology to compute natural frequencies of over-constrained closed-loop mechanism in the case of PKM. Thus, only the vibration behavior of the parallel mechanism composed of Leg 1, Leg 2, Leg 3 and the mobile platform (4) is studied in this paper (Figure 2). If a study of the complete PKM should be carried, a simple assembly of global mass and stiffness matrices of the parallel mechanism and the serial wrist can then be carried out.

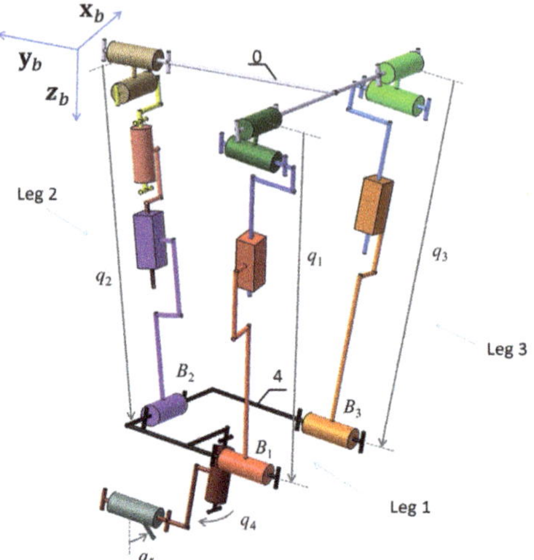

Figure 2. Kinematic model of Tripteor X7 PKM.

Each leg is connected to the mobile platform (4) by a revolute joint and to the fixed platform (0) by two revolute joints which form a universal joint (Figure 2). It is worth noting that Leg 2 has a supplementary revolute joint around its principal axis. Consequently, bending and traction-compression loads are prescribed to Leg 2 whereas Legs 1 and 3 are loaded in bending, traction-compression, and torsion [28]. The mobile platform is supposed to be rigid. Because stiffness and low frequencies are of primal importance in machining [11], an efficient and fast way to compute those frequencies is described in this paper. For this purpose, a local model of a PKM leg under bending, traction-compression, and torsion loads is introduced before the definition of constraint equations due to mechanism assembly.

3. Elasto-Dynamic Modeling of PKMs

This section introduces the novel elasto-dynamic modeling approach and its application to PKMs. For the sake of pedagogy, we introduce our method for PKMs with telescopic legs such as the Tripteor X7 (Figure 3). A_i and B_i are the leg extremities. A_i is attached to the PKM base and B_i is fixed to the mobile platform. The coordinate system associated with the fixed base is denoted as $\mathcal{R}_b = (O, \mathbf{x}_b, \mathbf{y}_b, \mathbf{z}_b)$ and with the mobile platform as $\mathcal{R}_m = (P, \mathbf{x}_m, \mathbf{y}_m, \mathbf{z}_m)$. $\mathbf{z}_i$ is the unit vector along leg i direction.

This method can be applied to any PKMs whose legs can be modeled by beams.

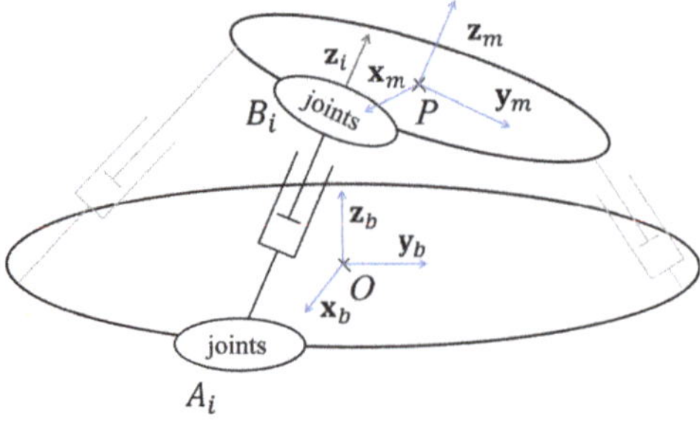

Figure 3. Definition of used parameters for PKM with telescopic legs.

3.1. Local Modeling of PKM Leg

First, each leg is modeled assuming the usual Euler-Bernoulli beam theory. It is worth noting that this model considers the usual mechanical loads, which are bending, traction-compression, and torsion conditions. Each leg is considered to be being connected at one of its ends via a given joint to a fixed part (Figure 4). This boundary condition depends on the joint type used for connecting the leg to the fixed base. In this sub-section, we first study the modeling and the parameter definition of a leg assuming generic boundary conditions to elaborate a leg stiffness matrix $\mathbf{K}_e$ and a leg mass matrix $\mathbf{M}_e$ in the leg coordinate system. With a generic beam model, leg kinematics are defined through six displacement functions, which correspond to the displacement of the neutral axis. $\mathbf{u}(x)$ is the vector that collects those displacement functions:

$$\mathbf{u}(x)^T = \begin{bmatrix} u(x) & v(x) & w(x) & \theta_\alpha(x) & \theta_\beta(x) & \theta_\gamma(x) \end{bmatrix}^T \quad (1)$$

$u(x)$ is the displacement along the neutral axis, $v(x)$ and $w(x)$ are the displacements along **fi**-direction and **fl**-direction. $\theta_\alpha(x)$, $\theta_\beta(x)$, and $\theta_\gamma(x)$ correspond to the small rotations about **ff**-axis, **fi**-axis and **fl**-axis, respectively. Using a Bernoulli beam theory, cross-sectional displacements and rotations satisfy the following equations:

$$\theta_\gamma(x) = \frac{dv}{dx}(x), \ \theta_\beta(x) = -\frac{dw}{dx}(x) \quad (2)$$

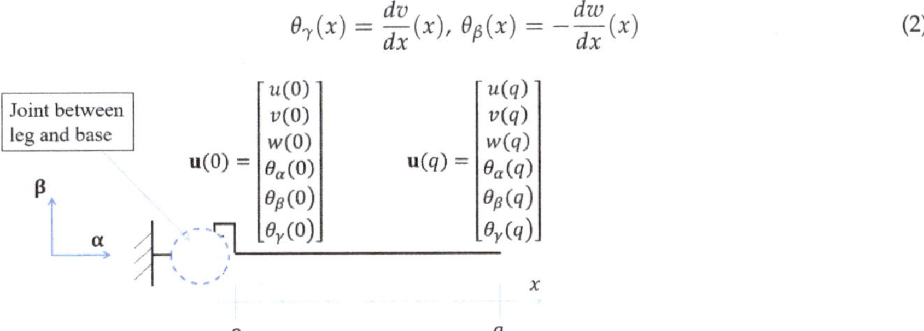

$\mathcal{R}(\alpha, \beta, \gamma)$ is the leg coordinate system.
x is the leg abscissa and q is the leg length.

Figure 4. Local leg model and displacement parameters.

Each kinematic function corresponds to a force function. In the following, $\mathbf{f}_{beam}$ denotes the force vector and $\mathbf{m}_{beam}$ is the moment vector such that:

$$\begin{cases} \mathbf{f}_{beam}(x) = N(x)\mathbf{ff} + T_\beta(x)\mathbf{fi} + T_\gamma(x)\mathbf{fl} \\ \mathbf{m}_{beam}(x) = M_t(x)\mathbf{ff} + M_\beta(x)\mathbf{fi} + M_\gamma(x)\mathbf{fl} \end{cases} \quad (3)$$

where $N(x)$ is the axial force, $T_\beta(x)$ and $T_\gamma(x)$ are the shear forces, $M_t(x)$ is the torsion moment, and $M_\beta(x)$ and $M_\gamma(x)$ are the bending moments. As there is no distributed load, the governing equilibrium equations give:

$$\begin{cases} \dfrac{dN}{dx}(x) = 0, \ \dfrac{dT_\beta}{dx}(x) = 0, \ \dfrac{dT_\gamma}{dx}(x) = 0 \\[2mm] \dfrac{dM_t}{dx}(x) = 0, \ \dfrac{dM_\beta}{dx}(x) - T_\gamma(x) = 0, \ \dfrac{dM_\gamma}{dx}(x) + T_\beta(x) = 0 \end{cases} \quad (4)$$

In addition, the kinematic and force functions are related to each other as follows:

$$N(x) = ES\frac{du}{dx}(x), \ M_t(x) = GI_G\frac{d\theta_\alpha}{dx}(x), \ M_\beta(x) = EI_\beta\frac{d\theta_\beta}{dx}(x), \ M_\gamma(x) = EI_\gamma\frac{d\theta_\gamma}{dx}(x) \quad (5)$$

where E is the material Young's modulus, S is the cross-sectional area, and I_G, I_β and I_γ are moments of area about ff-axis, fi-axis and fl-axis.

Combining Equations (2), (4) and (5), we can conclude that $u(x)$ and $\theta_\alpha(x)$ are linear functions, whereas $v(x)$ and $w(x)$ are polynomial functions of order 3. In a generic case, all kinematic functions can thus be defined through their nodal values expressed in Equation (6).

$$\mathbf{u}(x) = \begin{bmatrix} u(0)+\frac{x}{q}(u(q)-u(0)) \\[4pt] v(0)+\theta_\gamma(0)x-\left(\frac{2\theta_\gamma(0)+\theta_\gamma(q)}{q}-\frac{3(v(q)-v(0))}{q^2}\right)x^2+\left(\frac{\theta_\gamma(0)+\theta_\gamma(q)}{q^2}-\frac{2(v(q))-v(0))}{q^3}\right)x^3 \\[4pt] w(0)-\theta_\beta(0)x+\left(\frac{2\theta_\beta(0)+\theta_\beta(q)}{q}+\frac{3(w(q)-w(0))}{q^2}\right)x^2-\left(\frac{\theta_\beta(0)+\theta_\beta(q)}{q^2}+\frac{2(w(q)-w(0))}{q^3}\right)x^3 \\[4pt] \theta_\alpha(0)+\frac{x}{q}(\theta_\alpha(q)-\theta_\alpha(0)) \\[4pt] \theta_\beta(0)-2\left(\frac{2\theta_\beta(0)+\theta_\beta(q)}{q}+\frac{3(w(q)-w(0))}{q^2}\right)x+3\left(\frac{\theta_\beta(0)+\theta_\beta(q)}{q^2}+\frac{2(w(q)-w(0))}{q^3}\right)x^2 \\[4pt] \theta_\gamma(0)-2\left(\frac{2\theta_\gamma(0)+\theta_\gamma(q)}{q}-\frac{3(v(q)-v(0))}{q^2}\right)x+3\left(\frac{\theta_\gamma(0)+\theta_\gamma(q)}{q^2}-\frac{2(v(q)-v(0))}{q^3}\right)x^2 \end{bmatrix} \tag{6}$$

$\mathbf{u}_{nod}$ denotes the vector collecting the nodal values of the kinematic functions (Figure 4) and is expressed as a 1×12 column vector:

$$\mathbf{u}_{nod} = \begin{bmatrix} \mathbf{u}(0) \\ \mathbf{u}(q) \end{bmatrix} \tag{7}$$

Vectors $\mathbf{N}_u(x)$, $\mathbf{N}_v(x)$, $\mathbf{N}_w(x)$, $\mathbf{N}_{\theta_\alpha}(x)$, $\mathbf{N}_{\theta_\beta}(x)$ and $\mathbf{N}_{\theta_\gamma}(x)$ express the nodal values as a function of x coordinate bounded between 0 and q, i.e., $x \in [0, q]$:

$$\begin{cases} u(x) = \mathbf{N}_u(x)^T\mathbf{u}_{nod} \\ v(x) = \mathbf{N}_v(x)^T\mathbf{u}_{nod} \\ w(x) = \mathbf{N}_w(x)^T\mathbf{u}_{nod} \\ \theta_\alpha(x) = \mathbf{N}_{\theta_\alpha}(x)^T\mathbf{u}_{nod} \\ \theta_\beta(x) = \mathbf{N}_{\theta_\beta}(x)^T\mathbf{u}_{nod} \\ \theta_\gamma(x) = \mathbf{N}_{\theta_\gamma}(x)^T\mathbf{u}_{nod} \end{cases} \tag{8}$$

From [31], the beam potential energy E_p is written as:

$$E_p = \frac{1}{2}\int_0^q\left[ES\left(\frac{du(x)}{dx}\right)^2+GI_G\left(\frac{d\theta_\alpha(x)}{dx}\right)^2+EI_\beta\left(\frac{d\theta_\beta(x)}{dx}\right)^2+EI_\gamma\left(\frac{d\theta_\gamma(x)}{dx}\right)^2\right]dx \tag{9}$$

From Equations (8) and (9), E_p can be expressed in a compact form as a function of the beam stiffness matrix $\mathbf{K}_e$ and $\mathbf{u}_{nod}$ as follows:

$$E_p = \frac{1}{2}\mathbf{u}_{nod}^T\mathbf{K}_e\mathbf{u}_{nod} \tag{10}$$

With matrix $\mathbf{K}_e$ taking the form of Equation (11).

$$\mathbf{K}_e = \begin{bmatrix}
\frac{ES}{q} & 0 & 0 & 0 & 0 & 0 & -\frac{ES}{q} & 0 & 0 & 0 & 0 & 0 \\
0 & \frac{12EI_\gamma}{q^3} & 0 & 0 & 0 & \frac{6EI_\gamma}{q^2} & 0 & -\frac{12EI_\gamma}{q^3} & 0 & 0 & 0 & \frac{6EI_\gamma}{q^2} \\
0 & 0 & \frac{12EI_\beta}{q^3} & 0 & -\frac{6EI_\beta}{q^2} & 0 & 0 & 0 & -\frac{12EI_\beta}{q^3} & 0 & -\frac{6EI_\beta}{q^2} & 0 \\
0 & 0 & 0 & \frac{GI_G}{q} & 0 & 0 & 0 & 0 & 0 & -\frac{GI_G}{q} & 0 & 0 \\
0 & 0 & -\frac{6EI_\beta}{q^2} & 0 & \frac{4EI_\beta}{q} & 0 & 0 & 0 & \frac{6EI_\beta}{q^2} & 0 & \frac{2EI_\beta}{q} & 0 \\
0 & \frac{6EI_\gamma}{q^2} & 0 & 0 & 0 & \frac{4EI_\gamma}{q} & 0 & -\frac{6EI_\gamma}{q^2} & 0 & 0 & 0 & \frac{2EI_\gamma}{q} \\
-\frac{ES}{q} & 0 & 0 & 0 & 0 & 0 & \frac{ES}{q} & 0 & 0 & 0 & 0 & 0 \\
0 & -\frac{12EI_\gamma}{q^3} & 0 & 0 & 0 & -\frac{6EI_\gamma}{q^2} & 0 & \frac{12EI_\gamma}{q^3} & 0 & 0 & 0 & -\frac{6EI_\gamma}{q^2} \\
0 & 0 & -\frac{12EI_\beta}{q^3} & 0 & \frac{6EI_\beta}{q^2} & 0 & 0 & 0 & \frac{12EI_\beta}{q^3} & 0 & \frac{6EI_\beta}{q^2} & 0 \\
0 & 0 & 0 & -\frac{GI_G}{q} & 0 & 0 & 0 & 0 & 0 & \frac{GI_G}{q} & 0 & 0 \\
0 & 0 & -\frac{6EI_\beta}{q^2} & 0 & \frac{2EI_\beta}{q} & 0 & 0 & 0 & \frac{6EI_\beta}{q^2} & 0 & \frac{4EI_\beta}{q} & 0 \\
0 & \frac{6EI_\gamma}{q^2} & 0 & 0 & 0 & \frac{2EI_\gamma}{q} & 0 & -\frac{6EI_\gamma}{q^2} & 0 & 0 & 0 & \frac{4EI_\gamma}{q}
\end{bmatrix} \tag{11}$$

From [31], the beam kinetic energy E_k is defined as:

$$E_k = \frac{1}{2}\rho \int_0^q \iint_S \mathbf{v}_M^T \mathbf{v}_M \, dx\,dy\,dz \tag{12}$$

where ρ is the beam material density, and $\mathbf{v}_M$ is the point velocity along the beam expressed as:

$$\mathbf{v}_M = \left(\dot{u}(x) + z\dot{\theta}_\beta(x) - y\dot{\theta}_\gamma(x)\right)\mathbf{ff} + \left(\dot{v}(x) - z\dot{\theta}_\alpha(x)\right)\mathbf{fi} + \left(\dot{w}(x) + y\dot{\theta}_\alpha(x)\right)\mathbf{fl} \tag{13}$$

Combining Equations (12) and (13) with the hypothesis that **fi**-axis and **fl**-axis are principal axes of inertia, the kinematic energy E_k writes:

$$E_k = \frac{1}{2}\rho S \int_0^q \left(\dot{u}(x)^2 + \dot{v}(x)^2 + \dot{w}(x)^2\right)dx + \frac{1}{2}\rho I_{\beta\beta} \int_0^q \left(\dot{\theta}_\alpha(x)^2 + \dot{\theta}_\beta(x)^2\right)dx$$
$$+ \frac{1}{2}\rho I_{\gamma\gamma} \int_0^q \left(\dot{\theta}_\alpha(x)^2 + \dot{\theta}_\gamma(x)^2\right)dx \tag{14}$$

with $I_{\beta\beta}$ and $I_{\gamma\gamma}$ the beam inertia about **fi**-axis and **fl**-axis. Equation (14) takes the following compact form:

$$E_k = \frac{1}{2}\dot{\mathbf{u}}_{nod}^T \mathbf{M}_e \dot{\mathbf{u}}_{nod} \tag{15}$$

with $\dot{\mathbf{u}}_{nod}$ the time derivative of vector $\mathbf{u}_{nod}$ and $\mathbf{M}_e$ the mass matrix of beam expressed in Equation (16).

$$\mathbf{M}_e = \begin{bmatrix}
\frac{b}{3} & 0 & 0 & 0 & 0 & 0 & \frac{b}{6} & 0 & 0 & 0 & 0 & 0 \\
0 & \frac{13}{35}b + \frac{6\rho I_{\gamma\gamma}}{5q} & 0 & 0 & 0 & \frac{\rho I_{\gamma\gamma}}{10} + \frac{11c}{210} & 0 & \frac{9b}{70} - \frac{6\rho I_{\gamma\gamma}}{5q} & 0 & 0 & 0 & \frac{\rho I_{\gamma\gamma}}{10} - \frac{13c}{420} \\
0 & 0 & \frac{6\rho I_{\beta\beta}}{5q} + \frac{13}{35}b & 0 & -\frac{\rho I_{\beta\beta}}{10} - \frac{11c}{210} & 0 & 0 & 0 & \frac{9b}{70} - \frac{6\rho I_{\beta\beta}}{5q} & 0 & -\frac{\rho I_{\beta\beta}}{10} + \frac{13c}{420} & 0 \\
0 & 0 & 0 & a & 0 & 0 & 0 & 0 & 0 & \frac{a}{2} & 0 & 0 \\
0 & 0 & -\frac{\rho I_{\beta\beta}}{10} - \frac{11c}{210} & 0 & \frac{2\rho I_{\beta\beta}q}{15} + \frac{d}{105} & 0 & 0 & 0 & \frac{\rho I_{\beta\beta}}{10} - \frac{13c}{420} & 0 & -\frac{\rho I_{\beta\beta}q}{30} - \frac{d}{140} & 0 \\
0 & \frac{\rho I_{\gamma\gamma}}{10} + \frac{11c}{210} & 0 & 0 & 0 & \frac{2\rho I_{\gamma\gamma}q}{15} + \frac{d}{105} & 0 & -\frac{\rho I_{\gamma\gamma}}{10} + \frac{13c}{420} & 0 & 0 & 0 & -\frac{\rho I_{\gamma\gamma}q}{30} - \frac{d}{140} \\
\frac{b}{6} & 0 & 0 & 0 & 0 & 0 & \frac{b}{3} & 0 & 0 & 0 & 0 & 0 \\
0 & \frac{9b}{70} - \frac{6\rho I_{\gamma\gamma}}{5q} & 0 & 0 & 0 & -\frac{\rho I_{\gamma\gamma}}{10} + \frac{13c}{420} & 0 & \frac{13}{35}b + \frac{6\rho I_{\gamma\gamma}}{5q} & 0 & 0 & 0 & -\frac{\rho I_{\gamma\gamma}}{10} - \frac{11c}{210} \\
0 & 0 & \frac{9b}{70} - \frac{6\rho I_{\beta\beta}}{5q} & 0 & \frac{\rho I_{\beta\beta}}{10} - \frac{13c}{420} & 0 & 0 & 0 & \frac{6\rho I_{\beta\beta}}{5q} + \frac{13}{35}b & 0 & \frac{\rho I_{\beta\beta}}{10} + \frac{11c}{210} & 0 \\
0 & 0 & 0 & \frac{a}{2} & 0 & 0 & 0 & 0 & 0 & a & 0 & 0 \\
0 & 0 & -\frac{\rho I_{\beta\beta}}{10} + \frac{13c}{420} & 0 & -\frac{\rho I_{\beta\beta}q}{30} - \frac{d}{140} & 0 & 0 & 0 & \frac{\rho I_{\beta\beta}}{10} + \frac{11c}{210} & 0 & \frac{2\rho I_{\beta\beta}q}{15} + \frac{d}{105} & 0 \\
0 & \frac{\rho I_{\gamma\gamma}}{10} - \frac{13c}{420} & 0 & 0 & 0 & -\frac{\rho I_{\gamma\gamma}q}{30} - \frac{d}{140} & 0 & -\frac{\rho I_{zz}}{10} - \frac{11c}{210} & 0 & 0 & 0 & \frac{2\rho I_{\gamma\gamma}q}{15} + \frac{d}{105}
\end{bmatrix} \tag{16}$$

with $a = \frac{\rho(I_{\beta\beta} + I_{\gamma\gamma})q}{3}$, $b = \rho Sq$, $c = \rho Sq^2$ and $d = \rho Sq^3$.

To compute the global mass and stiffness matrices, the constraints between the displacements of the leg extremities are determined in the next section.

3.2. Constraint Equations

The mobile platform is considered to be rigid for the elaboration of the constraint equations between the leg extremities. We therefore express the movement of the mobile platform due to leg flexibility in stationary configurations of the PKM architecture using screw theory [26] (Figure 5).

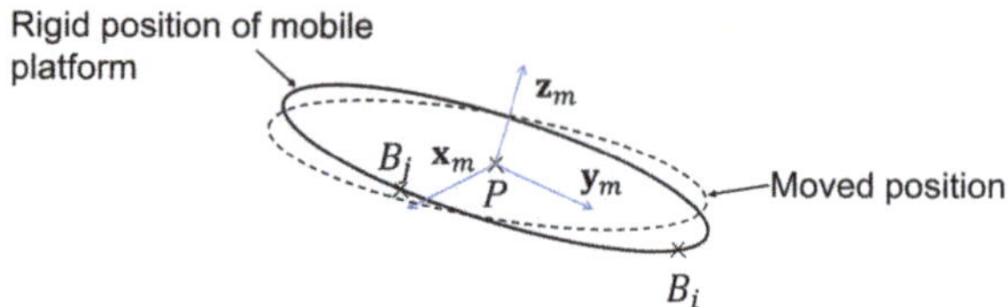

Figure 5. Parametrization of mobile platform small displacement.

The small displacement screw $\mathcal{T}_\mathbf{m}$ of the mobile platform is influenced by displacement and rotation $\mathbf{u}_i(q_i)$ of the leg i at point B_i regarding the movements of passive joints between leg i and mobile platform. A normalized screw is used to model passive joint movement [32]. Each movement between leg i and mobile platform is model with a one-DoF motion-screw:

$$
\mathcal{S}_{ik} = \begin{bmatrix} \mathcal{S}_{\mathbf{F}ik} \\ \mathcal{S}_{\mathbf{S}ik} \end{bmatrix} = \begin{cases} \begin{bmatrix} \mathbf{s}_{ik} \\ \mathbf{s}_{ik} \times \mathbf{r}_{ik} \end{bmatrix} \text{if it is a rotational movement} \\[2em] \begin{bmatrix} \mathbf{0} \\ \mathbf{s}_{ik} \end{bmatrix} \text{if it is a prismatic movement} \end{cases}
\tag{17}
$$

where $k = 1$ to n_i (n_i is the number of equivalent one-DoF joints between leg i and mobile platform, $n_i = 1$ for the case study), $\mathbf{s}_{ik}$ is a unit vector along the axis of the screw $\mathcal{S}_{ik}$, $\mathbf{r}_{ik}$ is a vector directed from any point on the screw axis to point B_i.

The computation of the movement transmitted by the leg to the mobile platform or vice versa can be done using the reciprocal screws of the motion-screws $\mathcal{S}_{ik}$. There is a unique normalized reciprocal screw system $\mathcal{T}_{ik}^\perp$ of order 5 of $\mathcal{S}_{ik}$ [32]. For a rotational joint, the reciprocal system is a 5-system which includes all rotational screws whose axes intersect the joint axis and all prismatic screws whose axes are perpendicular to joint axes and all the combinations of the above reciprocal screw [32]. The relation between screw $\mathcal{S}_{ik}$ and the five one-DoF reciprocal screws $\mathcal{S}_{ijk}^\perp$ is:

$$
\left(\begin{bmatrix} \mathbf{0} & \mathbf{I}_3 \\ \mathbf{I}_3 & \mathbf{0} \end{bmatrix} \mathcal{S}_{ik} \right)^T \mathcal{S}_{ijk}^\perp = 0
\tag{18}
$$

where $\mathbf{I}_3$ is the 3×3 identity matrix, $\mathbf{0}$ is the 3×3 zero matrix, $j = 1$ to 5. Then, $\mathcal{T}_{ik}^\perp = (\mathcal{S}_{ijk}^\perp$ with $j = 1$ to 5).

The small displacement screw is defined as $\mathcal{T}_\mathbf{m} = \begin{bmatrix} \boldsymbol{\Omega} \\ \mathbf{d}_{B_i} \end{bmatrix}$ with $\boldsymbol{\Omega}$ the rotation vector of the mobile platform and $\mathbf{d}_{B_i}$ the small displacement vector of point B_i ($\mathbf{d}_{B_i} = \mathbf{d}_{B_j} + \mathbf{b}_{ij} \times \boldsymbol{\Omega}$ with $\mathbf{b}_{ji}$ the vector from B_j to B_i). For leg i, the relation between $\mathcal{T}_m$ and $\mathbf{u}_i(q_i)$ is written:

$$
\prod_{k=1}^{n_i} \left(\sum_{j=1}^{5} \mathcal{S}_{ijk}^\perp \right) \left(\mathbf{u}_i(q_i) - \begin{bmatrix} \mathbf{R}_{leg_i \to beam} \mathbf{R}_{R_m \to leg_i} \mathbf{d}_{B_i} \\ \mathbf{R}_{leg_i \to beam} \mathbf{R}_{R_m \to leg_i} \boldsymbol{\Omega} \end{bmatrix} \right) = \mathbf{0}
\tag{19}
$$

where $\mathbf{R}_{leg_i \to beam}$ is the rotational matrix between the beam coordinate system and the leg i coordinate system, and $\mathbf{R}_{R_m \to leg_i}$ is the rotational matrix between the leg i coordinate system and the mobile platform coordinate system.

Constraint equations are defined by expressing Equation (19) for each leg and using small displacement screw properties.

In the same way, the nodal displacement $\mathbf{u}_i(0)$ is limited due to passive joints between the leg i and the fixed base. We note $\mathcal{S}'_{ik}$ the movement-screws between leg i and fixed base. Thus, the null nodal parameter is computed from:

$$\left(\prod_{k=1}^{n'_i} \left(\sum_{j=1}^{5} \mathcal{S}'_{ijk}{}^{\perp} \right) \right) \mathbf{u}_i(0) = \mathbf{0} \tag{20}$$

where n'_i is the number of equivalent one DoF joints between leg i and fixed base. For the case study, $n'_1 = n'_3 = 2$, $n'_2 = 3$.

From Equations (19) and (20), constraint equations are grouped together as:

$$\mathbf{L_D u_D} + \mathbf{L_I u_I} = 0 \Rightarrow \mathbf{u_D} = -\mathbf{L_D^{-1} L_I u_I} \tag{21}$$

where $\mathbf{u_D}$ is a vector composed of the set of beam nodal dependent parameters extracted from all $\mathbf{u}_i(q_i)$. Its size is equal to the rank of the equation system (19) for all legs. $\mathbf{u_I}$ is a vector composed of the set of beam nodal independent parameters, which is not included in $\mathbf{u_D}$. The choice of $\mathbf{u_D}$ and $\mathbf{u_I}$ is not unique. The components of matrices $\mathbf{L_D}$ and $\mathbf{L_I}$ are the factor of vectors $\mathbf{u_D}$ and $\mathbf{u_I}$ components in Equations (19) and (20).

Equation (21) corresponds to a simple formulation of the closed-loop geometric constraints for over-constrained PKM and is the major contribution of the paper. The determination of stiffness and vibratory behavior of a PKM is based on the definition of its global mass and stiffness matrices. These matrices are defined in the following section.

3.3. Mass Matrix Computation of Mobile Platform

The mobile platform and the serial wrist are considered to be a point-mass m located at its center of mass G. Thus, kinematic energy E_{km} writes:

$$E_{km} = \frac{1}{2} m \mathbf{v}_G^T \mathbf{v}_G \tag{22}$$

$\mathbf{v}_G$ being the linear velocity of point G. Under the assumption of small displacement $\mathbf{v}_G$ is expressed as:

$$\mathbf{v}_G = \dot{\mathbf{d}}_G = \dot{\mathbf{d}}_{B_1} + \mathbf{g}_1 \times \dot{!} \tag{23}$$

where $\mathbf{d}_G$ is the small displacement of point G, $\dot{\mathbf{d}}_G$ is its time derivative and $\mathbf{g}_1$ is the vector from G to B_1.

Equation (23) allows defining E_{km} according to beam nodal parameters:

$$E_{km} = \frac{1}{2} \begin{pmatrix} \dot{\mathbf{u}}_D \\ \dot{\mathbf{u}}_I \end{pmatrix}^T \mathbf{M}_{em} \begin{pmatrix} \dot{\mathbf{u}}_D \\ \dot{\mathbf{u}}_I \end{pmatrix} \tag{24}$$

where $\mathbf{M}_{em}$ is the mobile platform mass matrix. To compute natural frequency of the system, this mass matrix is added to the legs mass matrix.

3.4. Cartesian Stiffness Computation

$\mathbf{K}$ is the global stiffness matrix of the PKM. It is obtained from the assembly of stiffness matrices $\mathbf{K}_e$ of the three legs such that:

$$\mathbf{K} = \begin{bmatrix} \mathbf{K_D} & \mathbf{K_{DI}} \\ \mathbf{K_{DI}^T} & \mathbf{K_I} \end{bmatrix} \tag{25}$$

where $\mathbf{K_D}$ is the stiffness matrix corresponding to dependent beam nodal parameters $\mathbf{u_D}$, and $\mathbf{K_I}$ corresponds to the independent ones $\mathbf{u_I}$. $\mathbf{K_{DI}}$ is the coupling matrix.

From Equation (21), a decomposition between dependent and independent beam nodal parameters is used to express the potential energy E_p:

$$E_p = \frac{1}{2}\begin{bmatrix}\mathbf{u_D}\\\mathbf{u_I}\end{bmatrix}^T\begin{bmatrix}\mathbf{K_D} & \mathbf{K_{DI}}\\\mathbf{K_{DI}^T} & \mathbf{K_I}\end{bmatrix}\begin{bmatrix}\mathbf{u_D}\\\mathbf{u_I}\end{bmatrix} = \frac{1}{2}\mathbf{u_I}^T\widetilde{\mathbf{K}}\mathbf{u_I} \tag{26}$$

Consequently, this allows the introduction of the condensed matrix form $\widetilde{\mathbf{K}}$ of the global stiffness matrix of the studied PKM, which now only depends on the independent beam nodal parameters, such that:

$$\widetilde{\mathbf{K}} = \mathbf{L_I^T}\left(\mathbf{L_D^{-1}}\right)^T\mathbf{K_D}\mathbf{L_D^{-1}}\mathbf{L_I} - \left(\mathbf{K_{DI}^T}\mathbf{L_D^{-1}}\mathbf{L_I} + \left(\mathbf{K_{DI}^T}\mathbf{L_D^{-1}}\mathbf{L_I}\right)^T\right) + \mathbf{K_I} \tag{27}$$

The Cartesian stiffness matrix $\mathbf{K_c}$ can be introduced from the expression of the potential energy E_p according to:

$$E_p = \frac{1}{2}\mathbf{u_I}^T\widetilde{\mathbf{K}}\mathbf{u_I} = \frac{1}{2}\mathbf{ffix}^T\mathbf{K_c}\mathbf{ffix} \tag{28}$$

where $\mathbf{ffix}$ is the small displacement vector of point O_m expressed in the base frame. Thus, $\mathbf{ffix} = \mathbf{R}_m^{-1}\left(\mathbf{d}_{B_i} + \mathbf{b}_{mi} \times !\right) = \mathbf{L_{XI}}\mathbf{u_I}$ where $\mathbf{R}_m$ is the rotation matrix from the base frame to the mobile platform frame and $\mathbf{b}_{mi}$ is the vector from O_m to B_i. Finally, the Cartesian stiffness matrix of the PKM is computed from $\widetilde{\mathbf{K}}$:

$$\mathbf{K_c} = \left(\mathbf{L_{XI}}\mathbf{L_{XI}^T}\right)^{-1}\mathbf{L_{XI}}\widetilde{\mathbf{K}}\mathbf{L_{XI}^T}\left(\mathbf{L_{XI}}\mathbf{L_{XI}^T}\right)^{-1} \tag{29}$$

Note that $\mathbf{K_c}$ is a 3×3 matrix and its components are the stiffness along the PKM base frame vectors.

3.5. Natural Frequency Computation

The global stiffness matrix $\mathbf{M}$ of the PKM is obtained from the assembly of the mass matrices $\mathbf{M}_e$ and $\mathbf{M}_{em}$ of the three legs and mobile platform, respectively. It is expressed as:

$$\mathbf{M} = \begin{bmatrix}\mathbf{M_D} & \mathbf{M_{DI}}\\\mathbf{M_{DI}^T} & \mathbf{M_I}\end{bmatrix} \tag{30}$$

where $\mathbf{M_D}$ is the mass matrix corresponding to dependent beam nodal parameters $\mathbf{u_D}$, whereas $\mathbf{M_I}$ corresponds to the independent ones $\mathbf{u_I}$. $\mathbf{M_{DI}}$ is the coupling matrices.

The decomposition between dependent and independent parameters is used to express the kinetic energy E_k as follows:

$$E_k = \frac{1}{2}\begin{bmatrix}\mathbf{u_D}\\\mathbf{u_I}\end{bmatrix}^T\begin{bmatrix}\mathbf{M_D} & \mathbf{M_{DI}}\\\mathbf{M_{DI}^T} & \mathbf{M_I}\end{bmatrix}\begin{bmatrix}\mathbf{u_D}\\\mathbf{u_I}\end{bmatrix} = \frac{1}{2}\mathbf{u_I}^T\widetilde{\mathbf{M}}\mathbf{u_I} \tag{31}$$

Consequently, this allows the introduction of the condensed matrix form of the global mass matrix $\widetilde{\mathbf{M}}$, which now only depends on the independent beam nodal parameters:

$$\widetilde{\mathbf{M}} = \mathbf{L_I^T}\left(\mathbf{L_D^{-1}}\right)^T\mathbf{M_D}\mathbf{L_D^{-1}}\mathbf{L_I} - \left(\mathbf{M_{DI}^T}\mathbf{L_D^{-1}}\mathbf{L_I} + \left(\mathbf{M_{DI}^T}\mathbf{L_D^{-1}}\mathbf{L_I}\right)^T\right) + \mathbf{M_I} \tag{32}$$

The eigenvalues λ_{ev} and the associated eigenvector $\mathbf{u}_{ev}$ are determined from the spectral decomposition of matrix $\widetilde{\mathbf{M}}^{-1}\widetilde{\mathbf{K}}$. The eigenvalues are the solutions of the polynomial $\det\left(\widetilde{\mathbf{M}}^{-1}\widetilde{\mathbf{K}} - \lambda_{ev}\mathbf{I}\right) = 0$. The ith natural frequency f_{0i} is expressed as a function of the ith eignevalue λ_{evi} as follows: $f_{0i} = \frac{\sqrt{\lambda_{evi}}}{2\pi}$. The natural modes are the eigenvectors associated with those natural eigenvalues.

4. Natural Frequencies and Modes of Tripteor X7

The methodology described in Section 3 is applied in this section to calculate the natural frequencies and associated modes of the Tripteor X7 PKM.

4.1. Identification of Null Legs Nodal Parameters

The null nodal parameters of $\mathbf{u}_i(0)$ depend on the joint types between the legs and the base as shown in Figure 6. The first joint of Legs 1 and 3 with the base is a revolute joint about **fl**-axis, and the second joint is about **fi**-axis. Thus, for Legs 1 and 3, infinitesimal and reciprocal screw motion are the following:

$$\mathcal{S}'_{i1} = \begin{pmatrix} 0 \\ 0 \\ 1 \\ 0 \\ 0 \\ 0 \end{pmatrix} \text{ and } \mathcal{S}'_{i2} = \begin{pmatrix} 0 \\ 1 \\ 0 \\ 0 \\ 0 \\ 0 \end{pmatrix} \Rightarrow \sum_{j=1}^{5} \mathcal{S}'^{\perp}_{ij1} = \begin{pmatrix} 1 \\ 1 \\ 1 \\ 1 \\ 1 \\ 0 \end{pmatrix} \text{ and } \sum_{j=1}^{5} \mathcal{S}'^{\perp}_{ij2} = \begin{pmatrix} 1 \\ 1 \\ 1 \\ 1 \\ 0 \\ 1 \end{pmatrix} \tag{33}$$

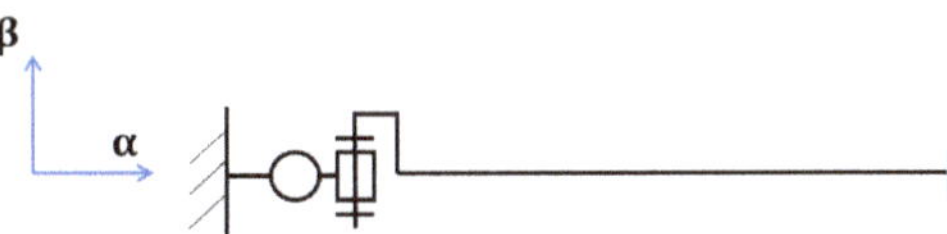

a) joint between legs 1 and 3, and the fixed base

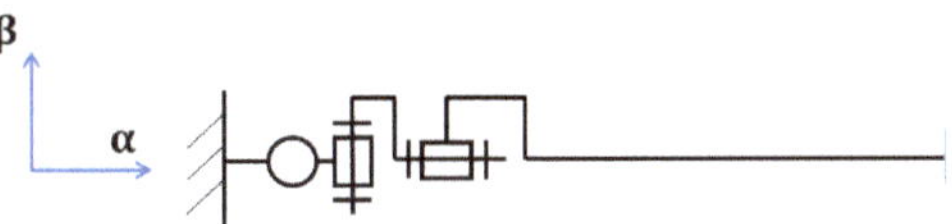

b) joints between leg 2 and the fixed base

Figure 6. Local parameters for Tripteor X7 legs.

Thus, from Equation (20), only $\theta_{\beta i}(0)$ and $\theta_{\gamma i}(0)$ are not null.

For Leg 2, with the same methodology, we obtain that only $\theta_{\alpha 2}(0)$, $\theta_{\beta 2}(0)$ and $\theta_{\gamma 2}(0)$ are not null (Figure 6b).

To compute the global mass and stiffness matrices, the motion constraints between the extremities of the three legs are explained hereafter.

4.2. Constraint Equations

The joints between Legs 1 and 3 and the mobile platform are revolute joints about $\mathbf{y}_m$-axis (**fi**-axis) and the joint between Legs 2 and the mobile platform is a revolute joint about $\mathbf{x}_m$-axis (**fl**-axis) (Figure 7). Thus, the reciprocal motion-screws in the leg coordinate system are:

$$\sum_{j=1}^{5} \mathcal{S}^{\perp}_{1j1} = \sum_{j=1}^{5} \mathcal{S}^{\perp}_{3j1} = \begin{pmatrix} 1 \\ 1 \\ 1 \\ 1 \\ 0 \\ 1 \end{pmatrix} \text{ and } \sum_{j=1}^{5} \mathcal{S}^{\perp}_{2j1} = \begin{pmatrix} 1 \\ 1 \\ 1 \\ 1 \\ 1 \\ 0 \end{pmatrix} \tag{34}$$

$$(u_2(q), v_2(q), w_2(q), \theta_{\alpha 2}(q), \theta_{\beta 2}(q), \theta_{\gamma 2}(q))$$

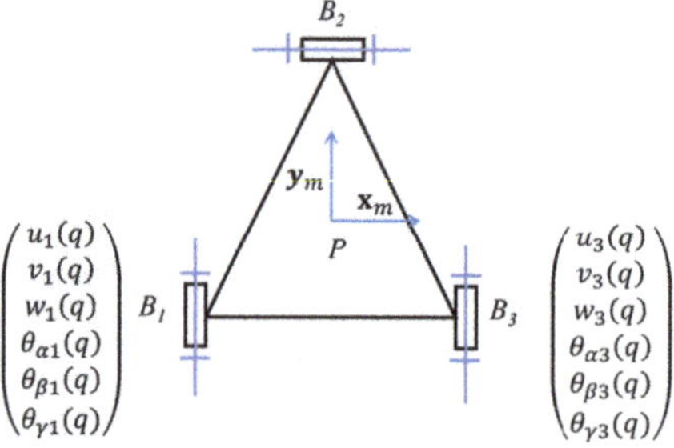

Figure 7. Parametrization of Tripteor mobile platform movement.

From Equation (19), the equation associated with the motion constraints between the mobile platform and the three legs of Tripteor X7 are the following:

$$\sum_{j=1}^{5} \mathcal{S}_{1j1}^{\perp} \left[\mathbf{u}_1(q_1) - \begin{bmatrix} \mathbf{R}_{leg_1 \to beam} \mathbf{R}_{R_m \to leg_1} \mathbf{d}_{B_1} \\ \mathbf{R}_{leg_1 \to beam} \mathbf{R}_{R_m \to leg_1}! \end{bmatrix} \right] = \mathbf{0} \tag{35}$$

$$\sum_{j=1}^{5} \mathcal{S}_{2j1}^{\perp} \left[\mathbf{u}_2(q_2) - \begin{bmatrix} \mathbf{R}_{leg_2 \to beam} \mathbf{R}_{R_m \to leg_2} \mathbf{d}_{B_2} \\ \mathbf{R}_{leg_2 \to beam} \mathbf{R}_{R_m \to leg_2}! \end{bmatrix} \right] = \mathbf{0} \tag{36}$$

$$\sum_{j=1}^{5} \mathcal{S}_{3j1}^{\perp} \left[\mathbf{u}_3(q_3) - \begin{bmatrix} \mathbf{R}_{leg_3 \to beam} \mathbf{R}_{R_m \to leg_3} \mathbf{d}_{B_3} \\ \mathbf{R}_{leg_3 \to beam} \mathbf{R}_{R_m \to leg_3}! \end{bmatrix} \right] = \mathbf{0} \tag{37}$$

$\mathbf{z}_1$ (resp. $\mathbf{z}_2$ and $\mathbf{z}_3$) is the unit vector along the direction of the ith prismatic joint, $i = 1, 2, 3$ as shown in Figure 8. The rotation matrix between the beam coordinate system and the leg coordinate system is expressed as:

$$\mathbf{R}_{beam \to leg} = \begin{bmatrix} 0 & 0 & 1 \\ 0 & 1 & 0 \\ -1 & 0 & 0 \end{bmatrix} \tag{38}$$

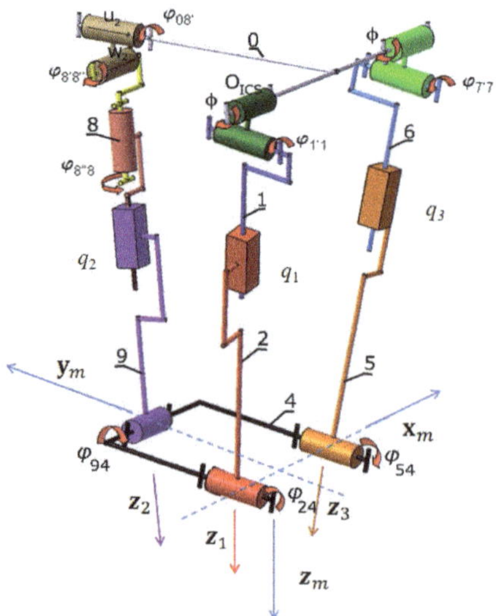

Figure 8. Leg and mobile platform coordinate systems and joint parametrization of Tripteor X7.

The angles between the legs and the mobile platform are $\varphi_{24} = (\mathbf{z}_m, \mathbf{z}_1)$, $\varphi_{94} = (\mathbf{z}_m, \mathbf{z}_2)$ and $\varphi_{54} = (\mathbf{z}_m, \mathbf{z}_3)$ (Figure 8). These angles can be measured with a CAD model or computed with the geometric model of the Tripteor PKM described in [33]. With this notation, the rotation matrices between Legs 1, 2 and 3 and the mobile platform coordinate system $\mathcal{R}_m$ are:

$$\mathbf{T}_{leg_1 \to R_m} = \begin{bmatrix} \cos(\varphi_{24}) & 0 & \sin(\varphi_{24}) \\ 0 & 1 & 0 \\ -\sin(\varphi_{24}) & 0 & \cos(\varphi_{24}) \end{bmatrix} \tag{39}$$

$$\mathbf{T}_{leg_2 \to R_m} = \begin{bmatrix} 1 & 0 & 0 \\ 0 & \cos(\varphi_{94}) & -\sin(\varphi_{94}) \\ 0 & \sin(\varphi_{94}) & \cos(\varphi_{94}) \end{bmatrix} \tag{40}$$

$$\mathbf{T}_{leg_3 \to R_m} = \begin{bmatrix} \cos(\varphi_{54}) & 0 & \sin(\varphi_{54}) \\ 0 & 1 & 0 \\ -\sin(\varphi_{54}) & 0 & \cos(\varphi_{54}) \end{bmatrix} \tag{41}$$

Consequently, in the case of Tripteor X7, by the application of a small-displacement screw relation at point B_1 and from Equations (35)–(37), nine constraint equations are obtained. Nine dependent parameters are chosen:

$$\mathbf{u}_\mathbf{D}^T = (u_2(q_2), v_2(q_2), w_2(q_2), \theta_{\beta 2}(q_2), u_3(q_3), v_3(q_3), w_3(q_3), \theta_{\alpha 3}(q_3), \theta_{\gamma 3}(q_3)) \tag{42}$$

Thus, 16 independent parameters are considered:

$$\mathbf{u}_\mathbf{I}^T = (\theta_{\beta 1}(0), \theta_{\gamma 1}(0), u_1(q1), v_1(q_1), w_1(q_1), \theta_{\alpha 1}(q_1), \theta_{\beta 1}(q_1), \theta_{\gamma 1}(q_1), \theta_{\alpha 2}(0),$$
$$\theta_{\beta 2}(0), \theta_{\gamma 2}(0), \theta_{\alpha 2}(q_2), \theta_{\gamma 2}(q_2), \theta_{\beta 3}(0), \theta_{\gamma 3}(0), \theta_{\beta 3}(q_3)) \tag{43}$$

The (9×9)-dimensional matrix $\mathbf{L_D}$ and the (9×16)-dimensional matrix $\mathbf{L_I}$, obtained from Equations (35)–(37), and the chosen vectors $\mathbf{u_D}$ and $\mathbf{u_I}$, are expressed in Equations (44) and (45).

$$\mathbf{L_D} = \begin{bmatrix} 0 & 0 & 0 & 0 & 0 & 0 & 0 & \sin(\varphi_{54}) & -\cos(\varphi_{54}) \\ 0 & 0 & 0 & 0 & 0 & 0 & 0 & \cos(\varphi_{54}) & \sin(\varphi_{54}) \\ 0 & 0 & 0 & -\sin(\varphi_{94}) & 0 & 0 & 0 & 0 & 0 \\ 0 & 0 & -1 & 0 & 0 & 0 & 0 & 0 & 0 \\ -\sin(\varphi_{94}) & -\cos(\varphi_{94}) & 0 & 0 & 0 & 0 & 0 & 0 & 0 \\ \cos(\varphi_{94}) & -\sin(\varphi_{94}) & 0 & -\cos(\varphi_{94})\mathbf{b}_{12}\cdot\mathbf{x}_m & 0 & 0 & 0 & 0 & 0 \\ 0 & 0 & 0 & 0 & \sin(\varphi_{54}) & 0 & -\cos(\varphi_{54}) & 0 & 0 \\ 0 & 0 & 0 & 0 & 0 & -1 & 0 & 0 & 0 \\ 0 & 0 & 0 & -\cos(\varphi_{94})\mathbf{b}_{13}\cdot\mathbf{x}_m & \cos(\varphi_{54}) & 0 & \sin(\varphi_{54}) & 0 & 0 \end{bmatrix} \tag{44}$$

$$\mathbf{L_I} = \begin{bmatrix} 0\,0\,0 & 0 & 0 & -\sin(\varphi_{24}) & 0 & \cos(\varphi_{24}) & 0\,0\,0 & 0 & 0 & 0\,0\,0 \\ 0\,0\,0 & 0 & 0 & -\cos(\varphi_{24}) & 0 & -\sin(\varphi_{24}) & 0\,0\,0 & 0 & 0 & 0\,0\,0 \\ 0\,0\,0 & 0 & 0 & -\cos(\varphi_{24}) & 0 & -\sin(\varphi_{24}) & 0\,0\,0\,\cos(\varphi_{94}) & 0 & 0\,0\,0 \\ 0\,0\,0 & -\sin(\varphi_{24}) & \cos(\varphi_{24}) & \cos(\varphi_{24})\mathbf{b}_{12}\cdot\mathbf{y}_m & 0 & \sin(\varphi_{24})\mathbf{b}_{12}\cdot\mathbf{y}_m & 0\,0\,0 & 0 & 0 & 0\,0\,0 \\ 0\,0\,0 & 1 & 0 & -\cos(\varphi_{24})\mathbf{b}_{12}\cdot\mathbf{x}_m & 0 & -\sin(\varphi_{24})\mathbf{b}_{12}\cdot\mathbf{x}_{MPS} & 0\,0\,0 & 0 & 0 & 0\,0\,0 \\ 0\,0\,0 & -\cos(\varphi_{24}) & -\sin(\varphi_{24}) & -\sin(\varphi_{24})\mathbf{b}_{12}\cdot\mathbf{y}_m & 0 & \cos(\varphi_{24})\mathbf{b}_{12}\cdot\mathbf{y}_m & 0\,0\,0 & 0 & -\sin(\varphi_{94})\mathbf{b}_{12}\cdot\mathbf{x}_m & 0\,0\,0 \\ 0\,0\,0 & -\sin(\varphi_{24}) & \cos(\varphi_{24}) & 0 & 0 & 0 & 0\,0\,0 & 0 & 0 & 0\,0\,0 \\ 0\,0\,0 & 1 & 0 & -\cos(\varphi_{24})\mathbf{b}_{13}\cdot\mathbf{x}_m & 0 & -\sin(\varphi_{24})\mathbf{b}_{13}\cdot\mathbf{x}_m & 0\,0\,0 & 0 & 0 & 0\,0\,0 \\ 0\,0\,0 & -\cos(\varphi_{24}) & -\sin(\varphi_{24}) & 0 & 0 & 0 & 0\,0\,0 & 0 & -\sin(\varphi_{94})\mathbf{b}_{13}\cdot\mathbf{x}_m & 0\,0\,0 \end{bmatrix} \tag{45}$$

4.3. Cartesian Stiffness Matrix

As discussed in Section 2, Tripteor X7 is modeled as a structural assembly of three legs, which link the rigid fixed platform to the rigid mobile platform. The three legs are considered to be steel beams of rectangular cross-section of size consistent with Tripteor X7 legs (Figure 9).

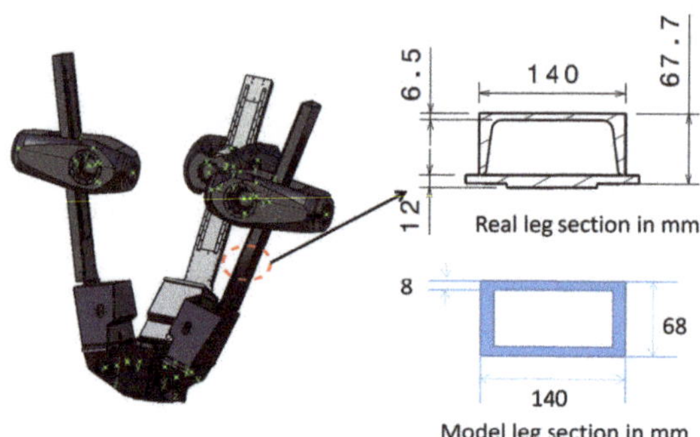

Figure 9. CAD model of the Tripteor X7 PKM and cross-section of its first leg.

To compute the Cartesian stiffness matrix of the robot at point O_m, the issue is to compute matrix $\mathbf{L_{XI}}$ (Equation (29)). Finally, we can compute the stiffness map at a level $z = 1.18$ m for example (Figure 10).

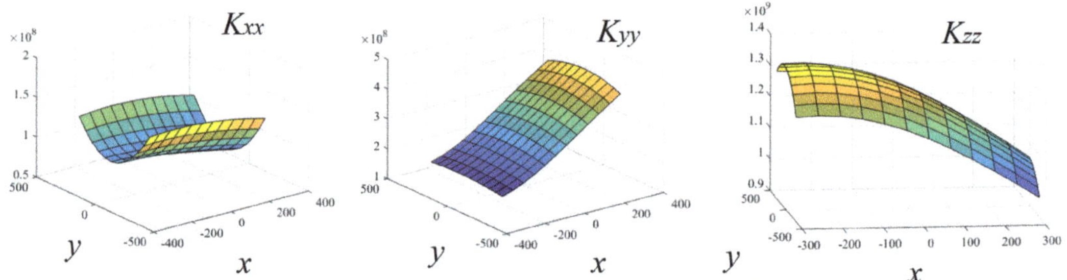

Figure 10. Principal stiffness coefficients of the Tripteor X7 along X, Y and Z axes in the XY−plane, $z = 1.18$ m.

To plot Figure 11, the method was implemented in Matlab®. To emphasize the benefit of this proposed approach, the calculation was performed on a tablet PC with a I5-7300U processor. The time taken to compute this map is 128.72 s and is mainly due to the time needed to calculate the Inverse Kinematic Model (IKM) of Tripteor X7.

The main advantage of the method is the fast calculation of the natural frequencies and associated modes of over-constrained parallel robots.

4.4. Natural Frequencies

The proposed method is used for the calculation, with a low computational cost, of the first natural frequencies and associated natural modes of the Tripteor X7 PKM.

For this first computation, the leg lengths are $q_1 = 1.212$ m, $q_2 = 1.314$ m and $q_3 = 1.22$ m, and the mass of the mobile platform is considered to be 8 kg. The approach introduced in Section 3 is applied that gives the following estimations for the first natural frequencies in 0.4 s:

$$\mathbf{f_0} = \begin{pmatrix} 46.37 \text{ Hz} \\ 51.26 \text{ Hz} \\ 52.35 \text{ Hz} \\ 90.20 \text{ Hz} \\ 140.11 \text{ Hz} \\ 249.16 \text{ Hz} \\ 297.70 \text{ Hz} \\ 328.35 \text{ Hz} \\ 350.00 \text{ Hz} \\ 389.74 \text{ Hz} \end{pmatrix} \qquad (46)$$

The associated modes of the first three natural frequencies given in Equation (46) are schematized in Figure 11. The variation of first natural frequency value at a level $z = 1.18$ m is illustrated in Figure 12.

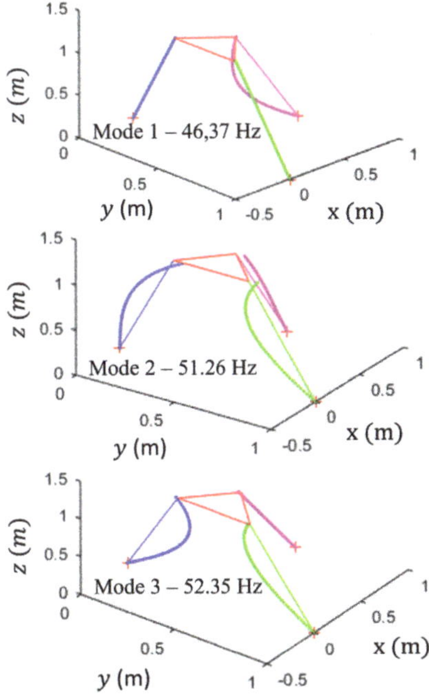

Figure 11. Modal shapes associated with the first tree natural frequencies of Tripteor X7.

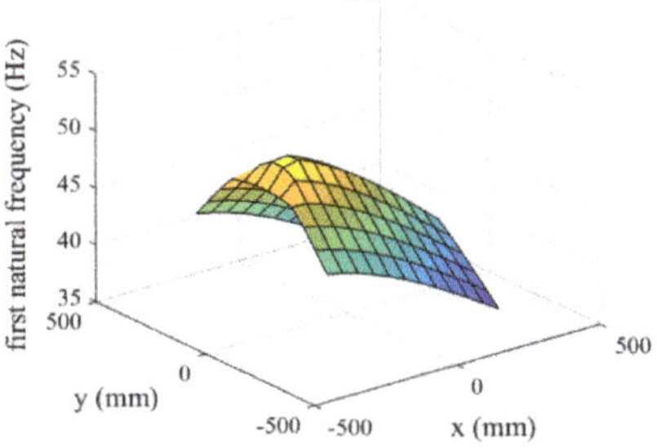

Figure 12. Variation of the first natural frequency.

4.5. Comparison with Experimental Results

In the literature, previous works studied the modal analysis of the Tripteor X7 PKM architecture using a shaker in three configurations [34] (Figure 13). The displacements of several points of the PKM architecture are measured with accelerometers located on the legs, and on the spindle. Nine DoF are considered for each leg, eight about **fi**-axis or **fl**-axis and one about **ff**-axis, and four DoF for the spindle (Figure 14). The obtained results show four natural frequencies close to 40, 50, 70 and 90 Hz.

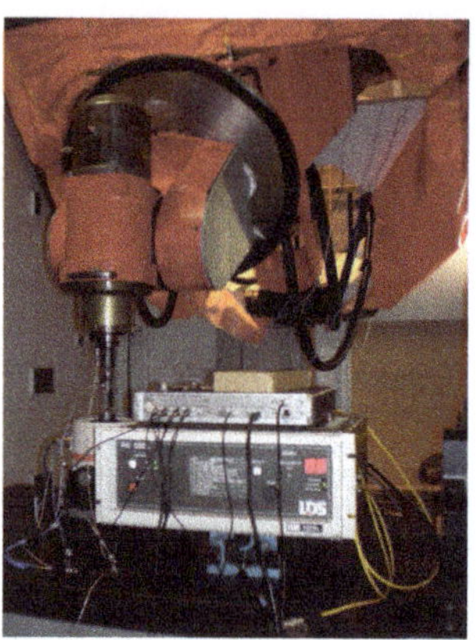

Figure 13. First natural frequency measurement.

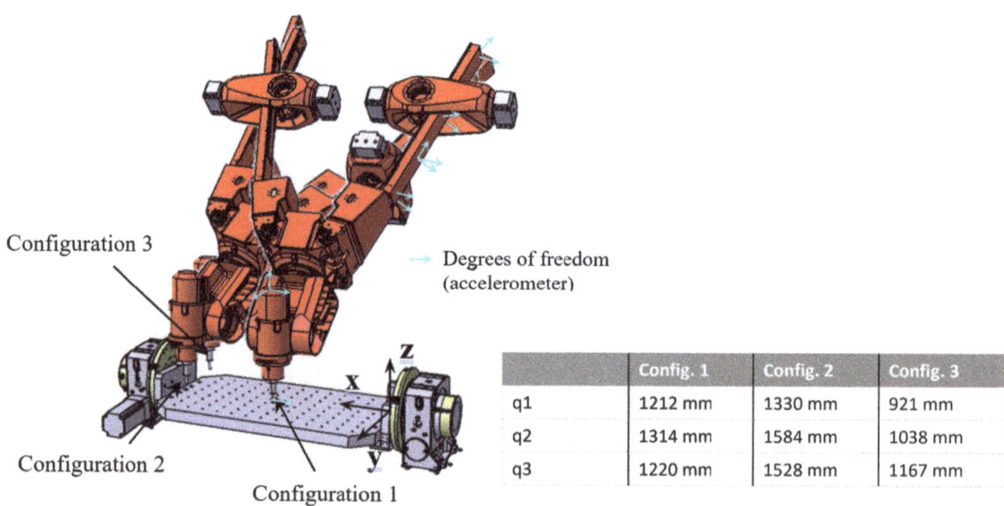

	Config. 1	Config. 2	Config. 3
q1	1212 mm	1330 mm	921 mm
q2	1314 mm	1584 mm	1038 mm
q3	1220 mm	1528 mm	1167 mm

Figure 14. Configuration of the Tripteor X7 for the experimental measurement.

The same configurations 1, 2 and 3 were considered with the proposed beam approach, and the associated results are given in Table 1. Experimental and simulated results are close for the first mode. For the other modes, the simplification of the legs modeling and the difference of mass distribution between the model and the real machine tool can explain the gap between estimated and measured natural frequencies. Indeed, the modeling of the leg under the assumption that their cross-section area is constant affects the leg stiffness, inertia and the position of its gravity center. This impact is different regarding the leg lengths and so to the studied configuration. It should be noted that this method is dedicated to the embodiment design of parallel manipulators. Therefore, such an error is acceptable, and the simplification of the leg shape allows the reduction of the calculation time of the elasto-dynamic model.

Table 1. Comparison of measured natural frequencies and calculated ones with the proposed method.

	Config. 1			Config. 2			Config. 3		
	Beam Model	**Measure**	**Error**	**Beam Model**	**Measure**	**Error**	**Beam Model**	**Measure**	**Error**
Mode 1	46.45 Hz	46.2 Hz	0.5%	34.84 Hz	39.6 Hz	12%	55.40 Hz	49.4 Hz	12.1%
Mode 2	50.98 Hz	55.3 Hz	7.8%	36.51 Hz	51.6 Hz	29.2%	66.66 Hz	50.2 Hz	32.8%
Mode 3	52.33 Hz	71.3 Hz	26.6%	44.46 Hz	58.4 Hz	23.9%	77.66 Hz	71.3 Hz	8.9%

5. Conclusions

This paper described a new methodology to compute the first natural frequencies of over-constrained PKMs based on the application of screw theory under the assumption of small displacements. Specifically designed to require few computing resources, the obtained model enables the estimation of stiffness maps, first natural frequencies and associated modes in a simple way without computing any Jacobian matrix. The proposed methodology was validated by comparing the obtained theoretical natural frequencies and associated modes of the Tripteor X7 to the measured experimental values. Errors between experimental and simulated results are less than 13% for the first mode and 33% for the second and third modes.

As a conclusion, the proposed methodology allows the mathematical expression of the simplified elasto-dynamic model of over-constrained parallel robots to reduce considerably the computation time of their natural frequencies. This low computation time allows the designer of over-constrained parallel robots to quickly estimate the natural frequencies of candidate robot architectures at the conceptual design stage and thus make the right choices of robot architecture with respect to required elasto-dynamic performance and a given task.

Later, variations in leg cross-sections will be implemented and joint stiffness will be added as an additional stiffness at beam extremities.

Author Contributions: Conceptualization, H.C., A.G., B.B. and S.C.; methodology, H.C. and S.C.; validation, H.C.; formal analysis, H.C., B.B. and S.C.; writing—original draft preparation, H.C.; writing—review and editing. All authors have read and agreed to the published version of the manuscript.

Funding: This research received no external funding.

Acknowledgments: This work was carried out within the Manufacturing 21 working group, which comprises 18 French research laboratories. The topics approached are: modeling of the manufacturing process, virtual machining, emergence of new manufacturing methods.

Conflicts of Interest: The authors declare no conflict of interest

Abbreviations

The following abbreviations are used in this manuscript:

q_i	*i*th joint variable
x	Beam abscissa
$\mathbf{u}(x)$	Displacement vector of the beam neutral axis
$\mathbf{u}_{nod}$	Vector collecting the nodal values of the kinematic functions
$u(x)$	Beam displacement along the neutral axis
$v(x)$	Beam displacement along **fi**-axis
$w(x)$	Beam displacement along **fl**-axis
$\theta_\alpha(x)$	Beam section rotation about neutral axis
$\theta_\beta(x)$	Beam section rotation about **fi**-axis
$\theta_\gamma(x)$	Beam section rotation about **fl**-axis

$\mathbf{f}_{beam}(x)$	Beam force vector at a given abscissa
$\mathbf{m}_{beam}(x)$	Beam moment vector at a given abscissa
$\mathbf{K}_e$	Beam stiffness matrix
$\mathbf{M}_e$	Beam mass matrix
$\mathbf{d}_{B_i}$	Vector of small displacements of point B_i
$\boldsymbol{!}$	Rotation vector of the mobile platform
$\mathbf{u}_D$	Vector of dependent parameters
$\mathbf{u}_I$	Vector of independent parameters
$\widetilde{\mathbf{K}}$	Global stiffness matrix
$\widetilde{\mathbf{M}}$	Global mass matrix
$\mathbf{f}_0$	First natural frequency vector

References

1. Patel, Y.; George, P. Parallel manipulators applications—A survey. *Mod. Mech. Eng.* **2012**, *2*, 57. [CrossRef]
2. Li, Y.; Ma, Y.; Liu, S.; Luo, Z.; Mei, J.; Huang, T.; Chetwynd, D. Integrated design of a 4-DOF high-speed pick-and-place parallel robot. *CIRP Ann.* **2014**, *63*, 185–188. [CrossRef]
3. Chanal, H.; Duc, E.; Ray, P. A study of the impact of machine tool structure on machining processes. *Int. J. Mach. Tools Manuf.* **2006**, *46*, 98–106. [CrossRef]
4. Terrier, M.; Dugas, A.; Hascoet, J.Y. Qualification of parallel kinematics machines in high-speed milling on free form surfaces. *Int. J. Mach. Tools Manuf.* **2004**, *44*, 865–877. [CrossRef]
5. Pritschow, G.; Eppler, C.; Garber, T. Influence of the dynamic stiffness on the accuracy of PKM. In Proceedings of the Chemnitz Parallel Kinematic Seminar, Chemnitz, Germany, 23–25 April 2002; pp. 313–333.
6. Pateloup, S.; Chanal, H.; Duc, E. Process parameter definition with respect to the behaviour of complex kinematic machine tools. *Int. J. Adv. Manuf. Technol.* **2013**, *69*, 1233–1248. [CrossRef]
7. Neumann, K. Exechon concept-parallel kinematic machines in research and practice. In Proceedings of the Chemnitz Parallel Kinematics Seminar, Chemnitz, Germany, 25–26 April 2006.
8. IFToMM. IFToMM Dictionnary—3. Dynamics. Available online: http://www.iftomm-terminology.antonkb.nl/2057_1036/frames.html (accessed on 6 March 2022).
9. Liu, W.L.; Xu, Y.D.; Yao, J.T.; Zhao, Y.S. Methods for force analysis of overconstrained parallel mechanisms: A review. *Chin. J. Mech. Eng.* **2017**, *30*, 1460–1472. [CrossRef]
10. Mei, B.; Xie, F.; Liu, X.J.; Yang, C. Elasto-geometrical error modeling and compensation of a five-axis parallel machining robot. *Precis. Eng.* **2021**, *69*, 48–61. [CrossRef]
11. Benti, A.; Durieux, S.; Chanal, H. Mesure du Comportement Vibratoire D'une Machine-Outil à Structure Parallèle. 8ème Assises MUGV 2014. Available online: https://tel.archives-ouvertes.fr/tel-00451081/document (accessed on 9 March 2022).
12. Pierrot, F.; Pierrot, F. Modelling and design issues of a 3-axis parallel machine-tool. *Mech. Mach. Theory* **2002**, *11*, 1325–1345.
13. Yan, S.; Ong, S.; Nee, A. Stiffness analysis of parallelogram-type parallel manipulators using a strain energy method. *Robot. Comput.-Integr. Manuf.* **2016**, *37*, 13–22. [CrossRef]
14. Wang, Y.; Huang, T.; Zhao, X.; Mei, J.; Chetwynd, D.G.; Hu, S.J. Finite element analysis and comparison of two hybrid robots-the tricept and the trivariant. In Proceedings of the 2006 IEEE/RSJ International Conference on Intelligent Robots and Systems, Beijing, China, 9–15 October 2006; pp. 490–495.
15. Klimchik, A.; Pashkevich, A.; Chablat, D. CAD-based approach for identification of elasto-static parameters of robotic manipulators. *Finite Elem. Anal. Des.* **2013**, *75*, 19–30. [CrossRef]
16. Germain, C.; Briot, S.; Caro, S.; Wenger, P. Natural frequency computation of parallel robots. *J. Comput. Nonlinear Dyn.* **2015**, *10*, 021004. [CrossRef]
17. Zhang, J.; Zhao, Y.Q.; Jin, Y. Elastodynamic modeling and analysis for an Exechon parallel kinematic machine. *J. Manuf. Sci. Eng.* **2016**, *138*, 031011. [CrossRef]
18. Kermanian, A.; Kamali, A.; Taghvaeipour, A. Dynamic analysis of flexible parallel robots via enhanced co-rotational and rigid finite element formulations. *Mech. Mach. Theory* **2019**, *139*, 144–173. [CrossRef]
19. Klimchik, A.; Pashkevich, A.; Chablat, D. Fundamentals of manipulator stiffness modeling using matrix structural analysis. *Mech. Mach. Theory* **2019**, *133*, 365–394. [CrossRef]
20. Guo, J.; Zhao, Y.; Chen, B.; Zhang, G.; Xu, Y.; Yao, J. Force Analysis of the Overconstrained Mechanisms Based on Equivalent Stiffness Considering Limb Axial Deformation. *Res. Sq.* **2021**. Available online: https://www.researchsquare.com/article/rs-634709/v1 (accessed on 6 March 2022). [CrossRef]
21. Yang, C.; Li, Q.; Chen, Q.; Xu, L. Elastostatic stiffness modeling of overconstrained parallel manipulators. *Mech. Mach. Theory* **2018**, *122*, 58–74. [CrossRef]
22. Bi, Z. Kinetostatic modeling of Exechon parallel kinematic machine for stiffness analysis. *Int. J. Adv. Manuf. Technol.* **2014**, *71*, 325–335. [CrossRef]

23. Bi, Z.M.; Jin, Y. Kinematic modeling of Exechon parallel kinematic machine. *Robot. Comput.-Integr. Manuf.* **2011**, *27*, 186–193. [CrossRef]
24. Amine, S.; Caro, S.; Wenger, P. Constraint and singularity analysis of the Exechon. *Appl. Mech. Mater.* **2012**, *162*, 141–150. [CrossRef]
25. Selig, J.; Ding, X. A screw theory of timoshenko beams. *J. Appl. Mech.* **2009**, *76*, 031003. [CrossRef]
26. Bourdet, P.; Mathieu, L.; Lartigue, C.; Ballu, A. The concept of small displacement torsor in metrology. *Ser. Adv. Math. Appl. Sci.* **1996**, *40*, 110–122.
27. Aydin, Y.; Kucuk, S. Quaternion based inverse kinematics for industrial robot manipulators with Euler wrist. In Proceedings of the 2006 IEEE International Conference on Mechatronics, Budapest, Hungary, 3–5 July 2006; pp. 581–586.
28. Bonnemains, T.; Chanal, H.; Bouzgarrou, B.C.; Ray, P. Stiffness computation and identification of parallel kinematic machine tools. *J. Manuf. Sci. Eng.* **2009**, *131*, 041013. [CrossRef]
29. Klimchik, A.; Pashkevich, A.; Caro, S.; Chablat, D. Stiffness modeling of robotic-manipulators under auxiliary loadings. International Design Engineering Technical Conferences and Computers and Information in Engineering Conference. *Am. Soc. Mech. Eng.* **2012**, *45035*, 469–476.
30. Kucuk, S. Dexterous workspace optimization for a new hybrid parallel robot manipulator. *J. Mech. Robot.* **2018**, *10*, 064503. [CrossRef]
31. Yoo, H.; Ryan, R.; Scott, R.A. Dynamic of flexible beams undergoing overall motions. *J. Sound Vib.* **1995**, *181*, 261–278. [CrossRef]
32. Kong, X.; Gosselin, C.M. *Type Synthesis of Parallel Mechanisms*; Springer: Berlin/Heidelberg, Germany, 2007; Volume 33.
33. Pateloup, S.; Chanal, H.; Duc, E. Geometric and kinematic modelling of a new Parallel Kinematic Machine Tool: The Tripteor X7 designed by PCI. *Adv. Mater. Res.* **2010**, *112*, 159–169. [CrossRef]
34. Bonnemains, T.; Chanal, H.; Bouzgarrou, B.C.; Ray, P. Dynamic analysis of the Tripteor X7: Model and experiments. *Robot. Comput.-Integr. Manuf.* **2010**, *29*, 180–191. [CrossRef]

MDPI

St. Alban-Anlage 66

4052 Basel

Switzerland

Tel. +41 61 683 77 34

Fax +41 61 302 89 18

www.mdpi.com

Machines Editorial Office

E-mail: machines@mdpi.com

www.mdpi.com/journal/machines